CREATION
"Behold, it was very good."

WORLDVIEWS and
a new SCIENTIFIC
AWAKENING

By Richard A. Schaefer

On the cover: *The discovery in 1953 of the double helix, the twisted-ladder structure of deoxyribonucleic acid (DNA) by James Watson and Francis Crick, marked a milestone in the history of science and gave rise to modern molecular biology, understanding how genes control chemical processes within cells.*

Researcher Rosalind Franklin captured the structure of DNA in her 51st X-ray diffraction image—before 1953. But a colleague, without Franklin's knowledge, gave the picture to researchers James Watson and Francis Crick in 1953. Watson said it was the missing information they needed; Franklin's photo of actual DNA structure showed that their theoretical model of DNA structure was correct. Watson and Crick won a Nobel Prize in 1962.

Copyright © 2019. Richard A. Schaefer
All rights reserved under United States of America and international copyright laws.

Edited by **H. I. Johnson**

Designed by **Larry Kidder, Loma Linda, California**
Published, printed, and available at **Lulu.com.** *Also available at* **BarnesandNoble.com** *and* **Amazon.com**

Basic cover graphic by **Design Cells (iStockphotos.com)**

Table of Contents

Preface iii
Pillars of Belief iii
New Style of Footnotes vi
Heads Up vi

I. Different Worldviews

Chapter 1 .... Origins and Evolution 1
Chapter 2 .... Creation 7
Chapter 3 .... The Minefield 13
Chapter 4 .... Different Kinds of Evolution 19
Chapter 5 .... Mutations 21
Chapter 6 .... Scientists Who Are Creationists 27
Chapter 7 .... What Did Darwin Say? 35
Chapter 8 .... Does Evolution Invest Chance with Intelligence? 39
Chapter 9 .... Refreshing Acknowledgments 43

II. Science and Life

Chapter 10 .... What Is Science? 49
Chapter 11 .... Why Is Creation an Important Issue? 53
Chapter 12 .... How Did Life Begin? 55
Chapter 13 .... Irreducible Complexity 63
Chapter 14 .... Are There Questions? Who's Asking? 67
Chapter 15 .... Where Did Humanity Come From? 71

III. Transitional Fossils—Intermediates

Chapter 16 .... Missing Links 83
Chapter 17 .... The Evidence Is Compelling 89
Chapter 18 .... The Cambrian Explosion 95
Chapter 19 .... Human Reproduction 99

IV. Hypothesis and Theory vs. Fact

Chapter 20 .... Neo-Darwinism—The Theory of Imagination 105
Chapter 21 .... Vestigial Organs 109
Chapter 22 .... Deoxyribonucleic Acid—DNA 113

V. Catastrophism

Chapter 23	The Biblical Flood	117
Chapter 24	The Geologic Column and Fossil Record	125
Chapter 25	Where Did the Floodwaters Come From, and Where Did They Go?	131
Chapter 26	Water, Water Everywhere	133
Chapter 27	The Worldwide Distribution of Coal and Other Water-laid Strata	139
Chapter 28	Fossil Evidence for a Worldwide Flood	145
Chapter 29	Evidence That Demands an Explanation	147
Chapter 30	Noah's Ark	151
Chapter 31	Volcanoes, Their Aftereffects, and Uniformitarianism	169

VI. Dating Techniques

Chapter 32	Radiocarbon Dating	175
Chapter 33	Radiometric Dating—Are the Dates Set in Stone?	179

VII. Dinosaurs and Humans

Chapter 34	Did Early Humans See Dinosaurs?	187
Chapter 35	Are Dinosaurs Mentioned in the Bible?	193
Chapter 36	Did Dinosaurs Evolve into Birds?	195

VIII. Prophecy and Astronomical Phenomena

Chapter 37	Evidences for Scripture's Reliability	201
Chapter 38	The Cosmos	203

IX. Conclusion

Chapter 39	Dark Edges	211
Chapter 40	Final Thoughts	215
Glossary		225
Index		230

Preface

This scientific book—*Creation: "Behold, it was very good"*—was written and compiled for non-scientist readers by a non-scientist author with collaboration from highly respected consultants, including those who have earned doctoral degrees in scientific disciplines from major universities. I say "compiled" because this book may document more quotations from other authors than any other book you have ever read. I documented statements from both creationists and evolutionists—scientists who are esteemed by their respective peers. I also quoted three attorneys, who have studied the subject comprehensively, and have applied logic from their informed perspective.

Why would anyone want to read a book on science written by someone who has absolutely no scientific credentials? Because my not being a scientist is really a strength. During my day job, I have spent more than 50 years employed at a highly respected health sciences university making scientific subjects understandable to the lay public.

I chose to have a variety of scientific consultants. If I'd consulted only a geologist to help me write the book, for example, he would not be able to write authoritatively about genetics, DNA, anatomy, physiology, biophysics, pharmacology, or cell chemistry. Some of my consultants did not feel qualified to critique the whole manuscript, an acknowledgment that only confirms my point. And one of them, a PhD, acknowledged that he had learned much from reading this manuscript.

My thanks goes to my editor, H. I. Johnson, whose writing and editing skills added clarity to the contents and their implications.

I pray that my readers will be as blessed from exposure to this material as I have been in compiling it.

Biblical texts are from the King James Version of the *Bible*, unless otherwise noted.

Pillars of Belief

What makes a concept believable? Are foundational ideas based on fact? Are these facts supported by unbiased science? In researching the creation/evolution controversy, I have discovered what might be called pillars of belief.

Pillars of Believing in Evolution
- Uniformitarianism (actualism: the present is key to the past)
- Abiogenesis (the spontaneous generation of life)
- The geologic column
- The fossil record (and the search for missing links)
- Macroevolution
- Radiometric dating
- DNA
- Mutations
- Natural selection
- Survival of the fittest
- The genetic code

Pillars of Believing in Creation
- The Bible and Nature, together, are the best key to the past
- Biogenesis (the origin of life)
- Irreducible complexity
- Intelligent design
- A global Flood
- The fossil record (lack of missing links)
- The Cambrian Explosion
- Microevolution (minor variation within a "kind")
- Cell complexity
- Radiometric dating

- DNA
- Mutations
- Natural selection
- Survival of the fittest
- The genetic code

Oddly, some beliefs appear on both lists.

In recent times some of the pillars and foundations of evolution have shown serious cracks. Some have crumbled. In this book are acknowledgments of some of the serious weaknesses in the worldview of mainstream science—candid and troubling statements which many have never read.

You will see evidence against some of the pillars of evolution by well-respected mainstream scientists:

Francis H. C. Crick, FRS
Richard Dawkins, PhD
Michael J. Denton, MD, PhD
Niles Eldredge, PhD
Richard B. Goldschmidt, MD, PhD
Stephen Jay Gould, PhD
Thomas Henry Huxley (also known as "Darwin's Bulldog")
Dean H. Kenyon, PhD
Mary Leakey
Richard C. Lewontin, PhD
Willard F. Libby, PhD
Ernst W. Mayr, PhD (hailed as the Darwin of the 20th century)
Stephen C. Meyer, PhD
Louis Trenchard More, PhD
Horatio H. Newman, PhD
David M. Raup, PhD
Michael Ruse, PhD
Carl Sagan, PhD
Allan Rex Sandage, PhD
John C. Sanford, PhD
Robert Shapiro, PhD
George Gaylord Simpson, PhD
William D. Stansfield, PhD

However—given evolution's presuppositions—evolution is still the mainstream science.

Subjects of special interest may be found in the Table of Contents or in the Index.

While working on this project for 10 years and after being exposed to the creation/evolution controversy all my life, I've collected what I trust is a unique compilation that will encourage each reader to take a closer look at his or her worldview. Most creationists agree that life started on Earth thousands of years ago, not hundreds of thousands, millions, or billions of years ago; and that whatever the age of the universe, solar system, and Earth, creation is a miracle.

Most statements in this book are quotations from others; some are strongly worded statements by people who have passionate feelings on both sides of this controversy. I am constantly amazed that seemingly very intelligent Democrats, Republicans, and Independents interviewed by the news media can with passion evaluate and describe an issue so differently. Likewise, I have lived and worked in the scientific community long enough to know that not all scientists agree with each other on every issue. Based on education and experience, very well educated and sincere scientists can have honest disagreements.

I am a believer in the Bible; that "holy men of God spake as [they were moved] by the Holy Ghost" [God's Spirit] (II Peter 1:21). I will acknowledge that my worldview is based on my lifetime experiences and education, including my five decades of service in a health sciences university. I have seen the scientific evidence. And I have seen various *interpretations* of the evidence.

However, where I also believe my approach is unique is that my focus will be on answering the question: Is it possible to find scientific evidence to support a conservative interpretation of the Genesis Creation Story? In particular, is it possible to look at the geologic column and its fossils, mountain building, plate tectonics, erosion rates, carbon-14 and radiometric dating, amino acids, DNA, "vestigial organs," petroglyphs, geology, archaeology, and paleontology, and (from evidence in these fields) not twist the data but see that it is consistent with creationism—often clearly more consistent than an evolutionary interpretation? Creation of the universe by a divine being is exactly the opposite of the materialistic worldview being taught in the public-supported educational system of America and promoted in the mainstream scientific literature as the scientific consensus.

One might think this approach would be unfeasible, unachievable, impossible. But on the contrary, I found a wealth of scientific documentation, much of it from noncreationists, to support the creationist worldview.

Scientists on both sides of this controversy look at the same laboratory or field data and interpret it as scientific evidence to support their conclusions. In the area of "origins" their differences are based on their different *interpretations* of the evidence (and that is based on something commonly known as presuppositional bias).

> Gareth J. Nelson and Norman Platnick: The mere compilation of data about the world in which we live, no matter in how highly ordered a fashion, is not sufficient for understanding the world. Data so compiled and ordered are still only data in search of interpretation.[1]

Scientific data are interpreted differently based on the assumptions, biases, prejudices, and faith (yes, *faith*) inherent in each worldview. Creation and evolution, their supporting "facts," and their proponents are poles apart. They are opposites.

So how should we respond?

While reading the voluminous documentation in this manuscript, one of my consultants acknowledged wondering how could *anybody* believe in evolution? Here's my answer. With some, it is an *a priori* (presumptive, taken for granted) *disbelief* in the Bible and its Creator. For some, it is really that simple. I recommend that readers who have such a worldview consider the possibility that this presupposition deserves the second look offered within these pages.

The following statement by evolutionary biologist and geneticist Richard C. Lewontin, PhD, a leader in developing the mathematical basis of population genetics and evolutionary theory, provides an "honest-moment" synopsis of his worldview:

> We take the side of science *in spite* of the patent absurdity of some of its constructs, *in spite* of its failure to fulfill many of its extravagant promises of health and life, *in spite* of the tolerance of the scientific community for unsubstantiated just-so stories,[2] because we have a prior commitment, *a commitment to materialism* [also known as naturalism]. It is not that the methods and institutions of science somehow compel us to accept a material explanation of the phenomenal world, but, on the contrary, that we are forced by our *a priori* [presumptive, taken for granted] adherence to *material causes* to create an apparatus of investigation and a set of concepts that produce material explanations, no matter how counterintuitive, no matter how mystifying to the uninitiated. Moreover, that materialism is an absolute, for **we cannot allow a Divine Foot in the door.**[3]

On the other hand, Lewis Thomas, MD, former chancellor of the Memorial Sloan-Kettering Cancer Center in New York, emphasizes a growing scientific dilemma. His acknowledgments, subsequent frustration, and almost hopelessness makes me want to gently suggest, "There is another, and, I believe, better explanation."

> Dr. Thomas: I cannot make my peace with the randomness doctrine: I cannot abide the notion of purposelessness and blind chance in nature. And yet I do not know what to put in its place for the quieting of my mind. It is absurd to say that a place like this place is absurd, when it contains, in front of our eyes, so many billions of different forms of life, each one in its way absolutely perfect.... We talk—some of us, anyway—about the absurdity of the human situation, but we do this because we do not know how we fit in, or what we are for. The stories we used to make up to explain ourselves do not make sense anymore, and we have run out of new stories, for the moment.[4]

What follows is my response.

Richard A. Schaefer
June 2019

---

[1] Gareth J. Nelson and Norman Platnick, "Comparative Biology: Space, Time, and Form," *Systematics and Biogeography, Cladistics and Vicariance*, Columbia University Press, New York, 1981, p. 5.

[2] In science, a *just-so story* is an unverifiable, fictional explanation, and there is no experiment that could be devised to prove it is false—it is not falsifiable.

[3] Richard C. Lewontin, PhD, quoted by Joe White et al., *Darwin's Demise*, Green Forest, AR: Master Books, 2001, p. 139, see also: Richard Lewontin, "Billions and Billions of Demons," *The New York Review*, January 9, 1997, p. 31, (emphasis added).

[4] Lewis Thomas, MD, "On the Uncertainty of Science," *Harvard Magazine*, Cambridge, MA, September-October 1980, p. 19-22.

New-style Footnotes

I have initiated a documentation and footnote style which I have never before seen. As you may have already noticed, quotations are preceded by the identity of the author, usually including any educational degree. I use this style so that the reader will not need to find the footnotes to determine who the author is, and his or her educational background. (Because this style is not used in the scientific literature, I was unable to locate some degrees.) In academic writing, the degree is not listed, it is assumed to be at the doctoral level. "Creation Science" is assumed by non-creationists to be unscientific. I have been careful mostly to quote scientists who have earned doctoral-level degrees in fields such as geology, paleontology, astronomy, biochemistry, biostatistics, etc. from standard, well recognized universities (most of which are not creation-oriented).

***Indented paragraphs* are quotations from primary and secondary sources, including evolutionists, atheists, agnostics, theistic evolutionists, and creationists.** When a footnote appears within such a paragraph, the author identified at the beginning of the statement is documenting his or her source. All other footnoted statements are either my condensations or my documented paraphrases of other authors.

I need to emphasize that these many sources have a large *variety of worldviews*, and that their interpretations of the evidence may vary considerably from mine and from each other's. I have chosen their quotations because they document my thesis, which is that a conservative interpretation of the Biblical Creation Story can be supported scientifically. To compare and contrast views, in some cases I have quoted authors with whom I generally disagree.

I have used the *ellipsis* (. . .) to indicate omitted words—in some cases just one word, in other cases many. Within quoted material, I may skip from the author's introductory summary statement to his or her conclusion using a simple 3-or-4-dot ellipsis (and not using the more formal ellipsis style to indicate a paragraph[s] deletion). I have provided more than 900 footnotes so that if there is any question the reader can verify the context of my condensations. They do not change the meaning of the statement or intent of the author; they simply condense the comment. I have added some italics for emphasis of thought.

Heads Up

Throughout this book, you will see numerous references to *millions of years for the emergence of life on earth*. Yet you also will see from respected sources convincing evidence that is cause for "reasonable doubt" about such assumptions. The evidences for *Recent Life* (Young Life) *Creationism* will be well documented with footnotes. Here is one to whet your appetite.

As we shall see in a future chapter, Carbon-14 is found in coal, dinosaurs, and even diamonds, (Old Life proponents claim these are too old to contain *any* carbon-14.)

These pages will offer many opportunities for the reader to question long-held assertions and assumptions of evolution—many counter-evidences which suggest "reasonable doubt."

The Profound Witness of Paraconformities

Ariel A. Roth, PhD: What is a paraconformity? It is a substantial gap in the sedimentary layers [of the geologic column], where the surfaces above and below are flat. You can tell you have a gap when a part of the geologic column is missing. What is the significance of paraconformities? Paraconformities challenge the geologic time scale. The usual lack of evidence at the underlayer for the long ages postulated for the gaps, especially the lack of erosion, suggests that the long geologic ages never occurred.[1]

According to Dr. Roth, these gaps at the Grand Canyon (which are common over the

---

[1] Ariel A. Roth, PhD, "Flat Gaps in the Rock Layers," www.sciencesandscriptures.com

whole world) represent from 6 million years of missing sediment (according to the traditional geological timescale), to 600 million years. Question: Why are they flat? Are we to believe that there was no weathering or erosion for 6 to 600 million years? According to Dr. Roth, based on conservative scientific measurements of the rates of erosion (one foot per thousand years), instead of being flat, these paraconformities in the Grand Canyon should exhibit from 1,000 to 6,000 feet of erosion. This missing erosion profoundly challenges long geological ages.[2]

Here is a sneak preview (from chapters 32 and 33) about some of evolution's dating problems: In 1987, a team at the University of California at Berkeley published a study comparing the mitochondrial DNA (mtDNA) of 147 people from five of the world's geographic locations.[3] They concluded that all 147 had the same female ancestor. She is now called "the mitochondrial Eve." Researchers concluded that she lived up to 200,000 years ago.

A greater surprise, even disbelief, occurred 11 years later, in 1998, when it was announced that mutations in mtDNA occur 20 times more rapidly than previously thought. Mutation rates can now be determined directly, by comparing the mtDNA of many mother-child pairs. Using the new, more accurate rate, mitochondrial Eve "would be a mere 6,000 years old."[4]

The highly respected sources of this story (*Nature* and *Science*) are mainstream science journals—not creation-science publications. The *Science* headline acknowledged that this "Research News" section raised "troubling questions about the dating of evolutionary events."[5]

Reasonable doubt is also suggested by NASA's Search for Extraterrestrial Intelligence (SETI) which inadvertently attributes DNA information to intelligence:

Alex Williams and John Hartnett, PhD: It is . . . no surprise to find a fundamental contradiction lying right at the very heart of the SETI program. SETI pioneer Professor Carl Sagan devised a set of four criteria that they could use to distinguish possible intelligent communications among the constant noise of radio static that comes from all parts of the sky. When those four criteria were applied to the *information on the DNA molecule* (to see if it comes from an intelligent source), it passed the test.[6] So, the criteria that would alert the SETI astronomers to the presence of "evolved" life in outer space would tell them, if they wanted to know, that life did not evolve—it came from an intelligent [source—i.e., it came from a] Creator.[7]

To illustrate the double standard and challenge many scientists face in today's mainstream scientific reality, the following conversation occurred between George V. Calor, a newspaper reporter for the *The Ledger* (Lynchberg, VA), and (name withheld by scientist), a molecular biologist involved in medical research on genetic diseases.

Reporter: Do you believe that the [DNA] information evolved?

Scientist: George, nobody I know in my profession believes it evolved. It was engineered by 'genius beyond genius,' and such information could not have been written any other way. The paper and ink did not write the book! Knowing what we know, it is ridiculous to think otherwise.

Reporter: Have you ever stated that in a public lecture, or in any public writings?

Scientist: No. It all just evolved.

Reporter: What? You just told me—

Scientist: Just stop right there. To be a molecular biologist requires one to hold on to two insanities at all times. One, it would be insane to believe in evolution when you can see truth for yourself. Two, it would be insane to say you don't believe in evolution. All government work, research grants, papers, big college lectures—everything would stop. I'd be out of a job, or relegated to the outer fringes where I couldn't earn a decent living.

---

[2] Ariel A. Roth, PhD, "Flat Gaps in the Rock Layers," www.scienceandscriptures.com
[3] Rebecca L. Cann et al., "Mitochondrial DNA and Human Evolution," *Nature*, Vol. 325, January 1, 1987, pp. 31-36.
[4] Ann Gibbons, "Calibrating the Mitochondrial Clock," *Science*, Vol. 279, January 2, 1998, p. 29; Walt Brown, PhD, In the Beginning, pp. 229-231.
[5] Ann Gibbons, "Calibrating the Mitochondrial Clock," *Science*, Vol. 279, January 2, 1998, p. 29.

[6] Walter I. Sivertsen, "SETI and DNA," *Creation Research Society Quarterly* 39 (3) (2000). Abstract at: http://www.creationresearch.org/crsq/abstracts/Abstracts39-3.htm, (emphasis added).
[7] Alex Williams and John Hartnett, PhD, "The Big-Bang Model," *Dismantling the Big Bang, God's Universe Rediscovered*, Master Books, Green Forest, AR, 2005, p. 161.

Reporter: I hate to say it, Jeff, but that sounds. . .

Scientist: The work I do in genetic research is honorable. We will find the cures to many of mankind's worst diseases. But in the meantime, we have to live with "the elephant in the living room."

Reporter: What elephant?

Scientist: Creation design. It's like an elephant in the living room. It moves around, takes up an enormous amount of space . . . and yet we have to swear it isn't there![8]

And what about missing links?

Stephen Jay Gould, PhD: All paleontologists know that the fossil record contains precious little in the way of intermediate forms; transitions between major groups are characteristically abrupt.[9]

Niles Eldredge, PhD: A smooth transition from one form of life to another which is implied in the theory is . . . not borne out by the facts. The search for "missing links" between various living creatures, like humans and apes, is probably fruitless . . . because they probably never existed as distinct transitional types. . . . No one has yet found any evidence of transitional creatures. This oddity has been attributed to gaps in the fossil record which gradualists expected to fill when rock strata of the proper age had been found. *In the last decade, however, geologists have found rock layers of all divisions of the last 500 million years and no transitional forms were contained in them.* If it is not the fossil record which is incomplete then it must be the theory.[10]

Is there an awakening?

I. L. Cohen, mathematician, researcher, author, member of New York Academy of Sciences, and officer of the Archaeological Institute of America: In a certain sense, the debate transcends the confrontation between evolutionists and creationists. We now have a debate within the scientific community itself; it is a confrontation between scientific objectivity and ingrained prejudice—between logic and emotion—between fact and fiction. . . . In the final analysis, objective scientific logic has to prevail—no matter what the final result is—no matter how many time-honored idols have to be discarded in the process.[11]

After all, it is not the duty of science to defend the theory of evolution, and stick by it to the bitter end—no matter what illogical and unsupported conclusions it offers. . . . If in the process of impartial scientific logic, they find a creation by outside super-intelligence is the solution to our quandary, then let's cut the umbilical cord that tied us down to Darwin for such a long time. It is choking us and holding us back.[12]

As we will see (in Chapter 7: What Did Darwin Say), Darwin believed man was more highly evolved than woman—that he had attained a "higher eminence, in whatever he takes up, than can a women," and that this "decided pre-eminence" showed that "the average of mental power in man must be above that of a woman."[13]

How many in today's mainstream scientific community would agree with this assertion by Charles Darwin?

---

[8]George V. Caylor, "The Biologists," *The Ledger*, Vol. 2, Issue 48, No. 92, Lynchburg, VA, February 17, 2000, p. 2; see also, Harold G. Coffin, "Summary: Is Creation a Viable Theory of Origins?" *Origin by Design*, Revised Edition, Review and Herald Publishing Association, Hagerstown, MD, 2005, pp. 435-436.

[9]Stephen Jay Gould, PhD, "Return of the Hopeful Monster," *The Panda's Thumb*, W.W. Norton & Company, NY, 1982, p. 189; "The Return of Hopeful Monsters," *Natural History*, Vol. 86, 1977, p. 22.

[10]Niles Eldredge, PhD, "Missing, Believed Nonexistent," *Manchester Guardian* (The Washington Post Weekly), Vol. 119, No. 22, November 26, 1978, p. 1, (emphasis added).

[11]I. L. Cohen, "Introduction," *Darwin Was Wrong: A Study in Probabilities*, New Research Publications, Inc., Greenvale, NY, 1984, pp. 6-8.

[12]I. L. Cohen, "In Retrospect," *Darwin Was Wrong: A Study in Probabilities*, New Research Publications, Inc, Greenvale, NY, 1984, pp. 214-215.

[13]Charles Darwin, "Sexual Selection in Relation to Man," *The Descent of Man and Selection in Relation to Sex*, Second Edition, A. L. Burt Company; New York, 1874, p. 643.

I
Different Worldviews

Chapter 1
Origins and Evolution

Sooner or later everyone wonders:
Who am I?
Where did I come from?
Why is there pain, suffering, and evil?
What is the solution to life's problems?
Why am I here?
Where am I going when this life is over?

The answers to these questions are based on one's personal experience, education, bias, prejudice, and faith, which eventually results in one's worldview. Ariel A. Roth, PhD, summarizes several:

> When it comes to the dominant question about the origin of life, the salient concepts being evaluated include: (a) life evolved by itself and no God was involved (naturalistic evolution); (b) there is some kind of designer (intelligent design); (c) God used the process of evolution (theistic evolution); (d) God created various forms of life during billions of years (progressive creation); (e) God created the various forms of life a few thousand years ago as implied in the Bible (recent creation) [Genesis 1:11-31].[14]

Naturalism or supernatural creation by God?

The following statement provides a preview of the focus of this book and supports the concept of reasonable doubt—is there good reason to doubt evolution?

> E. Theo Agard, PhD: The theory of evolution is not as scientifically sound as many people believe. In particular, the problem of the origin of life is well stated by the question, "Which came first, the chicken or the egg?" Every egg anyone has ever seen was laid by a chicken and every chicken was hatched from an egg. Hence, the first chicken or first egg which appeared on the scene in any other way would be unnatural,to say the least. The natural laws under which scientists work are adequate for explaining how the world functions, but are inadequate to explain its origin....[15]

In his book *Science Discovers God*, Dr. Roth reports that the *American Association of Petroleum Geologists Explorer* for January 2000 had an editorial discussion suggesting that geologists stay out of creation debates because of the politics involved and because "a scientist who goes and debates with these folks is going to get chewed up."[16]

The response from readers of the editorial was overwhelmingly that science should be more open to various ideas about creation or God.[17] I do not endorse "chewing up" those who may disagree with us; instead, I offer strong new ideas.

Believing that most people are open and even eager to know the issues, I will provide a summary of my 10 years of research on the subject. Unfortunately, I found that evolution is assumed in textbooks and educational lectures to be the only scientific position worthy of consideration. Sometimes harsh, forceful statements of belief in evolution are found in textbooks and supporting statements from scientists about what they think of anyone who doesn't share their opinions about the theory of evolution.

I believe in the benefits of honest scientific discovery. Our civilization today is the beneficiary. For example, discoveries in modern medical science are contributing to mankind's longevity and quality of life.

But I have also noticed how many scientific "facts," based on research, duplicated experiments, observation, peer-reviewed publications, and professional consensus, have been made obsolete by new scientific discoveries.

---

[14]Ariel A. Roth, PhD, "Can a Scientist Dare to Believe in God?" *Science Discovers God: Seven Convincing Lines of Evidence for His Existence*, Autumn House Publishing, a division of Review and Herald Publishing, Hagerstown, MD, 1998, p. 27.
[15]Edited by John Ashton, PhD, E. Theo Agard, PhD, "E. Theo Agard," *In Six Days: Why Fifty Scientists Choose to Believe in Creation*, Master Books, Inc., Green Forest, AR, 2000, p. 212, (emphasis added).
[16]Ariel A. Roth, PhD, "Can a Scientist Dare to Believe in God?" *Science Discovers God: Seven Convincing Lines of Evidence for His Existence*, Autumn House Publishing, a division of Review and Herald Publishing, Hagerstown, MD, 1998, p. 34; D. Brown, 2000, "Quiet agenda puts science on defensive: creation debate evolves into politics," *American Association of Petroleum Geologists Explorer* 21(1) pp. 20-22.
[17]See 10 letters in Readers' Forum, 2000, *American Association of Petroleum Geologists Explorer*, 21(3): pp. 32-37.

When we look at the bottom line of human knowledge, and compare it to how much we have learned in just the last few decades, and project how much we will learn in the future, how then can anyone be certain that the creation story is impossible or that evolution is the only possibility? Some of the evidence and conclusions I present in this book may in time be proven false. However, the reader is invited to consider evolution's major assumptions in the light of the many evidences in these pages that more clearly support creation science.

I once saw a bumper sticker that said: **"Question Authority."** I suggest this would be a good place to start.

Evolution

Eugenie Scott, PhD, an American physical anthropologist who was the executive director of the National Center for Science Education from 1987 to 2014, is a leading critic of creationism and intelligent design and is a supporter of the standard evolutionary model:

> The scientific evidence indicates that (1) earth is more than 4 billion years old; (2) life on earth began about 3.5 billion years ago; and (3) evolution has occurred, continues to occur, and is responsible for the great diversity of life on earth.[18]

Naturalistic/materialistic evolution assumes that everything began with a Big Bang 13.7 billion years ago.[19] Estimates in the past have varied from 7 to 20 billion years ago.[20] The Big Bang Theory is usually attributed to Georges Lemaître, a Belgian physicist. It was named by famous British cosmologist Sir Fred Hoyle, although he actually has been one of the theory's most vocal critics. He introduced the name Big Bang on the British Broadcasting Corporation's Third Programme broadcast at 1830 Greenwich Mean Time[21] on March 28, 1949—to compare it to the "steady state" cosmological model, idea, or hypothesis to which he subscribed—but the term stuck.[22]

The standard view is that about 4.6 billion years ago the earth cooled down and formed a rocky crust. Oceans formed and rain fell on the rocks for millions of years. Swirling in the waters of these oceans was a bubbling broth of complex chemicals—the chemicals of life. The first self-replicating systems supposedly emerged about three billion years ago from this "warm primordial soup." Life evolved from a simple cell into more complex organisms over millions of years. Some scientific evidence interpreted as supporting this theory includes the geologic column, radiometric dating, and paleontology (the study of fossil remains of plants and animals and their relationship to modern plants and animals). The presupposition of the standard secular, modern view of evolutionists is that these geological and astronomical events had to happen without a divine creator, so they would have had to have happened over billions of years by chance. This standard view has been created to interpret nature.

Because the evidence somewhat centers around fossils, we will start with them. David C. Read, JD, an attorney, defines fossilization as he might do for a jury:

> The term *fossilization* is sometimes used to describe a *chemical process in which organic material is replaced or fortified by hard minerals*. This process is more correctly called *permineralization* or *petrifaction*. (Petrifaction is the complete replacement of the original bone; permineralization is the [replacement of tissue or] filling of open spaces in the bone with hard minerals.) . . . Scientists also use the term fossil to describe any ancient plant, animal, or human trace, regardless whether or to what extent mineralization has occurred.[23]

It is the evidence from fossils that the standard model of evolutionary science relies on most

---

[18] Eugenie Scott, PhD, quoted by Cain, Yoon, Singh-Cundy, "Biology on the Job," *Discover Biology*, W. W. Norton & Company, NY, Fourth Edition, 2009, p. 338.
[19] "How old is the Universe?" 50 Greatest Mysteries of the Universe, *Astronomy Magazine*, Collector's Edition, 2007, p. 9.
[20] "The Origin of the Universe, Earth, and Life," *Science and Creationism: A View From the National Academy of Sciences*, National Academy of Sciences, Washington, DC, 1999, p. 4, see also: R. Cowen, "Further evidence of a youthful universe," *Science News*, September 9, 1995, p. 166.
[21] Solar time at the Royal Observatory in Greenwich, London, England.
[22] Fred Hoyle, from *Wikipedia, the free encyclopedia*.
[23] David C. Read, JD, "When Did the Dinosaurs Live? The 'Scientific' View," *Dinosaurs*, Clarion Call Books, Keene, TX, 76059, 2009, p. 46, [from American Heritage Dictionary], (Emphasis added).

heavily to support its claims. Yale paleontologist Carl O. Dunbar, PhD, says:

> Although the comparative study of living animals and plants may give very convincing circumstantial evidence, fossils [in the geologic column] provide the only historical, documentary evidence that life has evolved from simpler to more and more complex forms.[24]

Biology—Principles & Explorations: Most scientists think that life on earth had a spontaneous origin, developing by itself through natural chemical and physical processes. They hypothesize that molecules of non-living matter reacted chemically during the first 1 billion years of earth's history, forming a variety of simple organic molecules. They further hypothesize that complex organic molecules, some of which were capable of replicating themselves and other molecules, formed associations that became increasingly complex.[25]

Life Science: The first animals appeared in the seas about 540 million years ago. These animals included worms, sponges, and other invertebrates—animals without backbones. About 500 million years ago, fishes evolved. These early fishes were the first vertebrates—animals with backbones.

The first land plants, which were similar to mosses, evolved around 410 million years ago. Land plants gradually evolved strong stems that held them upright. These plants were similar to modern ferns and cone-bearing trees.[26]

Stanley Miller, a graduate student at the University of Chicago, attempted to simulate the earth's early conditions as hypothesized by Oparin, Urey, and other scientists. During his experiment [published in 1953],[27] he took four gases (methane, ammonia, water vapor, and hydrogen), ran them through some tubes and into a spark chamber supposed to simulate lightning. Because a red goo containing amino acids formed in the bottom of the apparatus, he concluded that he had created the beginning elements of life.

Evolution teaches that all plant and animal life forms advanced from simple to complex. What do the textbooks say?

Glencoe Biology—Living Systems: All the many forms of life on earth today are descended from a common ancestor, found in a population of primitive unicellular organisms. What were those first cells like? How do we know? What events led up to their formation? No traces of those events remain, and scientists can't travel backward in time to witness what happened. Instead they turn to scientific methods of observing data, forming hypotheses, making predictions, and constructing experiments to test their predictions.[28]

General Science: About 30 million years ago, larger primates, such as monkeys and apes, evolved. The earliest fossil apes that may be ancestral to both humans and modern apes date from about 15-20 million years ago. Primate fossils date from the past four million years. The humans, or *Homo sapiens*, are newcomers to the world. The earliest human fossils date from only about 125,000 years ago.[29]

The World of Biology: [Charles] Darwin speculated that all forms of life are related through descent with modification from earliest organisms. This speculation has been verified as we have learned more about molecular biology. . . . The genetic code has been passed along essentially unchanged through all the branches of the evolutionary tree since its origin in an extremely early form of life.[30]

Charles Darwin outlined this relationship in 1859. Although evolutionary ideas had existed since Antiquity, Darwin usually receives credit for the theory. Charles Darwin set sail on the HMS Beagle in 1831. On a five-year voyage, he was influenced by Charles Lyell's book, *Principles of Geology*. Darwin claims the book changed his life forever.

In a letter to his friend Alfred Russell Wallace in 1868, Darwin, who had studied theology, wrote, "Disbelief [in the Bible and its creation story] crept over me at a very slow rate. . . . The

---

[24] Carl O. Dunbar, PhD, *Historical Geology*, John Wiley and Sons, NY, 1949, p. 52.
[25] George B. Johnson, Peter H. Raven, "The Origin of Life," *Biology: Principles & Explorations*, Holt, Rinehart and Winston, NY, 1996, p. 226.
[26] "The Fossil Record," *Life Science*, Prentice-Hall, Inc., Upper Saddle River, New Jersey, 2002, p. 161.
[27] George B. Johnson, Peter H. Raven, "History of Life on Earth," *Biology: Principles & Explorations*, Holt, Rinehart and Winston, NY, 2001, p. 254.
[28] "Origin of Life," *Glencoe Biology, Living Systems*, Glencoe/McGraw-Hill, New York, NY; Columbus, Ohio: Mission Hills, California; Peoria, IL, 1998, p. 324.
[29] "The Cenozoic Era," *General Science*, Harcourt Brace Jovanovich, Inc., Chicago, IL, 1989, p. 385.
[30] P. William Davis, Eldra Pearl Solomon, Linda R. Berg, "Darwin and Natural Selection," *The World of Biology*, Saunders College Publishing, Philadelphia, PA, 1990, pp. 294-295.

rate was so slow that I felt no distress." One of the Beagle's stops was on the Galapagos Islands, where Darwin collected 14 different varieties of finches. Later, when an ornithologist pointed out that they had different beak shapes, Darwin concluded that the birds all had a common ancestor.

He then concluded that all things living are related:

> It is a truly wonderful fact . . . that all animals and all plants, throughout all time and space, should be related to each other. . . .[31]

"Evolution" means gradual modification of life into diverse forms. Sometimes referred to as "Darwinism" or the "neo-Darwinian synthesis," and "Universal Common Ancestry," evolution is thought to have been caused by random variation and natural selection.

Natural selection, also known as survival of the fittest, as defined by Charles Darwin, is his theory that plants and animals that *inherit superior genes and traits* and who adapt better to their environment and to changes in it, will pass on those genes and traits to the next generation. They are the fittest, so they will survive.

Charles Darwin's 1859 book, *On the Origin of Species By Means of Natural Selection, or the Preservation of Favored Races in the Struggle for Life*, made a major impact on the theory of evolution and on disbelief in the Bible's account of creation. *Time* magazine reported, "Charles Darwin didn't want to murder God, as he once put it. But he did."[32] His theory predicted that a steady progression of organisms, from simple to complex, would be found as fossils in the geologic column.

By the way, most references to Darwin's book today reduce the title to *The Origin of Species*, which may give the impression that his theory is fact. I have "A Facsimile of the First Edition," which shows the title to be **On** *the Origin of Species*, which tells me that at the time he published the book, he considered his theory to be just that—a theory. Of course, "theory" is a useful construct to define what will be subjected to testing and experimentation.

However, those who promote the theory of evolution today state it is no longer a theory—they say it is scientific fact.

> Carl Sagan, PhD: Evolution is a fact, not a theory.[33]

> Ernst W. Mayr, PhD, at Harvard University: No educated person any longer questions the validity of the so-called theory of evolution, which we now know to be a simple fact.[34]

How did "natural selection" and "survival of the fittest" fit into Darwin's theory?

> Ariel A. Roth, PhD: Darwin . . . proposed that life evolved from simple to advanced forms [complex], one minute step at a time, through a process he called natural selection. He reasoned that organisms constantly vary and that over-reproduction results in competition for food, space, and other resources. Under such conditions those organisms that have some advantage will survive over inferior ones. Thus we have evolutionary advancement by survival of the fittest.[35]

That is, those with superior traits will likely pass these genes on to their offspring. By the way, the word "species" is used by most scientists to depict a group of animals that can breed and produce fertile offspring. Evolutionists also claim that the very general advancement noted in the fossils of the stratified layers in the geologic column provides scientific evidence which substantiates evolution.

> *The World of Biology:* Macroevolution refers to evolution above the species level that might give rise to new genera or higher-level categories of organisms. Because of macroevolution, it follows that all forms of life developed over time from very different and often much simpler ancestors, and all lines of their descent can be traced back to a common ancestral organism.[36]

---

[31]Charles Darwin, "Natural Selection," *On the Origin of Species By Means of Natural Selection, or the Preservation of Favored Races in the Struggle for Life*, The World's Popular Classics, Books, Inc., New York, Boston, 19 [sic], p. 106.

[32]"Iconoclast of the Century: Charles Darwin, (1809-1882)," *Time*, December 31, 1999, p.186.

[33]Carl Sagan, PhD, "One Voice in the Cosmic Fugue," *Cosmos*, Random House Publishing Group, NY; Carl Sagan Productions, Inc., 1980, p. 17.

[34]Ernst W. Mayr, PhD, "Darwin's Influence on Modern Thought," *Scientific American*, July 2000, pp. 79-83.

[35]Ariel A. Roth, PhD, "The Perplexity of Complexity," *Science Discovers God: Seven Convincing Lines of Evidence for His Existence*, Autumn House Publishing, a division of Review and Herald Publishing, Hagerstown, MD, 2008, p. 100.

[36]P. William Davis, Eldra Pearl Solomon, Linda R. Berg, "Speciation and Macroevolution," *The World of Biology*, Fourth Edition, Saunders College Publishing, Philadelphia, PA, 1990, p. 316, (emphasis added).

Biology—Principles and Explorations: Fossils offer the most direct evidence that evolution takes place.... Fossils, therefore, provide an actual record of earth's past life-forms. Change over time (evolution) can be seen in the fossil record.[37]

And finally, it is said that radiometric dating provides solid scientific evidence which demonstrates that the earth and its fossils are billions of years old. Evolutionists in the past believed in the theory of *uniformitarianism*—that "the present is the key to the past." Although not applied as strictly today as it was in the past, the theory is sometimes termed "actualism." Therefore, they still do not believe in classical *catastrophism*—the biblical account that a global Flood destroyed the world, and that Noah, his family, and representative animals were saved on a huge ship he built called the Ark. They assert that is a myth.

A textbook outlines the two opposing concepts:

Introduction to Geology: Assuming that life did have its beginnings upon the earth, there remain two hypotheses in regard to its origin: that of special creation, and that of spontaneous generation, which merely implies the genesis of life from nonliving matter, through natural processes. Of course, only the latter appears amenable to scientific investigation.[38]

Not only do evolutionists believe that evolution is a demonstrated fact, but also, they believe that only religious fundamentalists doubt Darwinism; that in this age of science and technology it's simply irrational to believe ancient myths from the Bible, such as God created the world and shaped human beings in His own image. They conclude, modern man has split the atom, put astronauts on the moon, and uncovered ancient fossils in the geologic column that substantiate evolution beyond all reasonable doubt.

Then we must ask, Is the Bible just a collection of fairy tales and myths? Can "creation science" be shown to be "good science"—truly scientific? . . . and not simply scripture-based?

---

[37] George B. Johnson, Peter H. Raven, "Evidence of Evolution," *Biology: Principles and Explorations*, Holt, Rinehart, Winston, Austin, New York, Orlando, 2001, p. 283.

[38] Howard E. Brown, Victor E. Monnett, J. Willis Stoval, "Life on Earth," *Introduction to Geology*, Gin and Company, Chicago, New York, 1958, p. 406.

Chapter 2
Creation

On Christmas Eve, December 24, 1968, during NASA's first lunar flight, as the crew of Apollo 8 orbited the moon and approached a lunar sunrise they broadcast "for all the people back on earth" the first words of the Holy Bible.[39]

> ***William Anders:*** In the beginning God created the heaven and the earth. And the earth was without form, and void; and darkness was upon the face of the deep. And the Spirit of God moved upon the face of the waters. And God said, Let there be light: and there was light. And God saw the light, that it was good: and God divided the light from the darkness.
>
> ***Jim Lovell:*** And God called the light Day, and the darkness he called Night. And the evening and the morning were the first day. And God said, Let there be a firmament in the midst of the waters, and let it divide the waters from the waters. And God made the firmament, and divided the waters which were under the firmament from the waters which were above the firmament: and it was so. And God called the firmament Heaven. And the evening and the morning were the second day.
>
> ***Frank Borman:*** And God said, Let the waters under the heaven be gathered together unto one place, and let the dry land appear: and it was so. And God called the dry land Earth; and the gathering together of the waters called he Seas: and God saw that it was good (Genesis 1:1-10, *King James Version*).
>
> And from the crew of Apollo 8, we close with good night, good luck, a Merry Christmas . . . and God bless all of you, all of you on this good earth.[40]

Wernher von Braun, PhD, father of the United States space program, thought by many to be the greatest rocket scientist who ever lived, once remarked, "They (evolutionists) challenge science to prove the existence of God. *But must we really light a candle to see the sun?*"[41]

Creationism as taught in the first chapters of the book of Genesis is a description of how God created all life. "And God said, Let the waters bring forth abundantly the moving creatures that have life, and fowl that may fly above the earth in the open firmament of heaven" (Genesis 1:20). And how long did this take? "For in six days the Lord made heaven and earth, the sea, and all that in them is" (Exodus 20:11). And the Holy Bible teaches that this last statement, part of the Ten Commandments, was written in stone by the finger of God (Exodus 31:18), the most direct words from God in the Bible.

According to this worldview, about 6,000 years ago [. . . some believe as much as 10,000 years ago][42] God created life on our planet (and a habitat for life—its atmospheric heavens, the air we breathe) in six, literal evening-and-morning, 24-hour days, after which He rested on the seventh day—which is identified in Exodus 20 as the Sabbath.

Is the Genesis story of Creation prose? Or is it just poetry? Are the Bible's words to be taken literally or figuratively? Theistic evolution proposes that God is in charge of the biological process called evolution. He supposedly directs and guides the development of life forms over millions of years. Theistic evolution contends that there is no conflict between science and the biblical book of Genesis.

> Answers in Genesis: Since the early 1800s, many Christians have accepted the idea that [life on] the earth is billions of years old. This notion

---

[39] The Apollo 8 Christmas Eve Broadcast, December 24, 1968, NASA National Space Science Data Center.
[40] The Apollo 8 Christmas Eve Broadcast, December 24, 1968, NASA National Space Science Data Center.
[41] Wernher von Braun, letter read by Dr. John Ford to California State Board of Education, September 14, 1972, quoted in Ann Lamont, *Twenty-One Great Scientists Who Believed the Bible*, Creation Science Foundation, Acacia Ridge, Queensland, Australia, 1995, p. 47, (emphasis added).
[42] "Creationist Views of the Origin of the Universe, Earth, and Life," *Science and Creationism: A View From the National Academy of Sciences*, National Academy of Sciences, Washington, DC, 1999, p. 7.

contradicts a plain reading of the biblical text; so, many have searched for a way to harmonize the early chapters of Genesis with the idea of long ages. Many theories have been proposed, such as the Gap Theory, the Day-Age Theory, and Progressive Creationism.[43]

So, what is the evidence for a literal interpretation of Genesis?

David C. Read, JD: The Hebrew word for "day," *yôm*, can refer to something other than a literal twenty-four-hour period. In Genesis 1, however, each *yôm* of the creation week is modified by an ordinal number, as in "the first day," "the second day," etc. When so modified, *yôm* always means a literal twenty-four-hour day.[44]

Dr. John R. Howitt wrote to appropriate professors in nine leading universities, asking, "Do you consider that the Hebrew word *yôm* (day), as used in Genesis 1, accompanied by a numeral should properly be translated as (a) a day as commonly understood, (b) an age, (c) either a day or an age without preference?" Oxford and Cambridge did not reply but the professors at Harvard, Yale, Columbia, Toronto, London, McGill, and Manitoba universities replied unanimously that it should be translated as a day as commonly understood. Professor Robert H. Pfeiffer of Harvard added to his reply, "of twenty-four hours."[45]

> And on the seventh day God ended his work which he had made; and he rested on the seventh day from all his work which he had made (Genesis 2:2).

These words, confirmed by New Testament writers, tell us that God's work of creating life on earth and its life-supporting habitat did not span millions of years as believed by theistic evolutionists.

Jacques Monod, PhD, a Nobel Prize winner in biology, when commenting on theistic evolution (that God used the process of evolution) shared this profound insight:

> Natural selection is the blindest and most cruel way of evolving new species, and more and more complex and refined organisms. . . . The struggle for life and elimination of the weakest is a horrible process, against which our whole modern ethics revolts. An ideal society is a non-selective society, one where the weak is protected; which is exactly the reverse of the so-called natural law.
>
> I am surprised that a Christian would defend the idea that this is the process which God more or less set up in order to have evolution.[46]

Philosopher David L. Hull, PhD, of Northwestern University wrote regarding the implications of theistic evolution on the character of God: "Whatever the God [Who is] implied by evolutionary theory and the data of natural history may be like, He is not the Protestant God of waste not, want not. He is also not a loving God who cares about His productions. . . . The God of Galapagos is careless, wasteful, indifferent, almost diabolical. He is certainly not the sort of god to whom anyone would be inclined to pray."[47]

Warren L. Johns, JD: Dismissing the Genesis account of the creation week by suggesting the seven days are merely symbolic, allegory, or metaphor, mocks the majestic beauty of the event. Arbitrarily scrambling nature's time frame as symbolism, merely to accommodate evolution theory, diminishes the role of the Creator and mars the big picture of life's purpose.[48] . . . A fence-straddling theistic compromise that speculates a benign God used evolutionary process to "create" life in some vague series of gradual increments discounts the Scriptures' plain words. Christ, the Apostle Paul, and other Bible writers testify to a literal, seven-day creation week.

Gerhard F. Hasel, PhD: "The creation of vegetation with seed-bearing plants and fruit trees took place on the third day (Genesis 1:11-12). Much of this vegetation seems to need insects for pollination. Insects were created on the fifth day (vs. 20). If the survival of those types of plants which needed insects for pollination depended on them to generate seeds and to perpetuate themselves,

---

[43]Meredith G. Kline, "Space and Time in the Genesis Cosmogony," *Perspectives on Science and Christian Faith*, 48, March 1996, p. 48.
[44]David C. Read, *Dinosaurs*, Clarion Call Books, Keene, TX, 2009, p. 94. Randall W. Yonker, God's Creation, Pacific Press Publishing Association, Nampa, ID, 1999, p. 29.
[45]Dr. John R. Howitt, letter to the editor, *Journal of the American Scientific Affiliation*, 15:2:66, June 1962, p. 66.

[46]Jacques Monod, PhD, "The Secret of Life," Interview with Laurie John, Australian Broadcasting Co., June 10, 1976.
[47]David L. Hull, "The God of the Galapagos," review of Darwin on Trial by Phillip E. Johnson, JD, *Nature*, Vol. 352, August 8, 1991, p. 486.
[48]Warren L. Johns, JD, "Day One, Young Life, Old Earth," *Genesis File*, www.GenesisFile.com, Lightning Source, LaVergne, TN, 2010, p. 158.

then there would be a serious problem should the creation "day" consist of long ages or aeons. The type of plant life dependent on this type of pollination process without the presence of insects could not have survived for these long periods of time, if "day" were to mean "age" or "aeon." In addition, consistency of interpretation in the "day-age theory" would demand a long period of light and darkness during each of the ages. This would quickly be fatal both to plant and animal life.[49]

Some have postulated that the first two chapters of Genesis are poetry and not to be taken literally.

> Harold G. Coffin, PhD: Is the creation story written in poetry? Has the author used poetic license in recording this account? Various features are used to identify Hebrew poetry—each line starting with a new letter of the alphabet, similar thoughts repeated in different words, meter, poetic vocabulary, et cetera. Prose lacks these elements and makes frequent use of conjunctions to link sentences together. It is not difficult to determine that the first two chapters of Genesis are prose, not poetry.[50]

> Gerhard F. Hasel, PhD: Compared to the hymns in the Bible, the creation account is not a hymn; compared to the parables in the Bible, the creation account is not a parable; compared to the poetry in the Bible, the creation account is not a poem; compared to cultic liturgy, the creation account is not a cultic liturgy. Compared to various kinds of literary forms, the creation account is not a metaphor, a story, a parable, poetry, or the like.[51]

> Warren L. Johns, JD: Why would the infinite, all-powerful Creator set in motion an easy-to-read time calibrator tied to earth's orbit on its axis around the sun, and then inspire an account of origins that discarded real days, leaving human imaginations to decipher the abstract? It's less than logical to suggest the Creator put this obvious time measurement in place only to play a game of gotcha, confusing humans with a Genesis account referencing the *days of creation week* as mere *metaphors* or *symbols* of indeterminate time spans.[52]

Lexicographers of the Hebrew languages are among the most qualified of Hebrew scholars. They are expected to give great care in their definitions and also usually indicate alternative meanings, if there is warrant to do so in given instances. None of the lexicographers have departed from the meaning of the word "day" as a literal day of 24 hours for Genesis 1.

The most widely recognized Hebrew lexicons and dictionaries of the Hebrew language published in the twentieth and twenty-first centuries affirm that the designation "day" in Genesis 1 is meant to communicate a 24-hour day, that is, a solar day.

A prestigious [1994] lexicon refers to Genesis 1:5 as the first scriptural entry for the definition of "day of 24 hours" for the Hebrew term *yôm* (day).[53] Holladay's Hebrew-English lexicon refers to Genesis 1:5 as a "day of 24 hours."[54]

The words "evening and morning" accompanying each Genesis 1 day of creation is further evidence that these were the same kind of days we experience in the 21st Century. A Hebrew lexicographer agrees:

> Gerhard F. Hasel, PhD: The Brown-Driver-Briggs lexicon, the classical Hebrew-English lexicon, also defines the creation "day" of Genesis 1 as a regular "day as defined by evening and morning."[55]

---

[49] Lloyd R. Bailey, *Genesis, Creation, and Creationism*, Paulist Press, NY, Malwah, NJ, 1993, p. 126; quoted by Gerhard F. Hasel, PhD, "The 'Days' of Creation in Genesis 1: Literal 'Days' or Figurative 'Periods/Epochs' of Time?" *Origins*, a publication of the Geoscience Research Institute, Loma Linda, CA, Vol. 21, No. 1, 1994, p. 30.
[50] Harold G. Coffin, PhD, "Creation Week—The first three days," *Creation—Accident or Design?* Review and Herald Publishing Association, Washington, DC, 1969, p. 18.
[51] Gerhard F. Hasel, PhD, "The 'Days' of Creation in Genesis 1: Literal 'Days' or Figurative 'Periods/Epochs' of Time?" *Origins*, a publication of the Geoscience Research Institute, Loma Linda, CA, Vol. 21, No. 1, 1994, p. 19.
[52] Warren L. Johns, JD, "Day One, Young Life, Old Earth," *Genesis File*, www.GenesisFile.com, Lightning Source, LaVergne, TN, 2010, p. 158, (emphasis added).
[53] Benedikt Hartmann, Philippe Reymond, and Johann Jakob Stamm, *Hebräisches und Aramäisches Wörterbuch der Hebräischen Sprache*, Leiden: E. J. Brill, 1990, p. 382.
[54] William H. Holladay, *A Concise Hebrew and Aramaic Lexicon of the Old Testament*, William B. Eerdmans Publishing Co., Grand Rapids, MI, 1971, p. 130; quoted by Gerhard F. Hasel, PhD, "The 'Days' of Creation in Genesis 1: Literal 'Days' or Figurative 'Periods/Epochs' of Time?" *Origins*, a publication of the Geoscience Research Institute, Loma Linda, CA, Vol. 21, No. 1, 1994, p. 22.
[55] Francis Brown, S. R. Driver, and Charles A. Briggs, *A Hebrew and English Lexicon of the Old Testament*, Oxford: Clarendon Press, 1974, p. 398; quoted by Gerhard F. Hasel, PhD, "The 'Days' of Creation in Genesis 1: Literal 'Days' or Figurative 'Periods/Epochs' of Time?" *Origins*, a publication of the Geoscience Research Institute, Loma Linda, CA, Vol. 21, No. 1, 1994, p. 22.

At Mount Sinai, when God reminded the Children of Israel about the Sabbath, He said, "For in six days the Lord made heaven and earth, the sea and all that in them is. . . ." (Exodus 20:11). So as to emphasize how literal were the creation days, He also linked the weekly Sabbath with His supplying manna on six literal days; this allowed them to collect twice as much on the sixth day (Exodus 16:22) so they would have manna to eat on the seventh, a day of rest.

Creationists believe that instead of everything evolving from a common ancestor, they had a Common Designer. The creationist sees irreducible complexity and incredible design as well as a symbiotic relationship between plants, insects, animals, and man, and concludes that the Intelligent Designer is God. This worldview also concludes that since God created us, I owe Him my very life. I am very thankful. His "rules for living" or 10 commandments (Exodus 20) point to a Creator who longs to have me love Him (the first 4) and to love and care for others (the last 6 commandments) (Matthew 22:37-40).

> And God said, let us make man in our image, after our likeness: and let them have dominion over the fish of the sea, and over the fowl of the air, and over the cattle, and over all the earth, and over every creeping thing that creepeth upon the earth. So God created man in his own image, and in the image of God created he him; male and female created he them. And God blessed them, and God said unto them, Be fruitful, and multiply, and replenish the earth, and subdue it: and have dominion over the fish of the sea, and over the fowl of the air, and over every living thing that moveth upon the earth (Genesis 1:26-28).

The word "replenish" has misled some to believe there was an earlier creation during a "gap" between Genesis 1:1 and Genesis 1:2. The *"ruin and restoration"* theory is a false teaching that at some time in the past there was a series of creations and that the world was depopulated before the "Creation Week" of Genesis. But that interpretation is using today's understanding of the word replenish, not the understanding of those who wrote the King James Version and other translations of the Bible. According to a respected standard desk reference for theologians, in Genesis 1:28 and 9:1, replenish comes from the Hebrew word *male*, which Young's concordance tells us means literally, "to fill."[56]

- *Jesus Himself* endorsed the biblical story of creation by citing the book of Genesis 25 times. For example: "And he answered and said unto them, Have ye not read that he which made them *at the beginning* made them male and female?" (Matthew 19:4). He also mentioned the evil of Noah's time, and specifically refers to the day that Noah entered the ark (Matthew 24:37-38; Luke 17:26-27).
- *Moses* told us that on the seventh day God "rested from all his work that God created" (Genesis 2:3).
- *King David* wrote, "When I consider thy heavens, the work of thy fingers . . . what is man?" (Psalm 8:3); and "By the word of the Lord were the heavens made, and all the hosts" (Psalm 33:6).
- *King Solomon* wrote, "He hath made everything beautiful" (Ecclesiastes 3:11); and "Remember now thy creator in the days of thy youth" (Ecclesiastes 12:1).
- *Malachi* wrote, "Have we not all one father? Hath not one God created us?" (Malachi 2:10).
- *The Apostle John* wrote, "All things were made by him" (John 1:3).
- *The Apostle Paul* wrote, "By him were all things created that are in heaven and . . . earth" (Colossians 1:16); and "Adam was first formed, then Eve" (I Timothy 2:13).
- *John the Revelator* wrote, "For thou hast created all things" (Revelation 4:11).[57]

"And God saw every thing that he had made, and, behold, it was very good. And the evening and the morning were the sixth day" (Genesis 1:31).

However, science has not unraveled certain "problems." Many evolutionists today have been

---

[56]Robert Young, *Young's Analytical Concordance to the Bible*, Hendrickson Publishers, Peabody, MD, 2008, p. 808 and its Hebrew Lexicon, p. 24.
[57]Bernard Brandstater, MBBS, "The Centrality of Creation in Adventist Belief," paper presented to Adventist Theological Society Spring Symposium, Southwestern Adventist University, Keene, TX, April 17, 2010, p. 8.

taught, from K through 12 and for years beyond, that despite the unsolved, evolution explains the universe and the natural world, and therefore that the Bible account of origins is unscientific. They may have never met a creationist who was solidly trained in good science in a top secular university who is doing good creation science research published in reputable journals. They may have never read "good science" in creationist publications.

The apostle Peter stated (in II Peter 3:3) that "scoffers" in the last days of earth's history would be "willingly ignorant" of creation by God and destruction by the Flood. He also authenticated the account of Noah being saved by the ark during the Flood (I Peter 3:20; II Peter 2:5). In the Old Testament book of Isaiah God authenticates the Flood account and repeats His promise: "I have sworn that the waters of Noah should no more go over the earth" (Isaiah 54:9). Also, they would scoff at the thought that "long ago by God's word the [starry] heavens existed and the earth was formed" (II Peter 3:5 NIV).

Until recently, our reckoning of time was divided by the letters BC (Before Christ) and AD (*Anno Domini* or year of our Lord, referring to what was thought to be the year of Christ's birth). Today, these designations have changed to BCE (Before Common Era) and CE (Common Era), (perhaps for the same reason that it is no longer politically correct to say Merry Christmas, but simply Happy Holidays). Nevertheless, if one studies the background of this change, it will be clear that the Common Era starts at the time of Christ's birth. Changing BC/AD to BCE/CE did not change the event—the birth of Christ—that still marks the centerpoint of human history's time divide. Whenever anyone dates a letter, check, or footnote, this act indirectly acknowledges a major event and Person in earth's time span.

But one of the most compelling concepts derived from the Bible and used by virtually all civilizations for all time is the seven-day week. From astronomy we derive the length of the day, month, and year. But there is nothing in astronomy on which to base the length of the week. The seven-day week comes to us directly from the creation week found in the book of Genesis. Even though there have been changes in the identity of the year, the seven-day week ending in the Sabbath is unquestionably based on creation week. It also can be traced to the six-day distribution of manna (Exodus 16) and the 4th of 10 Commandments written on tables of stone by the finger of God (Exodus 31:18), given to the Children of Israel at Mount Sinai, and kept faithfully by the Jews and many Bible believers ever since.

While surveying the *Encyclopædia Britannica*, I was pleasantly surprised to see a very positive comment about God.

> Peter J. Wyllie: We all have a sense of awareness and appreciation of the Earth; we all admire the scenery. One of the rewards of studying and understanding the Earth is the development of this sense to a greater extent. This development brings us closer to nature, also to an awareness of some transcendental power, closer to God if we choose to define God in these terms. . . . The Earth provides all of our material needs and satisfies some of our spiritual needs: "I will lift up mine eyes unto the hills, from whence cometh my help." A day in the mountains, at the seashore, or in the countryside sharpens that sense of awareness of the Earth which was compared above with an awareness of God.[58]

Near the beginning of this book I want to suggest another important "heads up." As the creation/evolution controversy is explored here, I hope you will notice that throughout this book, the theory of evolution (including the science of radiometric dating) is built on the pillars of several major assumptions. The following quotation summarizes just one obvious and undeniable example, which supports one of my underlying themes—reasonable doubt.

---

[58] Peter J. Willie, "The Great Globe Itself," *Propaedia: Outline of Knowledge, Guide to the Britannica,* Encyclopædia Britannica, Inc., William Benton, Chicago, London, Toronto, Geneva, Sydney, Tokyo, Manila, Seoul, 1979, p. 76.

Henry M. Morris, PhD: In spite of the fact that [earth] history [as] documented by written records covers only the past few thousand years (since the first dynasty of Egypt, say, or the first king lists in Sumeria), evolutionists inevitably allege that the earth is several billion years old (the current "official" figure is 4.6 billion) and that the early *hominids* [humanlike beings] began to evolve from their nonhuman ancestors several million years ago. All such age estimates, of course, have to be based on a number of unprovable assumptions, since actual records have existed only since the invention of the calendar and some form of written language.[59]

---

[59] Henry M. Morris, PhD, "The Evolutionary Basis of Modern Thought," *The Long War Against God*, Master Books, Inc., Green Forest, AR, 2000, p. 25.

Chapter 3
The Minefield

Evolution is taught in the publicly funded educational system of America. Creation is rarely mentioned as a worldview even worthy of consideration. Information on the science of creation is not widely available (and is claimed by some not even to exist). Therefore, I have compiled such information.

Quotation mining

If you turn to the final footnote, you will notice that I have compiled almost a thousand quotations. These were discovered over a period of 10 years as I studied a variety of resources. Some enthusiastic evolutionists, in attempting to discredit creationists, criticize creationist writers by claiming that they practice *"quotation mining."*

However, when you think of "mining," do you think of prospectors? What were these prospectors searching for? Did it have value? Did these prospectors have to search far and wide before finding what they were looking for? Did they experience satisfaction if they found something of value? Did they employ effort to retrieve it? I have carefully searched for good, factual information.

To compile almost a thousand quotations, I "prospected" through numerous books, magazines, and scientific journals. I watched hundreds of hours of debates and educational videos, attended educational lectures, reviewed television specials on this topic, and visited numerous museums (including the Smithsonian Institution in Washington, DC).

I found numerous "gold nuggets." You will see here quotations from evolutionists, as well as from creationists. You will see quotations from Charles Darwin's book, *On the Origin of Species*. You will see quotations from two highly respected atheists who became believers in God by following their science and their consciences.

You will have to decide if identifying a person as a creationist (the messenger) is good enough reason to reject the information presented in the quotation (the message). In my research, I have seen authors reject—outright—good information, scientific facts, not because of what was said, but because of who said it. This is a position that is at least questionable; we can learn from everyone, even those with whom we do not always agree.

Reasonable doubt

Football coach Vince Lombardi reportedly once said, "The best defense is a good offense." Therefore, I am going to use reasonable doubt near the beginning of this book as a form of persuasion against the theory of evolution and its presupposition of millions of years of life on Earth. Reasonable doubt is a legal concept used every day not only in America's judicial system, but also in the court of public opinion. My line of reasoning begins with three subjects which I believe awaken "reasonable doubt." (Many more will follow.)

1. *Spontaneous generation of life*—the materialistic belief that life can evolve from non-life.
2. *Rates of erosion*—and therefore the impossibility that the geologic column and its fossil record could continue to exist through "deep time" (millions of years).
3. *Macroevolution*—the idea that one "kind" of organism can progress to a higher "kind" of organism by natural selection and survival of the fittest, given enough "deep time."

Spontaneous generation of life

Harold G. Coffin, PhD: The general theory of evolution is contrary to some of the basic laws of science. . . . In all recorded history, in all the experience of man, past and present, there has never once been a documented observation or laboratory experiment of the change of nonliving matter into a living organism. Yet the *modus operandi* [its way of working] of science involves observation and experimentation. One of the most fundamental laws of biology is that life must come from life. True, the beliefs in spontaneous generation were common during the [Middle] Ages, but none today dare suggest that such ideas resulted from reliable observations or controlled experiments. In fact, it was just such observations and experiments that laid the beliefs to rest.[60]

Rates of erosion

Dr. Ariel Roth documents from scientific authorities that *erosion rates* are faster in high mountains and slower in regions of less relief.[61] In the Hydrographers Range in Papua New Guinea, field researchers noted erosion rates of 80 millimeters *[3.15 inches]* per *1,000 years* near sea level and 520 millimeters *[20 inches]* per 1,000 years at an altitude of 975 meters *[3,199 feet]*.[62] Investigators report rates of 920 millimeters *[36 inches]* per 1,000 years for mountains along the Guatemala-Mexico border,[63] while in the Himalayas, rates of 1,000 millimeters *[39 inches]* per 1,000 years are noted.[64] In the Mount Rainier region of Washington rates can reach up to 8,000 millimeters *[315 inches or 26 feet]* per 1,000 years.[65] Probably the fastest recorded regional rate is 19,000 millimeters *[748 inches or 63 feet]* per 1,000 years from a volcano in Papua, New Guinea.[66] The above-listed current rates of geologic change obviously challenge the validity of the standard geologic timescale, because, at present rates of erosion, the continents and their fossils should have been eroded away many times during the long ages postulated for the geologic past. How many times? *Continents* average 623 meters *[2,044 feet] above sea level. At an average rate of only 1 millimeter per 1,000 years* (unless they were higher in the past), *they would have been eroded to sea level in 623 million years.*[67]

Using an estimated average erosion rate of 61 millimeters *[2.4 inches]* per 1,000 years,[68] a number of geologists point out that North America could be leveled in "a mere 10 million years."[69] In other words, at present rates of erosion, the North American continent would have been eroded away about 250 times in 2.5 billion years.[70] Correcting for the erosional effects of agriculture, etc., it seems that our continents could have been eroded away at least 100 times during the estimated billions of years of their existence.

I was amazed to see similar scientific facts (because these facts contradict the evolutionary theory) reported in the *Encyclopædia Britannica*.

> Peter J. Wyllie: The force of gravity and the rivers together carry the products of weathering downhill to the ocean reservoir. The average rate at which the surface of the land is being worn down in the land dispersed into the oceans is a trivial 1.5 inches per 1,000 years, but the dimension of geological time gives significance to small numbers. At this rate, all the continents would be worn down to sea level within 20,000,000 years. This means that during the 4,600,000,000 years since the earth was formed, the continents would have been worn down to sea level at least 200 times. By now there should be no land rising above sea level, but we still see high mountains.
>
> The mountains exist and persist because the effects of the hydrologic cycle are offset by the mountain-building cycle. Forces within the earth cause large regions of the surface to rise very slowly, imperceptibly in human terms.[71]

---

[60] Harold G. Coffin, PhD, "Summary—Is Creation a Viable Theory of Origins?" *Origin By Design*, Review and Herald Publishing Association, Washington, DC; Hagerstown, MD, 1983, pp. 422-423.
[61] Ariel A. Roth, PhD, "Some Geologic Questions About Geologic Time," *Origins: Linking Science and Scripture*, Review and Herald Publishing Association, Hagerstown, MD, 1998, pp. 262-264; F. Ahnert, "Functional relationships between denudation, relief, and uplift in large mid-lateral drainage basins," *American Journal of Science*, 1970, 268:243-263.
[62] B.P. Ruxton, I. McDougall, "Denudation rates in northeast Papua from potassium-argon dating of lavas," *American Journal of Science*, 1967, 265:545-561.
[63] J. Corbel, "Vitesse de L'erosion," *Zeitschrift für Geomorphologie*, 1959, 3:1-28.
[64] H.W. Menard, "Some rates of regional erosion," *Journal of Geology*, 1961, 69:154-161.
[65] H.H. Mills, "Estimated erosion rates on Mount Rainier, Washington," *Geology*, 1976, 4:401-406.
[66] C.D. Ollier, M.J.F. Bown, "Erosion of a young volcano in New Guinea," *Zeitschrift für Geomorphologie*, 1971, 15:12-28.
[67] Ariel A. Roth, PhD, "Some geologic questions about geologic time," *Origins: Linking Science and Scripture*, Review and Herald Publishing Association, Hagerstown, MD, 1998, pp. 262-274.
[68] S. Judson, D.F. Ritter, "Rates of regional denudation in the United States," *Journal of Geophysical Research*, 1964, 69:3395-3401.
[69] R.H. Dott Jr, R.L. Batten, *Evolution of the Earth*, 4th ed. New York, St. Louis, and San Francisco: McGraw-Hill Book Co., 1988, p. 155.
[70] Ariel A. Roth, PhD, "Some geologic questions about geologic time," *Origins: Linking Science and Scripture*, Review and Herald Publishing Association, Hagerstown, MD, 1998, p. 263.
[71] Peter J. Wyllie, "The Great Globe Itself," *Propaedia: Outline of Knowledge, Guide to the Britannica, Encyclopaedia Britannica*, Ency-

Here you can see that the "scientific" answer as to why the continents and mountains have not eroded away is mountain building. But this answer fails to acknowledge the scientific fact that mountain building comes from the bottom up. And that *didn't replace* the geologic column and its fossils, which should have eroded from the top down. The column is *still there*.

Thus, the existence of the scarcely eroded geologic column, with all of its fossils, is positive evidence for the age of life on earth being thousands, not millions, of years old.

Megaevolution

Harold G. Coffin, PhD: The theory of evolution requires gradual change from simple to complex, from one kind of organism to another. Here again, humanity's total experience as grower of crops, breeder of animals, as reproducing organisms themselves, contradicts such continuous transformation.

Organisms do change, as witnessed by the many modern breeds of dogs, the new varieties of roses that appear regularly, DDT-resistant strains of flies, and strains of bacteria impervious to antibiotics [microevolution]. But to insist that such minor changes accumulate into major differences across the vast boundaries between basic kinds of organisms [megaevolution] is an extrapolation not based on empirical evidence and contrary to another basic law of life—that offspring resemble their parents. The dogs are still dogs, the roses are still roses, the resistant flies are very much pestiferous flies, and unfortunately the bacteria still produce the same diseases despite the presence of antibiotics.[72]

The theory of evolution stands only by contradicting two of the most fundamental laws of life that we know—that life begets life and like begets like.[73]

And there is no compelling evidence to document macroevolution over time.

Richard B. Goldschmidt, MD, PhD: Microevolution does not lead beyond the confines of the species. . . . Microevolution by accumulation of micro-mutations . . . leads to diversification strictly within the species. . . . The decisive step in evolution, the first step toward macroevolution, the step from one species to another, requires another evolutionary method than that of sheer accumulation of micro-mutations.[74]

Frank Lewis Marsh, PhD: Two very influential books in recent years have been the beautifully colored Life Nature Library volume, *Evolution*, by Ruth Moore and the Editors of Life, and the even more beautifully colored and produced volume *Atlas of Evolution*, by Sir Gavin de Beer. The impressive demonstrable evidence which fills these volumes is micro-evolution only![75]

Leonard R. Brand, PhD: If such evidence exists, we might expect a good evolution textbook to provide it. Two of the most well-respected evolution textbooks—Douglas J. Futuyma's *Evolution* and Mark Ridley's *Evolution*—provide abundant evidence supporting the reality of microevolution and speciation . . . [But they] do not provide any convincing evidence for a genetic process of megaevolution. This lack of evidence in evolution texts is significant. It suggests that evolutionary science merely assumes the existence of a genetic process that can evolve new structures or gene complexes, but that there is no convincing evidence of such a process. The evidence they use depends on the assumption of naturalism.[76]

If evolution were true we should find a mechanism for change:

Bernard Brandstater, MBBS: Dr. Jerry Fodor, Rutgers University philosopher of cognitive sciences and an atheist, published [in 2008] an extraordinary paper, "Why Pigs Don't Fly," that raised a storm amongst evolutionists. In it Fodor **demolished Darwinian natural selection as an effective basis for species evolution.** The commenting blogs in the science community went wild. In February 2010 Fodor co-authored a book, *What Darwin Got Wrong*. It is described by a reviewer

---

clopaedia Britannica, Inc., William Benton, Chicago, IL, 1979, p. 77.
[72]Harold G. Coffin, PhD, with Robert H. Brown, PhD, and R. James Gibson, PhD: "Summary—Is Creation a Viable Theory of Origins?" *Origin By Design,* Review and Herald Publishing Association, Hagerstown, MD, 2005, p. 431.
[73]Harold G. Coffin, PhD, "Summary—Is Creation a Viable Theory of Origins?" *Origin By Design,* Review and Herald Publishing Association, Washington, DC; 1983, p. 423.

[74]Richard B. Goldschmidt, MD, PhD, *The Material Basis of Evolution,* Pageant Books, Inc., Paterson, NJ, 1960, p. 183.
[75]Frank L. Marsh, "The Form and Structure of Living Things," *Creation Research Society Quarterly,* June 1969, p. 21.
[76]Leonard R. Brand, PhD, "Modern biology and the challenge to Darwinism," *Beginnings: Are Science and Scripture Partners in the Search for Origins?* Pacific Press Publishing Association, Nampa, ID, 2006, pp. 83-84, (emphasis added); Dr. Brand can also be found in journals such as *Nature, PLOS,* and *Geology.*

This once horizontal, water-laid strata, at an elevation of about 5,000 feet, is in Southern California, on top of the Yucaipa Ridge on the southern edge of Mill Creek Canyon. Its existence, at this elevation, where science says erosion rates are faster, illustrates the fact that the continents haven't been eroded even once.

as, "A new book that dares to attack the theory of natural selection by using . . . science." Fodor has an immense reputation.[77]

His degrees were from Columbia University, Princeton, and Oxford. He was a "longtime faculty member of Rutgers" and was described as "the world's leading philosopher of mind."[78]

> This celebrated academic declares that natural selection will not do what Darwin claimed.
> . . . And natural selection is the essential core of Darwin's theory, the only "discovery" for which he is acclaimed. . . .
>
> In 2008, sixteen of the world's top evolutionary scientists convened privately behind closed doors in a castle in Altenburg, Austria. Jerry Fodor was one of them. Preliminary reports from evolutionary journalist Susan Mazur point to a widely shared dissatisfaction with the inadequacies of the Neo-Darwinian synthesis that science has relied on for decades. Previously that synthesis was triumphant, unchallengeable; but in the light of new science it is now tottering. If mechanisms for speciation remain a mystery, too complex to be explained, and if natural selection is rejected, Darwin's whole theory is dead. But what can replace it? There's nothing except God.[79]

The age of the earth

How old is the earth? In 1770, George Buffon said the earth was 70,000 years old.[80] By 1905, the official age of the earth was said to be 2 billion years old.[81] By 1927, British geologist Arthur Holmes[82] determined that the earth was 3 billion years old. By 1969, the age of the earth and moon had increased to 3.5 billion years.[83] Today the earth is thought to be 4.6 billion years old.[84]

---

[77]Bernard Brandstater, MBBS, "The Centrality of Creation in Adventist Belief," personal paper presented to Adventist Theological Society Spring Symposium, Southwestern Adventist University, Keene, TX, April 17, 2010, pp. 5-6, (emphasis in the original).
[78]ruccs.rutgers.edu/jerry; https://www.nytimes.com/2017/11/30
[79]Bernard Brandstater, MBBS, "The Centrality of Creation in Adventist Belief," personal paper presented to Adventist Theological Society Spring Symposium, Southwestern Adventist University, Keene, TX, April 17, 2010, pp. 5-6.
[80]"Old Earth, Ancient Life: Georges-Louis Leclerc, Comte de Buffon," evolution.berkeley.edu
[81]Sharon Begley, "Science finds God," Newsweek, July 20, 1998, p. 50.
[82]Arthur Holmes, Wikipedia
[83]The Minneapolis Tribune, August 25, 1969.
[84]Henry M. Morris, PhD, "The Evolutionary Basis of Modern Thought," The Long War Against God, Master Books, Inc., Green

Since 1770, the earth has "aged" more than 19 million years per year.

In a process called *desertification*, the Sahara Desert, the largest desert on the planet, is estimated to be about 4,000 years old. What is believed to be the oldest living thing on the planet is a Bristlecone Pine in the Owens Valley of California. It is said to be 4,300 years old. Question: If the earth is billions of years old, why does the Sahara Desert, and the oldest living thing, point to a beginning near the time of Noah's Flood?

According to *National Geographic*, the oldest writing systems in the world started about 3,200 BC.[85] The year 2000 was 4700 on the Chinese calendar. Our year 2000 was 5760 on the Hebrew calendar. These easily verified facts lead to another profound question: If life on earth is billions of years old, why are the records which predate the theory of evolution less than 6,000 years old?

Other vital considerations

In addition to erosion rates indicating a young earth, even carbon-14 points to life on earth being thousands of years old, not millions.

Genesis 1:2 states that in the beginning "the earth was without form, and void." That may mean that the earth was without form and void, *but was* (i.e., that it existed for possibly eons—the "old earth" model). Different interpretations are possible.

But, as you will see later in greater detail, Willard F. Libby, PhD, the evolutionary scientist who developed carbon-14 dating methods, determined that if a new earth were to be created, it would take about 30,000 years for carbon-14 to reach equilibrium; where the amount entering from the atmosphere matched the decay rate.[86]

In 1947, Libby acknowledged that carbon-14 on earth had not yet reached equilibrium. He attributed this problem to his margin of error.

But, the figures are even worse than Libby acknowledged:

American Antiquity: Radiocarbon is forming 28-37 percent faster than it is decaying.[87]

Conclusion: If carbon-14 is still forming that much faster than it is decaying, according to Libby's estimate, the earth's atmosphere, and thus life on earth, is significantly less than 30,000 years old.

I will end this chapter with a quotation which summarizes the focus of this book:

David C. Read, JD: Why do *humans* have art, architecture, literature, theater, music, fashion, religion, commerce, law, politics, history, paleontology, science, technology, medicine, engineering, manufacturing, agriculture, animal husbandry, sports, cuisine, formal education, marriage ceremonies, funerals, charity benefits, and all the other innumerable things that separate humans from animals? Creationists have an explanation. We believe that we were created in the image of God to commune with God and to be conscious and self-conscious creatures with free will. The animals were not created for this purpose. That is why there is such an enormous gap between humans and animals.[88]

Forest, AR, 2000, p. 25.
[85]Joel Swerolow, "The Power of Writing," *National Geographic*, August 1999, p. 110.
[86]W.F. Libby, Radiocarbon Dating, University of Chicago Press, Chicago, IL, 1955, p. 7.
[87]R.E. Taylor, et al, "Major Revisions in the Pleistocene Age Assignments for North American Human Skeletons by C-14 Accelerator Mass Spectrometry," *American Antiquity*, Vol. 50. No. 1, 1985, pp. 136-140.
[88]David C. Read, JD, *Dinosaurs*, Clarion Call Books, Keene, TX, 2009, pp. 367-377, (emphasis added).

Chapter 4
Different Kinds of Evolution

Here are four different concepts that relate to evolution:
1. The origin of higher elements from hydrogen and helium (sometimes used to describe the origin of the first life from chemicals).
2. The origin of life from non-living material (spontaneous generation).
3. An organism changing from one major kind into another. (None of these first three, the basis of the evolutionary theory, have ever been observed.)
4. Minor changes (or variation) within biblical "kinds" (the only form of evolution that has actually been observed). These minor changes are "microevolution," like changes from one variety of bird beak to another. But not the changes from one major kind of plant or animal to another, or macroevolution.

Micro- versus macroevolution

Most creationists believe in microevolution. The evidences evolutionists present for evolution are mostly evidences for microevolution. They assert that macroevolution is just microevolution over longer periods of time. And that is an assumption; a major leap in logic. School children are being taught to believe that microevolution demonstrates that all evolution is true.

> *The World of Biology:* Based on current knowledge, it seems likely that, given enough time, the mechanisms of microevolution lead to speciation and macroevolution.[89]

That is philosophy, not observable science. Here you will see some mainstream scientists question this general theory of evolution:

> *Science:* The central question of the Chicago conference was whether the mechanisms underlying microevolution can be extrapolated to explain the phenomena of macroevolution. At the risk of doing violence to the positions of some of the people at the meeting, the answer can be given as a clear, "No."[90]

From the date on the following quotations (1903 and 1940) and others, one can see that this problem has been recognized for a very long time.

> Richard B. Goldschmidt, MD, PhD: We started this chapter with the conviction, gained from an unbiased analysis of all pertinent facts, that microevolution by means of micromutation leads only to diversification within the species, and that the large step from species to species is neither demonstrated nor conceivable on the basis of accumulated micromutations.[91]

Similarity does not mean evolution

I will use the automobile as a metaphor to make a point about macroevolution—the concept that everything descended from a common ancestor because they have similar structures.

Henry Ford's first Model T had four wheels, a steering mechanism, and an engine. Over more than 100 years, the Ford, by intelligent design, has "evolved" into an automobile that still has four wheels, a steering mechanism, and an engine. Cadillac also developed a car with four wheels, a steering mechanism, and an engine. It, too, has evolved into a modern automobile with the same components.

Actually, the Model T Ford evolved from earlier Fords. And they might be seen to have evolved and improved from a stage coach and an earlier ox cart.

Are Fords and Cadillacs related? They have the same components. No, man's intelligence determined that, in order to be efficient and reliable, all "kinds" of the "fittest" cars needed

---

[89] P. William Davis, Eldra Pearl Solomon, Linda R. Berg, "Speciation and Macroevolution," *The World of Biology*, Saunders College Publishing, Philadelphia, PA, 1990, p. 311, (emphasis added).

[90] Roger Lewin, "Evolution theory under fire," *Science*, November 21, 1980, p. 883, see also: Jerry Adler, John Carey, "Is Man a Subtle Accident?" *Newsweek*, November 3, 1980, p. 95.

[91] Richard B. Goldschmidt, MD, PhD, "Macroevolution," *The Material Basis of Evolution*, Yale University Press, New Haven and London, 1940, 1982, p. 199.

to have four wheels, a steering mechanism, and an engine. Also, man determined that wheels, a steering mechanism, and an engine would be worthless by themselves, if they were not connected to something. So, by intelligent design, man brought these components together, simultaneously, so that they all could work together for a common purpose.

Likewise, an Intelligent Designer—God—in his infinite wisdom, decided that all mammals needed to have a circulatory system, a central nervous system, a skeletal system, a digestive system, lymphatic system, endocrine system, limbic system, reproductive system, and in addition to these systems, various similar organs. He created all of the integrated, complementary components of the mammal's body, which would have no survival capability if they somehow each evolved separately from simple organisms to complex.

We cannot conclude that various forms of animals, from simple to complex, evolved from one to the other any more than we can claim that an ox cart, stage coach, and Model T Ford eventually evolved into a Cadillac Escalade. They are unrelated, yet they have similar components which all work together in order to accomplish their purpose.

Intelligent design

The Moody Institute of Science determined that the red blood cell is an excellent example of intelligent design.

> Dr. Irwin A. Moon: Have you ever wondered about the peculiar shape of the red blood cell? Someone has described it as a cross between a donut and a pancake. Now if you or I had been designing the red blood cell, we probably would have made it spherical, for a sphere is the simplest of all compact shapes, and in many ways the most efficient. It has much greater volume than the bi-concave disk of the red blood cell. It's the strongest of all shapes. And it's the shape that would pass through the intricate maze of the blood vessels with greatest ease. But it has one fatal weakness. It wouldn't work.[92]

The Moody Institute of Science made models of several possible shapes of the red blood cell, cut them in half in order to observe their cross section in a colored liquid, and observed the rate at which the liquid was absorbed. The absorption rate with the sphere was rapid at first. But near the center the process "slowed to a snail's pace." It became obvious that a spherical cell would absorb oxygen much too slowly.[93]

Of course, the most obvious solution would be to flatten the sphere into a disk. And in actual tests it became clear that the disk did solve the problem of rapid absorption. The disk had adequate speed, but not enough volume.

The ideal red-blood-cell shape would be one that would combine volume, speed of absorption, and durability.

Starting with the laws of gas infusion, and then applying the principles of advanced calculus over a period of several weeks, the Moody Bible Institute's staff scientifically and mathematically developed a formula that would provide maximum volume with maximum speed in absorbing gasses. Their formula allowed for variables of the laws of gas infusion, volume, surface area, and time.

For more complete and reliable results, they submitted their formula for an ideal red blood cell to the applied science department of the International Business Machines Corporation. Dr. Edgar Smith, an IBM mathematician, programmed the Moody formula into one of the giant IBM research computers. What was the result? On a computer terminal screen appeared the shape of a red blood cell.

> Dr. Moon: The fact that the red blood cell turns out to be the one, perfect, ideal shape demands an explanation. And to me the only adequate explanation is intelligent design.[94]

---

[92]Dr. Irwin A. Moon, *Red River of Life*, Moody Science Classics, A Moody Institute of Science film presentation, Reviewed by the American Scientific Affiliation, 1957, 1968, 1998.
[93]Dr. Irwin A. Moon, *Red River of Life*, Moody Science Classics, A Moody Institute of Science film presentation, Reviewed by the American Scientific Affiliation, 1957, 1968, 1998.
[94]Dr. Irwin A. Moon, *Red River of Life*, Moody Science Classics, A Moody Institute of Science film presentation, Chicago, IL, Reviewed by the American Scientific Affiliation, 1957, 1968, 1998.

Chapter 5
Mutations

Darwinism was not looking good because evolution theorists had not yet come up with some kind of mechanism that would explain how Darwin's theory of evolution could really work. Scientists have since acknowledged that mutations are not that mechanism and do not contribute to macroevolution. So how did the problem evolve?

> David C. Read, JD: Around [the year] 1900, many scientists were beginning to doubt Darwin's theory of evolution. Especially among geneticists, the opinion was widespread that Darwinism was dead.[95] In order to save the theory someone had to find a way that the gene pool could be deepened.
>
> A Dutch botanist named Hugo de Vries (1848-1935) was working on the problem, not to rescue Darwin's theory, but to replace it with a theory of evolution by large spontaneous genetic changes. [He discovered] a change he termed "mutation." A mutation is a new genetic characteristic or trait that does not come from either of the parents, but it appears suddenly, in a manner that does not follow the rules of Mendelian genetics. . . .
>
> The modern theory of evolution, known as the "Neo-Darwinian synthesis," rests upon the foundation of inheritable genetic mutations.[96]

Is there scientific evidence to support mutations as a cause of macroevolution? Or does scientific opinion continue to undermine and challenge the concept?

Sir Ernst B. Chain, PhD, Biochemist and Nobel Prize winner: "To postulate that the development and survival of the fittest is entirely a consequence of chance mutations, or even that nature carries out experiments by trial and error through mutations in order to create living systems better fitted to survive, seems to me a hypothesis based on no evidence and irreconcilable with the facts."

British author Francis Hitching: On the face of it, then, the prime function of the genetic system would seem to be to resist change: to perpetuate the species in a minimally adapted form in response to altered conditions, and if at all possible to get things back to normal. The role of natural selection is usually a negative one: to destroy the few mutant individuals that threaten the stability of the species.[97]

So, can mutations, which have been one of the supporting pillars of evolution be relied upon?

> Ernst Mayr, PhD: We now believe that mutations do not guide evolution; the effect of a mutation is very often far too small to be visible.[98]
>
> Arthur Koestler: You cannot have a mutation A occurring alone, preserve it by natural selection, and then wait a few thousand or million years until mutation B joins it, and so on, to C and D. Each mutation occurring alone would be wiped out before it could be combined with the others. They are all interdependent. The doctrine that their coming together was due to a series of blind coincidences is an affront not only to common sense but to the basic principles of scientific explanation.[99]
>
> I. L. Cohen, after a study of probabilities: To propose and argue that mutations even in tandem with "natural selection" are the root-causes for 6,000,000 viable, enormously complex species, is to mock logic, deny the weight of evidence, and reject the fundamentals of mathematical probability.[100]
>
> Gary E. Parker, EdD: *Mutations point back to creation.* . . . After all, mutations are *only changes in already existing genes.* . . . The gene has to be there before it can mutate. All you get as a result of mutation is just a varied form (allele) of an already existing gene, i.e., variation within type.[101]

---

[95]Ernst W. Mayr, PhD, *The Growth of Biological Thought*, Harvard University Press, Cambridge, MA, 1982, 1985, p. 547.
[96]Hugo de Vries, *Die Mutations-Theorie*, 1901-1903.
[97]David C. Read, JD, "The Copying Error Theory of Evolution," *Dinosaurs*, Clarion Call Books, Keene, TX, 2009, pp. 246-247.
[98]Ernst Mayr, PhD, *Animal Species and Evolution*, Harvard University Press, Cambridge, MA, 1963, p. 7.
[99]Arthur Koestler, *The Ghost in the Machine*, Hutchinson, London, 1967, p. 129.
[100]I. L. Cohen, *Darwin Was Wrong: A Study in Probabilities*, New Research Publications, Inc., Greenvale, NY, 1984, p. 81.
[101]Gary E. Parker, EdD, "Darwin and the Nature of Biologic Change," *What is Creation Science?* Master Books, El Cajon, CA,

There are three recognized kinds of mutations: detrimental, neutral, and beneficial. Detrimental and neutral kinds of mutations are recognized by everybody as unhelpful to the theory of evolution. Are there any known beneficial mutations that have produced new types of organisms or increased their complexity? Have any mutations added information and caused a morphological change in the creature? What do the experts say?

> Respected evolutionist Pierre-Paul Grassé, MD, a French zoologist, author of over 300 publications, including the influential 35-volume *Traité de zoologie*, once president of the French Academy of Sciences: No matter how numerous they may be, mutations do not produce any kind of evolution.[102]

> Lee Spetner, PhD: But in all the reading I've done in the life-sciences literature, I've never found a mutation that added information. . . . All point mutations that have been studied on the molecular level turn out to reduce the genetic information and not increase it.[103]

> Maxim D. Frank-Kamenetskii, PhD in molecular biology (professor at Brown University Center for Advanced Biotechnology and Biomedical Engineering): Mutations are rare phenomena, and a simultaneous change of even two amino acid residues in one protein is totally unlikely. . . . One could think, for instance, that by constantly changing amino acids one by one, it will eventually be possible to change the entire sequence substantially. . . .
>
> These minor changes, however, are bound to result eventually in a situation in which the enzyme has ceased to perform its previous function but has not yet begun its "new duties." It is at this point it will be destroyed—together with the organism carrying it.[104]

> Ray Bohlin, PhD in molecular and cell biology: We see the apparent inability of mutations truly to contribute to the origin of new structures. The theory of gene duplication in its present form is unable to account for the origin of new genetic information—a must for any theory of evolutionary mechanism.[105]

Science vs. Evolution (2016): Hundreds of thousands of mutation experiments have been done, in a determined effort to prove the possibility of evolution by mutation. And this is what they learned: **NOT ONCE has there ever been a recorded instance of a truly beneficial mutation** (one which is a known mutation, and not merely a reshuffling of latent characteristics in the genes) **nor such a mutation that was permanent, passing on from one generation to another!**[106]

Drug resistant bacteria—evidence for evolution?

To advance their view, evolutionists have long pointed to mutations which are thought to have beneficial effects. The most common example given: mutations sometimes make bacteria resistant to germ-killing drugs called antibiotics. The conclusion then is that if mutations can make bacteria stronger, they must be able to do the same for other creatures.

This conclusion is based on a misunderstanding: that mutations cause antibiotic resistance. No, they cause a loss of genetic information:

> Lee Spetner, PhD: For example, to destroy a bacterium, the antibiotic streptomycin attaches to a part of the bacterial cell called ribosomes. Mutations sometimes cause a structural deformity in ribosomes. Since the antibiotic cannot connect with the misshapen ribosome, the bacterium is resistant. But even though this mutation turns out to be beneficial [for the moment], it still constitutes a loss of genetic information, not a gain. No "evolution" has taken place; the bacteria are not "stronger." In fact, under normal conditions, with no antibiotic present, they are weaker than their non-mutated cousins.[107]

If you were born without feet, you will not get

1987, p. 108, (emphasis in the original).
[102]Pierre-Paul Grassé, MD, *Evolution of Living Organisms*, Academic Press, New York, NY, 1977, pp. 88-103.
[103]Lee Spetner, PhD, *Not By Chance: Shattering the Modern Theory of Evolution*, Judaica Press, Brooklyn, NY, 1997, pp. 131, 138.
[104]Maxim D. Frank-Kamenetskii, PhD, "Where do Genes Come From?" *Unraveling DNA*, VCH Publishers, Inc., New York, NY, 1993, p. 76.
[105]Ray Bohlin, PhD, "The Natural Limits to Biological Change," *Creation, Evolution, and Modern Science*, Kregel Publications, Grand Rapids, MI, 2000, p. 41.
[106]Mutations," Chapter 10, Why Mutations cannot Produce Cross-species Change, *Science vs. Evolution*, pathlights.com, March 6, 2016, (emphasis in the original).
[107]James Perloff, "Evidence Against the Theory of Evolution," *The Case Against Darwin: Why the Evidence Should Be Examined*, Refuge Books, Burlington, Massachusetts, 2002, p. 24. See also: Lee Spetner, PhD, "Can Random Variation Build Information?" *Not By Chance: Shattering the Modern Theory of Evolution*, Judaica Press, Brooklyn, NY, 1997, p. 141. See also: Daniel Criswell, "The 'Evolution' of Antibiotic Resistance," Impact #378, Institute for Creation Research, El Cajon, CA, December 2004.

athlete's foot. Is that a beneficial mutation to not have feet? In all examples of bacteria becoming resistant to drugs, there is a loss of genetic information, not a gain.

> David C. Read, JD: Moreover, resistance to insecticides and antibiotics, and the sickle cell mutation, are obviously mutations of the type that do not change the physical appearance of the organism. But the theory of evolution depends upon mutations that change the shape, or morphology, of the organism. Evolution posits innumerable instances of changes in the shape of the bones and overall form of animals, as they evolved from one species into another. Trillions of mutations changing some aspect of animal morphology must have occurred if life on this planet evolved as Darwin believed.[108]

Mutations are the only known means by which new genetic material could possibly become available for evolution.[109] No known mutation has ever produced a form of life having greater complexity and viability than its ancestors.[110]

> Henry M. Morris, PhD: Mutations take place, but they are either reversible, deteriorative, or neutral. . . . If one must depend on mutations and natural selection to produce new species—let alone, the families, orders and phyla as evolutionists assume—then not even billions of years would suffice.[111]

> Well-known evolutionist Stephen Jay Gould, PhD: You don't make new species by mutating the species. . . . A mutation is not the cause of evolutionary change.[112]

> David C. Read, JD: Accumulated degenerative mutations can never lead to evolution, as it is commonly understood to mean progress toward more complex forms of life. For that type of evolution to occur, there must be many mutations that are not accidentally beneficial within the context of degeneration, but that are always and merely beneficial, but not degenerative.[113]

In discussing the many mutations needed to produce a new organ, a number of different scientists give their perspective:

> Ariel A. Roth, PhD: One of the most severe challenges the evolutionary model faces is its inadequacy to explain how complex organs and organisms with interdependent parts ever evolved. The basic problem is that random mutations cannot plan ahead to gradually design intricate systems, and the appearance of a multitude of the right kind of mutations, all emerging at the same time, to produce a new organ is implausible. If you are going to produce such complex things gradually, the very process of natural selection by survival of the fittest proposed by Darwin would tend to prevent their evolution. Until all the necessary parts of a complex system have assembled so that the system can actually work [later in this book identified as "irreducible complexity"], it has no survival value. Before that, the functionless extra parts of an incomplete developing system are useless—a cumbersome impediment. One would expect natural selection to get rid of them. As an example, what survival value would a newly evolving skeletal muscle have without a nerve to stimulate it to contract, and what purpose would a nerve have without an intricate control mechanism to provide the necessary stimulus?

> Biologist L. James Gibson, PhD: For a species to become more complex, it requires more than simply a mutation in a gene; it requires new genes. But simply adding a new gene would not work. Genes do not work in isolation. Rather, an organism's set of genes work together to produce the organism. A new gene must work properly with all the other genes in order for the organism to survive.

> Furthermore, several new genes would be required in order to produce a new structure and a more complex organism. Each new gene would require a control (regulatory) gene. In addition, each new gene would have to operate at the right time in development for the new structure to develop

---

[108] David C. Read, JD., "The Copying Error Theory of Evolution," *Dinosaurs*, Clarion Call Books, Keene, TX, 76059, 2009, p. 256, (emphasis in the original); Richard Milton, "Of Cabbages and Kings," *Shattering the Myths of Darwinism*, Park Street Press, Rochester, VT, 1997, p. 157.
[109] Ernst W. Mayr, PhD, "Evolutionary Challenges to the Mathematical Interpretation of Evolution," *Mathematical Challenges to the Neo-Darwinian Interpretation of Evolution*, proceedings of a symposium held at the Wistar Institute of Anatomy and Biology, April 25 and 26, 1966, the Wistar Institute Press, 1967, p. 50.
[110] Nils Heribert-Nilsson, PhD, *Synthetische Artbildung*, Lund, Sweden: Verlag CWK Gleerup, 1956; reprint edition, p. 1157.
[111] Henry M. Morris, PhD, "What they Say," *Back to Genesis*, March 1999.
[112] Stephen Jay Gould, Speech at Hobart College, February 14, 1980, cited by Luther Sunderland, *Darwin's Enigma*, El Cajon, CA, Master Books, 1984, p. 106, cited by Bert Thompson and Brad Harrub in "*National Geographic* Shoots Itself in the Foot Again," ApologeticsPress.Org online report, 2004, p. 36.

[113] David C. Read, JD, "The Copying Error Theory of Evolution," *Dinosaurs*, Clarion Call Books, Keene, TX, 76059, 2009, p. 256.

correctly. It does not seem reasonable to expect even one new gene to appear by chance, much less several highly coordinated genes working together to produce a new structure.[114]

Arthur Koestler: Each mutation occurring alone would be wiped out before it could be combined with the others. They are all interdependent. The doctrine that their coming together was due to a series of blind coincidences is an affront not only to common sense but to the basic principles of scientific explanation.[115]

Evolutionist George Gaylord Simpson, PhD: Simultaneous appearance of several gene mutations in one individual has never been observed, as far as I know, and any theoretical assertion that this is an important factor in evolution can be dismissed. . . . The probability that five simultaneous mutations would occur in any one individual would be about .0000000000000000000001. This means that if the population averaged 100,000,000 individuals with an average length of generation of only one day, such an event could be expected only once in about 274,000,000,000 years—a period of about one hundred times as long as the age of the earth.[116]

Richard Milner: Alfred Russell Wallace and Charles Darwin had insisted that through gradual, continuous change, species could (in Wallace's phrase) "depart indefinitely from the original type." Around [the year] 1900 came the first direct test of that proposition: the "pure line of research" of Wilhelm Ludwig Johannsen (1857-1927). What would happen, Johannsen wondered, if the largest numbers of a population were always bred with the largest, the smallest with the smallest? How big or how small would they continue to get after a few generations? Would they "depart indefinitely" from the original type, or are there built-in limits and constraints?

Experimenting on self-fertilizing beans, Johannsen selected and bred the extremes in sizes over several generations. But instead of a steady, continuous growth or shrinkage as Darwin's theory seemed to predict, he produced two stabilized populations (or "pure lines") of large and small beans. After a few generations, they had reached a specific size and remained there, unable to vary further in either direction. Continued selection had no effect.

Johannsen's work stimulated many others to conduct similar experiments [a vital concept in science]. One of the earliest was Herbert Spencer Jennings (1868-1947) of the Museum of Comparative Zoology at Harvard, the world authority on the behavior of microscopic organisms. He selected for body size in Paramecium and found that after a few generations selection had no effect. . . . Even after hundreds of generations, his pure lines remained constrained within fixed limits, "as unyielding as iron."[117]

Behold the common fruit fly

BBC Television: It is a striking, but not much mentioned fact that, though geneticists have been breeding fruit-flies for [80] years or more in labs around the world—flies which produce a new generation every eleven days—they have never yet seen the emergence of a new species or even a new enzyme.[118]

Pierre-Paul Grassé, MD: The fruit fly *(Drosophila melanogaster)* ["black-bellied lover of dew"], a favorite pet insect of the geneticists, whose geographical, biotopical, urban, and rural genotypes are now known inside out, seems not to have changed since the remotest times.[119]

Clyde L. Webster Jr., PhD: Over the past several decades this species has been the workhorse of the geneticist. Fruit flies have been exposed to essentially every known mutagen in existence, be it radiological, chemical, or biological. And yet this species still looks like a fruit fly! In addition, thousands of generations of fruit flies have been bred and examined for new gene combinations. . . . Nothing new has been created; only new genetic combinations have been expressed![120]

Maurice Caullery: Out of the 400 mutations that

---

[114]James L. Gibson, PhD, written communication, to Clyde L. Webster, Jr., PhD, "Evolutionary Processes," *A Scientist's Perspective on Creation and the Flood,* Geoscience Research Institute, Loma Linda, CA, 1995, pp. 12-13.
[115]Arthur Koestler, "Evolution: Theme and Variations," *The Ghost in the Machine,* The Macmillan Company, NY, 1968, p. 129.
[116]George Gaylord Simpson, PhD, *Tempo and Mode in Evolution,* Columbia University Press, NY, 1944, pp. 54-55, (emphasis added).
[117]Richard Milner, "Pure Line Research, Limits of Variability," *Encyclopedia of Evolution,* Facts on File, New York, NY, 1990, p. 376.
[118]Gordon Rattray Taylor, former Chief Science Adviser, BBC television, "The Problem of the Giraffe," *The Great Evolution Mystery,* Harper & Row Publishers, NY, 1983, p. 48.
[119]Pierre-Paul Grassé, MD, *Evolution of Living Organisms,* Academic Press, New York, NY, 1977, p. 130.
[120]Clyde L. Webster, Jr., PhD, "Evolutionary Processes," *A Scientist's Perspective on Creation and the Flood,* Geoscience Research Institute, Loma Linda, CA, 1995, p. 11.

have been provided by *Drosophila melanogaster*, there is not one that can be called a new species. It does not seem, therefore, that the central problem of evolution can be solved by mutations.[121]

Richard B. Goldschmidt, MD, PhD: In the best-known organisms, like Drosophila, innumerable mutants are known. If we were able to combine a thousand or more of such mutants in a single individual, this still would have no resemblance whatsoever to any type known as a species in nature.[122]

And what about the *Escherichia coli* microbe?

Warren L. Johns, JD: Modifications have been observed in 44,000 *E. coli* bacteria generations in the course of a twenty-year Michigan State University lab experiment launched in 1988. . . . Putting *E. coli* bacteria through its laboratory paces, 44,000 generations after the fact, descendants continue reproducing thousands of generations of unmistakable *E. coli*. Survival of modified *E. coli* bacteria represents the opposite of evolutionary transition to some new and different genome.[123]

The *ability* of a genome to adjust assures *variety*. But this intragenomic change cannot be extrapolated as evidence of evolution. Finches can modify beaks; bacteria develop immunity; and English tree moths can change from gray to white and back to gray. . . . But bottom line: after these adjustments (possibly thanks to pre-existing genes in a genome) the finches, bacteria, and moths remain finches, bacteria, and moths.[124]

What Darwin observed is called "microevolution." Again, a better word would be "variation." Dogs produce a large variety of dogs that have short and long fur, short and long legs, straight and floppy ears. But they are all dogs. Farmers today produce a variety of corn. But it is all corn. They can try as much as they want to get something else, but all they get is a variety of corn. Darwin saw finches with a variety of beaks. They not only were all birds, they also were all finches.

Warren L. Johns, JD: According to the Westminster Kennel Club, the estimated four-hundred species of dogs claim descent from common canine ancestry. Courtesy of selective gene mixing, collies and poodles look different, but continue as dogs, man's favorite animal companion. . . . All dog breeds, recognized and registered by the American Kennel Club, belong to the same *Canis familiaris* species, fully capable of interbreeding to produce lovable "mutts," treasured by their caretakers.[125]

Clyde L. Webster Jr., PhD: Excluding the wild members of the family *Canidae*, there are over 400 breeds of dogs! And yet they are still recognizable as dogs. Regardless of the amount of cross-breeding and selection applied to this species, a dog is still a dog. Not once has a dog breeder been able to produce . . . any animal that even closely resembles anything other than a dog. The same statement can be made for the cat family *Felidae*. In fact, these statements are true for any kind of animal![126]

In reality natural selection only gives us variety. The variety of dogs we see today came from the gene pool of the dogs on Noah's Ark. The same can be said about the variety of cats, horses, cattle, elephants, etc., and people we see today.

This observation fits well with creation theory. Variation only appears within the biblical kinds. The information in the new variety had to already be in the original gene code. No new information has been added. The gene pool of the new variety is always more limited. Genetic information is always lost, not added as demanded by the theory of evolution. To become a scientific reality, macroevolution would need an increase in genetic complexity.

David C. Read, JD: But a more fundamental flaw in Darwin's reasoning was in his premise that artificial selection can accomplish limitless change. It cannot. Mankind has been selectively breeding animals for millennia without ever having bred a new animal species.[127] Breeders have always

---

[121] Maurice Caullery, *Genetics and Heredity*, Walker and Company, NY, 1964, p. 119.
[122] Richard B. Goldschmidt, MD, PhD, "Evolution, as viewed by one geneticist," *American Scientist*, Yale University, Vol. 40, January 1952, No. 1, p. 94.
[123] Warren L. Johns, JD, "An Inconvenient Truth, Counterfeit Evolution," *Genesis File*, www.GenesisFile.com, Lightning Source, LaVergne, TN, 2010, pp. 54-56.
[124] Warren L. Johns, JD, "An Inconvenient Truth, Counterfeit Evolution," *Genesis File*, www.GenesisFile.com, Lightning Source, LaVergne, TN, 2010, pp. 52 (emphasis added).
[125] Warren L. Johns, JD, "An Inconvenient Truth, Counterfeit Evolution," *Genesis File*, www.GenesisFile.com, Lightning Source, LaVergne, TN, 2010, pp. 62-63.
[126] Clyde L. Webster Jr., PhD, "Evolutionary Processes," *A Scientist's Perspective on Creation and the Flood*, Geoscience Research Institute, Loma Linda, CA, 1995, p. 11.
[127] Richard Milton, "Green Mice and Blue Genes," *Shattering the Myths of Darwinism*, Park Street Press, Rochester, VT, 1997, p. 134.

known that there is a limit to the amount of change achievable by selective breeding.

As more was learned about genetics it became clear that breeders and horticulturists were not generating new variety, as Darwin had thought, but were merely influencing the visible manifestation of genes already found within a given species. In order for breeders to select for a visible trait, it had to be in the "gene pool" of that species to start with.[128]

Human consciousness and intellect

It may seem strange to include the subject of human consciousness and intellect in a chapter on mutations (when we commonly think of mutations as new or modified physical features, like feathers, scales, etc.). However, when we consider the complexity of human embryos and mind and consciousness, could Chance mutations plus Time plus Survival of the fittest really have enough brains to create ours?

In the pages you are reading is the testimony of many geologists and other scientists who have found evidence that corroborates the claims of the Judeo-Christian Scriptures from first verse to last, claims that the true God, revealed in its pages, is the Creator, always existing, uncreated. "God said, Let us make man in our image. . . ." (Genesis 1:26-31).

Science has few answers to questions about the origin of love and hate, joy and sadness, truth, beauty, appreciation of beauty, consciousness and many other human characteristics:

Evolutionist Michael Ruse, PhD: Why should a bunch of atoms have thinking ability? Why should I, even as I write now, be able to reflect on what I am doing and why should you, even as you read now, be able to ponder my points, agreeing or disagreeing, with pleasure or pain, deciding to refute me by deciding that I am just not worth the effort? No one, certainly not the Darwinian as such, seems to have any answer to this. . . . The point is that there is no scientific answer.[129]

Dr. Ariel Roth: Others have wondered why evolution should persist when so few points support it. Phillip Johnson, professor of law at the University of California at Berkeley, echoes some of these concerns[130] as he examines the tenets of evolution from a trial lawyer's perspective. Given the shaky case for evolution, he wonders why the experts can be so blind.[131] . . . Such turmoil is symptomatic of the absence of a workable model for evolution and the limited explanatory value of a naturalistic philosophy. Despite this, scientific thought shies away from such alternatives as creation, since the concept of a God is unacceptable in naturalistic scientific explanations.[132]

One of the most commonly heard beliefs is that "since there can be no God, there must be evolution," which often leads to circular reasoning: "since there is evolution, there can be no God."

---

[128] David C. Read, JD, "The Copying Error Theory of Evolution," *Dinosaurs*, Clarion Call Books, Keene, TX, 2009, pp. 242-244.

[129] Darwinist philosopher Michael Ruse, PhD, *Can a Darwinian Be a Christian?*, Cambridge: Oxford University Press, 2001, p. 73.

[130] Phillip E. Johnson, JD, *Darwin on Trial*, 2nd edition, 1993, Downers Grove, IL. InterVarsity Press; Phillip E. Johnson, JD, *Reason in the Balance: the Case Against Naturalism in Science, Law, and Education*, Downers Grove, IL, 1995, InterVarsity Press.

[131] Ariel A. Roth, PhD, "Is Science in Trouble?" *Origins: Linking Science and Scripture*, Review and Herald Publishing Association, Hagerstown, MD, 1998, p. 333.

[132] Ariel A. Roth, PhD, "Is Science in Trouble?" *Origins: Linking Science and Scripture*, Review and Herald Publishing Association, Hagerstown, MD, 1998, p. 333.

Chapter 6
Scientists Who Are Creationists

Strictly speaking, "creation" is not something that can be studied scientifically, because creation happened only once, through Supernatural means, and with no observers except God and His other created beings. Nevertheless, scientists who believe the biblical record of origins are able to do scientific studies of the natural world and the universe. *Faith does not make science impossible; it broadens the search area.*

Is science done by creation scientists valid? This is a topic I feel I must address early because some critics today claim that scientists who believe in creation are not real scientists. Even though many creationists have earned legitimate post-graduate degrees in scientific disciplines from major public and private universities (some with two or more doctoral degrees), hold responsible scientific positions, and have published numerous peer-reviewed scientific articles and books, they can't be true scientists? Critics claim that these scientists have abandoned the scientific method. Yet, it is interesting to note that Sir Francis Bacon, a scientist who believed in creation, established and developed the scientific method.

So, is there such a thing as scientific work by scientists in a creationist context?

One recent and telling example of real science in action—by creationists—is found at a fossil dig site at the Hanson Research Station in Eastern Wyoming. The dig, begun in 1997, has set a new standard of excellence in paleontology: Using the latest global positioning system (GPS) technology, Arthur Chadwick, PhD, and Larry Turner, PhD, professors at Southwestern Adventist University, in Keene, Texas, and a team of students and faculty with as many as 100 participants, conduct the ongoing dig—annual scientific studies of animal and plant remains and how they were buried and preserved, considered within a biblical framework. They have uncovered an astonishing record of mass burial by rapidly moving water.[133]

The university group chose to do research on a 247-acre area of the Hanson Ranch, which has an estimated 25,000 buried animals—a fossil-rich sedimentary layer located in the Western United States. The Lance Formation with its extensive outcrops is one of the largest bone beds in the world. The digs have continued every year since 1997. *Answers* magazine documented the 2009 dig:

> *Answers* magazine: When a fossil is discovered at Camp Cretaceous, researchers call in the GPS survey crew to record the find. The fossil is described in the researcher's notebook. The fossil is photographed digitally. A series of points is plotted across the surface of the bone. The photo's background is digitally removed, and the plot points are added. The image is added to the computer data bank, which displays the relative positions of all the bones recorded at the site.[134]

Popular GPS technology provides information from orbiting satellites with an accuracy of about *20-30 feet*. However, with special differential RTK (real-time-kinematic) equipment, these scientists can document their fossil finds to within *less than one quarter of an inch*. This sophisticated procedure can determine each fossil's position and orientation *in three dimensions.*

With many fossils now documented over the years, this mapping system has allowed the team to create an extraordinary reconstruction of the entire bone bed in 3-D. After more than 20 years of fieldwork these scientists have determined that this bone bed is not the result of slow accumulation.

> *Answers:* Based on GPS data collected so far, the more likely scenario is that an enormous herd of dinosaurs, mostly the duck billed *Edmontosaurus*, were overtaken by fast flowing water

---

[133] John Upchurch, "Mapping Out the Truth," *Answers*, Vol. 5, No. 1, January-March, 2010, pp. 56-59.
[134] John Upchurch, "Mapping Out the Truth," *Answers*, Vol. 5, No. 1, January-March, 2010, pp. 56-59.

and mud, died quickly, rotted in a freshwater environment, and then were swept into deeper water by another catastrophe. This mass carnage and devastation fits well with the creationist view of [geologic] upheavals caused during and after the Flood.[135]

The 2009 digging season located a record 1,223 specimens. Each member of the team has witnessed firsthand how science and technology confirm the biblical account. Dr. Chadwick sees that as the most important aspect of the dig. "We continue to feel that the benefits of this project to participants outweigh even the remarkable scientific value of the bones."[136]

Anyone can sift through the amazing collection of bones from the Hanson Research Station online (http://fossil.swau.edu/). Users can sort bones by type and other factors, view bones in context in relation to the entire quarry, and even examine 360° views of some of the fossils.[137]

Chadwick and Turner's belief in a biblical model of creation and the Flood has given them more degrees of freedom from which to draw their hypotheses; they are free to allow a biblical model to inform their theories—something a secular paleontologist cannot do.

Must "science" be materialistic? Does the philosophical perspective of the scientist affect his or her ability to analyze evidence? In this chapter you will see that science done by creationists is real, and when they actively investigate areas whose results support a biblical view, that evidence is often ignored by noncreationists and is considered unscientific simply because it doesn't support the current, mainstream "scientific" interpretations. Evolution is based on major assumptions that limit what interpretations will be considered scientific.

Public figures and their opposite worldviews

Henry M. Morris, PhD: It is worth noting that almost none of the [early] leaders of this evolutionary revival had been trained as scientists in the modern sense. None were educated as physicists or chemists or biologists or geologists or astronomers or other "natural" scientists:

Charles Darwin himself had been an apostate divinity student whose only degree was in theology. Charles Lyell was a lawyer, William Smith a surveyor, James Hutton an agriculturalist, John Playfair a mathematician, Robert Chambers a journalist. Alfred Russell Wallace had little formal education of any kind, with only a brief apprenticeship in surveying. Thomas Huxley had an indifferent education in medicine. Herbert Spencer received practically no formal education except some practical experience in railroad engineering. Thomas Malthus was a theologian and economist, while Erasmus Darwin was a medical doctor and poet. Of all the chief contributors to the *revival of evolutionism commonly associated with Charles Darwin*, only Jean Lamarck in France and Ernst Haeckel in Germany seemed to have had a *bona fide* education in the branch of evolutionary "science" that they pursued, and they also had their own particular anti-Christian agendas to promote.[138]

As you can see, most of the above had no education in biology. And as you will see, Haeckel (1834-1899) was convicted of fraud by his own university[139] more than 140 years ago for his main contribution to the evolution model (which, unfortunately, is still being published in some standard science textbooks to this day).

As Jesus' apostle John said, "And ye shall know the truth and the truth shall make you free" (John 8:32). History shows that scientists of the past were Christians who believed in the inspiration and authority of the Bible, the deity of Jesus Christ, and the supernatural creation of everything. In fact, their research established the scientific breakthroughs and the very laws and concepts of science which introduced our modern scientific age. These pioneers were not only scientists who believed in the biblical account of origins and the Flood, they were intellectual giants in fields from A to Z:

Charles Babbage (1792-1871)—Actuarial tables
Joseph Lister (1827-1912)—Antiseptic surgery

---

[135] John Upchurch, "Mapping Out the Truth," *Answers*, Vol. 5, No. 1, January-March, 2010, pp. 56-59.
[136] Arthur Chadwick, PhD, "Mapping Out the Truth," *Answers*, Vol. 5, No. 1, January-March, 2010, p. 59.
[137] John Upchurch, "Mapping Out the Truth," *Answers*, Vol. 5, No. 1, January-March, 2010, p. 58.

[138] Henry M. Morris, PhD, "The Dark Nursery of Darwinism," *The Long War Against God*, Master Books, Inc., Green Forest, AR, 2005, pp. 161-162, (emphasis added).
[139] Records from University of Jena trial in 1875. Dr. Edward Blick, Norman, OK.

Louis Pasteur (1822-1895)—Bacteriology, fermentation control, vaccination and immunization, Biogenesis Law [life comes only from life], Pasteurization

Blaise Pascal (1623-1662)—Barometer, Statistics

Isaac Newton (1642-1727)—Calculus, dynamics, law of gravity, reflecting telescope

Johannes Kepler (1571-1630)—Celestial mechanics, physical astronomy

Robert Boyle (1627-1691)—Chemistry, gas dynamics

Georges Cuvier (1769-1832)—Comparative anatomy, vertebrate paleontology

Michael Faraday (1791-1867)—Electric generator

Joseph Henry (1797-1878)—Electric motor, galvanometer

John Ambrose Fleming (1849-1945)—Electronics

James Clerk Maxwell (1831-1879—Electrodynamics

Lord Kelvin (William Thomson) (1824-1907)— Energetics, thermodynamics, absolute temperature scale, transatlantic cable

Henri Fabre (1823-1915)—Entomology of living insects

George Stokes (1819-1903)—Fluid dynamics

William Herschel (1738-1822)—Galactic astronomy, discoverer of Uranus

Gregor Mendel (1822-1844)—Genetics

Louis Agassiz (1807-1873)—Glacial geology, ichthyology

John Herschel (1792-1871)—Global star catalog

James Simpson (1811-1870)—Gynecology, chloroform

William Ramsey (1852-1916)—Isotopic chemistry

Matthew Murray (1806-1873)—Oceanography

John Woodward (1665-1728)—Paleontology

Rudolph Virchow (1821-1902)—Pathology

Sir Francis Bacon (1561-1626)—Scientific method

Nicholas Steno (1631-1686)—Stratigraphy

Carolus Linnaeus (1707-1778)—Systematic biology, classification system

Humphrey Davy (1778-1829)—Thermokinetics

Samuel F. B. Morse (1791-1872)—Telegraph, Morse code

This list, of course, is not comprehensive. But it does reveal the fact that our scientific forefathers were not living in the "Dark Ages." These true scientist-pioneers were Bible-believing Christians.[140]

Famed French mathematician and cosmologist Joseph Lagrange suggested that Isaac Newton's *Principia* was assured for all time "a preeminence above all other productions of the human intellect."[141]

At the end of his second edition of *Principia*, Newton gives credit to God as Creator: "This most beautiful system of the sun, planets, and comets could only proceed from the counsel and dominion of an intelligent and powerful Being."[142]

In 1999, *Time* magazine selected Newton as the most influential person of the seventeenth century.[143] Voltaire, one of the great French leaders in the burgeoning thought and reasoning movement of the time, lauded the scientist with "if all the geniuses of the universe were assembled, he should lead the band."[144]

Also, Johannes Kepler ranks among the leading scientists of all time. He demonstrated that planets move around the sun in an elliptical (oval-shape) rather than a circular orbit. In cosmology he developed three principles, known as Kepler's laws, which describe planetary motion. Like the famous Italian astronomer Galileo, Kepler saw God's nature in his rigorous study of the exacting mathematics of nature.[145]

Historians consider Robert Boyle to be the father of chemistry. Chemistry students know him for Boyle's law, which explains the inverse

---

[140] Henry M. Morris, PhD, "Bible-Believing Scientists of the Past," *Impact*, No. 103, Institute For Creation Research, El Cajon, CA, 1982, pp. 1-4.

[141] As quoted in: D. C. Miller, 1928, "Newton and optics," *The History of Science Society: Sir Isaac Newton, 1727: a bicentenary evaluation of his work*, Williams and Wilkins Co, Baltimore, p. 15.

[142] I. Newton, 1686, 1934. *Mathematical Principles of Natural Philosophy and His System of the World*. Translated into English by Andrew Motte in 1729, revised translation by Florian Cajori. Los Angeles: University of California Press, p. 544.

[143] P. Gray, "The most important people of the millennium," *Time* 154 (27): 1999, pp. 139-195.

[144] As quoted in: D. C. Miller, 1928, "Newton and optics," *The History of Science Society: Sir Isaac Newton, 1727: a bicentenary evaluation of his work*, Williams and Wilkins Co, Baltimore, 1928, p. 15.

[145] W.C. Dampier, *A History of Science: Its Relations with Philosophy and Religion*, 4th ed. NY: Macmillan Co., 1949, p. 127.

relationship of pressure and volume in gases. Boyle believed that one glorifies God by explaining His creation, and that God not only created the world but also is continually needed to keep it going.[146]

But don't all scientists believe in evolution today?

> Russell Humphreys, PhD: Using a simple statistical approach, I would conservatively estimate that in the United States alone, there are around 10,000 practicing professional scientists who openly believe in six-day recent creation.[147]

The Wright brothers were creationists who studied God's design of birds to develop the airplane.[148]

> Wernher von Braun, PhD, father of the United States space program and at one time director of NASA's Marshall Space Flight Center, where he helped develop the Saturn V rocket that carried Neil Armstrong to the moon: The vast mysteries of the universe should only confirm our belief in the certainty of its Creator. I find it as difficult to understand a scientist who does not acknowledge the presence of a superior rationality behind the existence of the universe as it is to comprehend a theologian who would deny the advances of science.[149] . . . *To be forced to believe only one conclusion—that everything in the universe happened by chance—would violate the very objectivity of science itself.* . . . What random process could produce the brains of a man or the system of the human eye? . . . They (evolutionists) challenge science to prove the existence of God. *But must we really light a candle to see the sun?*[150]

Raymond Damadian, MD, a scientist with scores of patented ideas, including Magnetic Resonance Imaging, believes in creation.[151]

However, Harvard Professor Stephen Jay Gould, PhD, a famous evolutionary scientist, claims that creationists are "religious fundamentalists, not scientists," and that "professionally trained scientists, virtually to a person, understand the factual basis of evolution and don't dispute it."[152]

Many of today's scientific professionals obviously dispute Gould's claim: In the year 2000, Master Books, Inc., of Green Forest, Arkansas, published a 384-page book entitled *In Six Days: Why Fifty Scientists Choose to Believe in Creation*, edited by John F. Ashton, PhD. In his preface, Dr. Ashton reported hearing a university lecturer several years ago state that he did not believe that any scientist with a PhD would advocate a literal interpretation of the six days of creation. The experience motivated Dr. Ashton to publish the book.[153]

Each chapter begins with a brief summary of the author's education and place of employment. Authors—one of them with two PhD degrees—graduated from:
Pennsylvania State University
Wayne State University
Columbia Pacific University
The University of Minnesota
Loma Linda University
University of Michigan
University of California, Irvine
George Mason University
　(University of Virginia)
University of Melbourne
University of Western Australia
University of Sydney
University of Adelaide (Australia)
University of Newcastle (Australia)
Victoria University of Wellington, New Zealand
University of Edinburgh, Scotland
Columbia University
California Institute of Technology
Tulane University

---

[146]E. Nordenskiöld, *The History of Biology: A Survey*, Eyre L. B., trans. Tudor Pub. Co., NY, 1928, 1942, pp. 206-207; Ariel A. Roth, PhD, "Can a Scientist Dare to Believe in God?" *Science Discovers God: Seven Convincing Lines of Evidence for His Existence*, Autumn House Publishing, a division of Review and Herald Publishing, Hagerstown, MD, 2008, p. 22.
[147]D. Russell Humphreys, PhD, a nuclear physicist who worked for years at Sandia National Laboratories. Answers in Genesis website
[148]"Celebrate the success of the Wright Brothers," WrightStories.com., and many other entries on Google.
[149]Quoted in Cal Thomas, "Gone Bananas," *World*, September 7, 2002.
[150]Wernher von Braun, letter read by Dr. John Ford to California State Board of Education, September 14, 1972, quoted in Ann Lamont, *Twenty-One Great Scientists Who Believed the Bible*, Creation Science Foundation, Acacia Ridge, Queensland, Australia, 1995, p. 47, (emphasis added).

[151]James Perloff, "Conclusion," *The Case Against Darwin*, Refuge Books, Burlington, MA, 2002, p. 71.
[152]Edited by John Ashton, PhD, *In Six Days: Why Fifty Scientists Choose to Believe in Creation*, Master Books, Inc., Green Forest, AR, 2000, cover.
[153]Edited by John Ashton, PhD, "Preface," *In Six Days: Why Fifty Scientists Choose to Believe in Creation*, Master Books, Inc., Green Forest, AR, 2000, p. 5.

University of Toronto
University of California, Los Angeles (UCLA)
University of Birmingham
Ohio State University
University of Liverpool
University of Wales
University College of Wales
Cambridge University
University of London
University of Cape Town
Indiana University
University of New York at Albany
University of Oklahoma
University of Southern California
University of Illinois
Colorado State University
Rutgers University
The State University of New Jersey
Iowa State University
Harvard University.

Scientists who utterly reject evolution may be one of the fastest-growing controversial minorities. In his book *The Case for a Creator*, Lee Strobel, a former evolutionist, and investigative reporter for the *Chicago Tribune*, identified others who went public: "despite the specter of professional persecution, they broached the politically incorrect opinion that the emperor of evolution has no clothes."

> Lee Strobel: There were one hundred of them—biologists, chemists, zoologists, physicists, anthropologists, molecular and cell biologists, bioengineers, organic chemists, geologists, astrophysicists, and other scientists. Their doctorates came from such prestigious universities as Cambridge, Stanford, Cornell, Yale, Rutgers, Chicago, Princeton, Purdue, Duke, Michigan, Syracuse, Temple, and Berkeley.
>
> They included professors from Yale Graduate School, the Massachusetts Institute of Technology, Tulane, Rice, Emory, George Mason, Lehigh, and the Universities of California, Washington, Texas, Florida, North Carolina, Wisconsin, Ohio, Colorado, Nebraska, Missouri, Iowa, Georgia, New Mexico, Utah, Pennsylvania, and elsewhere.
>
> Among them was the director of the Center for Computational Quantum Chemistry; and scientists at the Plasma Physics Lab at Princeton, the National Museum of Natural History at the Smithsonian Institution, the Los Alamos National Laboratory, and the Lawrence Livermore Laboratories. . . . And they wanted the world to know one thing: *they are skeptical*.

After spokespersons for the Public Broadcasting System's seven-part television series *Evolution* asserted that "all known scientific evidence supports [Darwinian] evolution" as does "virtually every reputable scientist in the world," these professors, laboratory researchers, and other scientists published a two-page advertisement in a national magazine under the banner: "A Scientific Dissent from Darwinism."

Their statement was direct and defiant. "We are skeptical of claims for the ability of random mutation and natural selection to account for the complexity of life," they said. "Careful examination of the evidence for Darwinian theory should be encouraged."[154]

These were not narrow-minded fundamentalists . . . or rabid religious fanatics—just respected, world-class scientists like Nobel nominee Henry F. Schaefer, the third most-cited chemist in the world; James Tour of Rice University's Center for Nanoscale Science and Technology; and Fred Figworth, professor of cellular and molecular physiology at Yale Graduate School.

Viewers of the popular 2001 PBS series weren't told that [credible scientists harbored significant skepticism toward Darwinian theory.] In fact, its one-sided depiction of evolution spurred a backlash from many scientists. A detailed, 151-page critique claimed it "failed to present accurately and fairly the scientific problems with the evidence for Darwinian evolution" and even systematically ignored "disagreements among evolutionary biologists themselves."[155]

Discover Institute Chair Bruce Chapman, who assembled the list of 100 doctoral-level scientists, stated, "Realizing that there were likely more scientists worldwide who share some skepticism of Darwinian evolution and were willing to go on

---
[154]Lee Strobel, "Doubts About Darwinism," *The Case for a Creator*, Zondervan, Grand Rapids, MI, 2004, pp. 31-32; "A Scientific Dissent from Darwinism" *The Weekly Standard*, October 1, 2001, pp. 20-21.
[155]Lee Strobel, "Doubts About Darwinism," *The Case for a Creator*, Zondervan, Grand Rapids, MI, 2004, pp. 31-32, (emphasis in the original).

record, the Institute has maintained the list and added to it continually since its inception. The list has now grown from 100 to 1,000.[156]

As my research progressed, I felt that more people needed to be exposed to just such currently increasing "reasonable doubt." Although most evolutionists and creationists believe in natural selection and survival of the fittest, *this simply means the ability to adapt to an environment. The difference* is that creationists don't believe this process adds genetic information. It cannot cause one kind of creature to become a new kind.

Darwin could not provide good examples of natural selection in action, and so he had to rely heavily on analogy:

> Evolutionist Douglas J. Futuyma, PhD: When Darwin wrote *[On] The Origin of Species*, he could offer no good cases of natural selection because no one had looked for them. He drew instead an analogy with the artificial selection that animal and plant breeders use to improve domesticated varieties of animals and plants. By breeding only from the woolliest sheep, the most fertile chickens, . . . breeders have been spectacularly successful in altering almost every imaginable characteristic of our domesticated animals and plants to the point where most of the time they differ from their wild ancestors far more than related species differ from them.[157]
>
> Loren Eiseley, PhD: It would appear that careful domestic breeding, whatever it may do to improve the quality of race horses or cabbages, is not actually in itself the road to the endless biological deviation which is evolution. There is great irony in this situation, for more than almost any other single factor, domestic breeding has been used as an argument for the reality of evolution.[158]
>
> Creationist Clyde L. Webster, Jr., PhD: Many species have been subjected to artificial selection such as: cats, dogs, cattle, horses, sheep, corn, roses, grasses, and numerous others.

The results of artificial selection suggest that changes in species could not result in evolutionary progress. This evidence is consistent with the creation account, which states that many different *kinds* of animals were created at the beginning.[159]

Here is the perspective of the Intelligent Design Movement: life is too complex to have evolved without the intervention of a Supernatural being. Phillip E. Johnson, JD, the "father" of the Intelligent Design Movement: "The analogy to artificial selection is misleading. Plant and animal breeders employ intelligence and specialized knowledge to select breeding stock and to protect their charges from natural dangers. The point of Darwin's theory, however, was to establish that purposeless natural processes can substitute for intelligent design. That he made that point by citing the accomplishments of intelligent designers proves only that the receptive audience for his theory was *highly uncritical.*"

> Phillip E. Johnson, JD: Artificial selection is not basically the same sort of thing as natural selection, but rather is something fundamentally different. Human breeders produce variations among sheep or pigeons for purposes absent in nature, including sheer delight in seeing how much variation can be achieved. If the breeders were interested only in having animals capable of surviving in the wild, the extremes of variation would not exist. When domesticated animals return to the wild state, the most highly specialized breeds quickly perish and the survivors revert to the original wild type. Natural selection is a conservative force that prevents the appearance of the extremes of variation that human breeders like to encourage.
>
> What artificial selection actually shows is that there are definite limits to the amount of variation that even the most highly skilled breeders can achieve. Breeding of domestic animals has produced no new species, in the commonly accepted sense of new breeding communities that are infertile when crossed with the parent group.[160]

Warren L. Johns, JD: Evolutionists routinely reject "Intelligent Design" theory as less than true

---

[156] WND, "1,000 Scientists Go Public With Doubts on Evolution," *Breaking News*, WND.con, February 9, 2019.
[157] Douglas J. Futuyma, PhD, quoted by Phillip E. Johnson, JD, "Natural Selection," *Darwin On Trial*, Regnery Gateway, Washington, DC, 1991, p. 17. (Johnson was a graduate of Harvard University and a law professor at the University of Berkeley).
[158] Loren Eiseley, PhD, *The Immense Journey*, Vintage Books, NY, 1958, p. 223.
[159] Clyde L. Webster Jr., PhD, "Evolutionary Processes," *A Scientist's Perspective on Creation and the Flood*, Geoscience Research Institute, Loma Linda, CA, 1995, p. 12, (emphasis added).
[160] Phillip E. Johnson, JD, "Natural Selection," *Darwin On Trial*, Regnery Gateway, Washington, DC, 1991, pp. 17-18, (emphasis added).

science on the grounds it is not readily falsifiable. The transparent inconsistency of claiming the design inference is less than scientific because it can't be readily tested, understandably exposes evolution's own fragile feet of "scientific" clay. . . .

The modern Evolutionary Synthesis *has yet to be verified in laboratory tests.*[161]

Gary E. Parker, EdD: . . . 24 years before Darwin's publication, a scientist named Edward Blyth published the concept of natural selection in the context of creation. He saw it as a process that adapted varieties of the created types to changing environments.[162]

Michael Pitman, PhD: But Blyth's idea may well have been closer to the truth than Darwin's development of it, for natural selection can only reduce rather than increase genetic variability. It operates in nature solely as a conservative mechanism, a sieve to weed out the weak, malformed or sick and maintain a healthy stock. It is indeed a force counteracting the tendency for mutation to cause a degeneration in the quality of living organisms—but it cannot be creative.[163]

But what about the Peppered Moth?

Dr. Parker, a former evolutionist, continues: "Perhaps the best example of Darwinian selection is the one that's in all the biology textbooks: the peppered moths. . . . The darker moth stood out, but the lighter one was camouflaged against the mottled gray lichen that encrusted the trees back then [in the 1850s]. As a result, birds ate mostly dark moths and light moths made up over 98 percent of the population.

But then pollution killed the lichen on the trees, revealing the dark color of the bark. As a result, the dark moths were more camouflaged than the light ones. The moths themselves didn't change; there were always dark moths and always light moths from the earliest observations. . . . The dark ones had a better chance of surviving and leaving more offspring that grew into dark moths in succeeding generations. Sure enough, just as Darwin would have predicted, the population shifted. . . . By the 1950s the population was over 98 percent dark colored—proof positive of "evolution going on today." At least that's the way it's stated in many biology books, and that's what I used to tell my biology students.

The question is, what kind of change do we see: change only within type (creation) or change also from one type to others (evolution)? . . . What did we start with? Dark and light varieties of the peppered moth, species *Biston betularia*. After 100 years of natural selection, what did we end up with? Dark and light varieties of the peppered moth, species *Biston betularia*. All that changed was the percentage of moths in the two categories—that is, just variation within type.[164]

In 2016, some researchers claimed that melanism (the change to dark color) in the peppered moth is due to a "jumping gene." But this would not alter the related broader problem for evolution described in *Icons of Evolution* by Jonathan Wells, PhD (whose doctorate in biology is from Berkeley). Dr. Wells' 10-chapter book quotes published scientific evidence that reveals serious misrepresentations (found in biology textbooks) that are commonly used to support belief in evolution. His and related evidence tells us Parker's statement is still valid: the famous peppered moths do not support evolution.[165]

Daniel Brooks in *Science: Natural selection may have a stabilizing effect, but it does not promote speciation. It is not a creative force as many people have suggested.*[166]

The dog, whatever the new variety, is still a dog.

---

[161] Warren L. Johns, JD, "Imagination, Living Cell," *Genesis File*, www.GenesisFile.com, Lightning Source, LaVergne, TN, 2010, p. 100, (emphasis added).
[162] Gary E. Parker, EdD, "Darwin and the Nature of Biologic Change," *What is Creation Science?* Master Books, El Cajon, CA, 1987, p. 82, (emphasis in the original).
[163] Michael Pitman, PhD, "Sports [mutations], Survival and the Hone," *Adam and Evolution,* Rider & Company, London, Melbourne, Sydney, Auckland, Johannesburg, 1984, p. 76.

[164] Gary E. Parker, EdD, "Darwin and the Nature of Biologic Change," *What is Creation Science?* Master Books, El Cajon, CA, 1987, pp. 78-82. *See also:* William D. Stansfield, PhD, "Natural Selection," *The Science of Evolution,* MacMillan Publishing Co., Inc. New York; Collier MacMillan Publishers, London, 1977, pp. 371-372. *See also:* J. B. Birdsell, "The Forces of Evolution," *Human Evolution: An Introduction to the New Physical Anthropology,* Rand McNally College Publishing Company, Chicago, 1975, pp. 406-409.
[165] Jonathan Wells, PhD, *Icons of Evolution: Science or Myth?,* Regnery Publishing Washington, DC, 2000.
[166] Daniel Brooks, "A Downward Slope to Greater Diversity," *Science,* Vol. 217, September 24, 1982, p. 1240, (emphasis added).

Chapter 7
What Did Darwin Say?

Charles Darwin: "The chief distinction in the intellectual powers of the two sexes is shown by man's attaining to a higher eminence, in whatever he takes up, than can woman—whether requiring deep thought, reason, or imagination, or merely the use of the senses and hands. If two lists were made of the most eminent men and women in poetry, painting, sculpture, music (inclusive of both composition and performance), history, science and philosophy, with half a dozen names under each subject, the two lists would not bear comparison. We may also infer . . . that if men are capable of decided pre-eminence over women in many subjects, the average of mental power in man must be above that of a woman."[167] "Man has ultimately become superior to woman."[168]

Few today would agree.

Although Darwin's family all denied that he made any "death bed" confessions, apparently at sometime before he died, to Darwin's credit he acknowledged flawed thinking and even expressed serious doubts about his theory:

> Warren L. Johns, JD: In a burst of candor, Darwin recognized the feeble fabric of his idea, acknowledging, "I am conscious that my speculations run beyond the bounds of true science."[169] He admitted that his theory seemed to be "a mere rag of an hypothesis with as many flaw[s] and holes as sound parts." . . . Before his 1882 death, the English naturalist verbalized uneasy equivocation in a whiff of prescience, confessing doubts about his thesis. He fretted he may "have devoted my life to a phantasy [sic]."[170]

Before Darwin died, as time went on, and he became more enlightened, he revised some of his writings. For example, Darwin had imagined that (given enough time) a bear might evolve into a marine mammal "as monstrous as a whale." In the first edition of his book he stated, "I can see no difficulty in a race of bears being rendered, by natural selection, more and more aquatic in their structure and habits . . . till a creature was produced as monstrous as a whale."[171] Then he seems to have rejected this obvious conjecture, as it was removed from all subsequent editions.

As you can see, he also recognized the lack of missing links:

> Charles Darwin: But, as by this theory innumerable transitional forms must have existed, why do we not find them embedded in countless numbers in the crust of the earth?

> The number of intermediate varieties, which have formerly existed on earth [must] truly be enormous. Why then is not every geological formation and every stratum full of such intermediate links? Geology assuredly does not reveal any such finely-graduated organic chain; and this, perhaps, is the most obvious and gravest objection which can be urged against my theory.[172]

Isn't Darwin stating here that no adequate physical evidence exists to support his theory? It seems to me that he just acknowledged the difficulty of finding intermediates—missing links. But Darwin continues and even emphasizes his predicament by quoting scientists of his time:

> Darwin: The abrupt manner in which whole groups of species suddenly appear in certain formations has been urged by several paleontologists, for instance by Agassiz, Pictet and none more forcibly than by Professor Sedgwick, as a

---

[167] Charles Darwin, "Sexual Selection in Relation to Man," *The Descent of Man and Selection in Relation to Sex*, Second Edition, A. L. Burt Company; New York, 1874, p. 643.
168 Charles Darwin, "Sexual Selection in Relation to Man," *The Descent of Man and Selection in Relation to Sex*, National Library Association, Chicago, IL, 1874, p. 588.
[169] Charles Darwin, to Asa Gray, cited by Adrian Desmond and James Moore, Darwin, W.W. Norton and Company, NY, 1991, p. 456.
[170] Charles Darwin, *Life and Letters*, 1887, Vol 2, p. 229; Desmond & Moore, Darwin, pp. 475, 477; Warren LeRoy Johns, JD, *Genesis File*, "Fact-free 'Science,'" www.GenesisFile.com, 2009, p. 32.
[171] Charles Darwin, *On the Origin of Species by Means of Natural Selection, or the Preservation of Favored Races in the Struggle For Life*, John Murray, Albemarle Street, London, 1859, p. 184.
[172] Charles Darwin, "On the Imperfection of the Geological Record," *On The Origin of Species by Means of Natural Selection or the Preservation of Favoured Races in the Struggle for Life*, John Murray, Albemarle Street, London, 1859, p. 280 ([bracketed] added).

fatal objection to the belief in the transmutation of species. If numerous species, belonging to the same genera or families, have really started into life at once, the fact would be fatal to the theory of descent with slow modification through natural selection. . . . There is another and allied difficulty, which is much more serious. I allude to the manner in which numbers of species of the same group suddenly appear in the lowest known fossiliferous rocks [since identified as "the Cambrian Explosion"]. . . . To the question why do we not find rich fossiliferous deposits belonging to these assumed earliest periods prior to the Cambrian system, I can give no satisfactory answer. . . . The case at present must remain inexplicable, and may be truly urged as a valid argument against the views here entertained.[173]

This acknowledgment by Darwin himself should have ended the theory. But it didn't.

Horse evolution to the rescue

Ernst Mayr, PhD: The most complete transition between an early primitive type and its modern descendant that has been described is that between *Eohippus*, the ancestral horse, and *Equus*, the modern horse.[174]

Textbooks often picture the famous fossil series that illustrates the so-called gradual evolution of the horse.[175] A 2009 biology textbook illustrated horse evolution as fossil evidence for evolution with the following caption: "The modern horse is the descendant of a long lineage of horses. . . . Over time, this lineage underwent evolutionary changes in size, in the number and shape of the toes, and in other anatomical features. Many fossils that are intermediate in form between the modern horse and its earliest known ancestor refute claims, made by some people who don't believe in evolution, that the fossil record fails to document evolution."[176]

According to Dr. David M. Raup, quoted below, this textbook is at least 30 years out of date. So is the highly acclaimed Smithsonian Institution in Washington, DC, where in 2009 I saw its display of horse evolution. Any lay person would think it is educational and excellent, but according to Dr. Raup's statement, it should be discarded or modified.

The Field Museum of Natural History in Chicago has one of the largest collections of fossils in the world. Consequently, in 1979, its Dean, Dr. David Raup, was highly qualified to summarize the situation regarding transitions that should be observed in the fossil record, and specifically mentioned the horse:

David M. Raup, PhD: Well, we are now about 120 years after Darwin and the knowledge of the fossil record has been greatly expanded. We now have a quarter of a million fossil species but the situation hasn't changed much. The record of evolution is still surprisingly jerky and, ironically, we have even fewer examples of evolutionary transition than we had in Darwin's time. By this I mean that some of the classic cases of *Darwinian change* in the fossil record, *such as the evolution of the horse* in North America, have had to be *discarded* or modified as a result of more detailed information—what appeared to be a nice simple progression when relatively few data were available now appears to be much more complex and much less gradualistic. So, Darwin's problem has not been alleviated in the last 120 years, and we still have a record which does show change but one that can hardly be looked upon as the most reasonable consequence of natural selection.[177]

Boyce Rensberger, senior editor of *Science 80*, adds his understanding and generalizes the horse example to apply to all other fossils:

Boyce Rensberger: *The popularly told example of horse evolution*, suggesting a gradual sequence of changes from four-toed, fox-sized creatures, living nearly 50 million years ago, to today's much larger one-toed horse, *has long been known to be wrong*. Instead of gradual change, fossils of each intermediate species appear fully distinct, persist unchanged, and then become extinct. Transitional forms are unknown.[178]

---

[173]Charles Darwin, "On the Imperfection of the Geological Record," *On The Origin of Species by Means of Natural Selection or the Preservation of Favoured Races in the Struggle for Life*, The World's Popular Classics, Books, Inc. New York, Boston, 19--, pp. 288, 291-293.
[174]Ernst Mayr, PhD, "What is the Evidence for Evolution?" *What Evolution Is*, Basic Books, New York, NY, 2001, p. 16.
[175]George B. Johnson, Peter H. Raven, "Evidence of Macroevolution," Biology: *Principles & Explorations*, Holt, Rinehart and Winston, Austin, New York, Orlando, Atlanta, San Francisco, Boston, Dallas, Toronto, London, 1996, p. 254.
[176]Cain, Yoon, Singh-Cundy, "Evolution is strongly supported by the fossil record," *Discover Biology*, W.W. Norton & Company, NY,
Fourth Edition, 2009, p. 333.
[177]David M. Raup, PhD, "Conflicts between Darwin and Paleontology," *Field Museum of Natural History Bulletin*, Vol. 50, No. 1, January 1979, p. 25, (emphasis added).
[178]Boyce Rensberger, "Ideas on Evolution Going Through a

Researchers are considering newer ideas about the evolution of the horse.[179] A recent opinion is that the whole issue requires more study. Evolutionists are now questioning the validity of the traditional arrangement of the horse as worked out by O. C. Marsh. Even as far back as 1953, George Gaylord Simpson, PhD, of Harvard University, perhaps the most influential paleontologist of the twentieth century, stated, "The most famous of all *equid* [horse] trends, 'gradual reduction of the side toes,' is flatly fictitious."[180] "Many examples commonly cited, such as the evolution of the horse family or of saber-tooth 'tigers' can be readily shown to have been unintentionally falsified and not to be really orthogenetic."[181] "The uniform continuous transformation of *Hyracotherium* into *Equus*, so dear to the hearts of generations of textbook writers, never happened in nature."[182]

Biology—The Unity and Diversity of Life: According to the traditional view, termed the gradualistic model, the branchings on family trees diverged gradually, with each new species emerging through many small changes in form over long spans of time. Some cases of gradualism are well documented. . . . Other examples, including the much-repeated "gradual" evolution of the modern horse, have not held up under close examination.[183]

Luther D. Sunderland: Nowhere in the world are the fossils of the horse series found in successive strata. When they are found on the same continent, like the John Day formation in Oregon, the three-toed and one-toed are found in the same geologic horizon (stratum). In South America, the one-toed is even found below the three-toed creature. And when other structures besides toes are considered, the picture does not look so impressive. For example, the four-toed Hyracotherium has 18 pairs of ribs, the next [horse-like] creature has 19, then there is a jump to 15, and finally to 18 for *Equus*, the modern horse.[184]

Science News Letter: The early classical evolutionary tree of the horse, beginning in the small, dog-sized Eohippus and tracing directly to our present day Equinus, was all wrong.[185]

Jonathan D. Sarfati, PhD: As the biologist Heribert-Nilsson said, "The family tree of the horse is beautiful and continuous only in the textbooks,"[186] and the famous paleontologist Niles Eldredge called the textbook picture "lamentable" and "a classical case of paleontologic *museology.*"[187]

Luther D. Sunderland: When scientists speak in their offices or behind closed doors, they sometimes make candid statements that sharply conflict with statements they make for public consumption before the media. For example, after Dr. Eldredge made the statement [in 1979] about the horse series being the best example of a lamentable imaginary story being presented as though it were literal truth, he then contradicted himself.

[On February 14, 1981] in California he was on a network television program. The host asked him to comment on the creationist claim that there were no examples of transitional forms to be found in the fossil record. Dr. Eldredge turned to the horse series display at the American Museum [of Natural History] and stated that it was the best available example of a transitional sequence.[188]

Revolution Among Scientists," *Houston Chronicle*, November 5, 1980, Section 4, p. 15, quoted in Duane T. Gish, *Creation Scientists Answer Their Critics*, Institute for Creation Research, El Cajon, CA, 1993, p. 80, (emphasis added).
[179]MacFadden, B. J., *Fossil Horses: Systematics, Paleobiology, and Evolution of the Family Equidae*, Cambridge and New York: Cambridge University Press, 1992, p. 330.
[180]Simpson, George Gaylord, "Trends and Orientation," *The Major Features of Evolution*, Columbia University Press, New York and London, 1953, p. 263.
[181]Simpson, George Gaylord, "Evolutionary Determinism and the Fossil Record," *Scientific Monthly*, Vol. 71, October 1950, p. 264.
[182]Simpson, George Gaylord, PhD, *Life of the Past: Introduction to Paleontology*, Yale University Press, New Haven, CT, 1953, p. 119.
[183]Cecie Starr, Ralph Taggart, "Life's Origins and Macroevolutionary Trends," *Biology: The Unity and Diversity of Life*, Wadsworth Publishing Company, Inc., Belmont, CA, 1992, pp. 303-304, (emphasis in the original).

[184]Luther D. Sunderland, "The Fossil Record—Reptile to Man," *Darwin's Enigma: Ebbing the Tide of Naturalism*, Master Books, Green Forest, AR, 1988, 2002, p. 94.
[185]"Little Eohippus Not Direct Ancestor of the Horse," *Science News Letter*, August 25, 1951, p. 118.
[186]Heribert-Nilsson, *Synthetische Artbildung*, Gleerup, Sweden, Lund University, 1954; cited in Luther Sunderland, *Darwin's Enigma: Fossils and Other Problems*, 4th edition, Master Books, Santee, CA, 1988, p. 81.
[187]Jonathan D. Sarfati, PhD, "The non-evolution of the horse," *Creation Ex Nihilo*, Vol. 21, No. 3, June 1999, pp. 28-31, (emphasis added), see also: Niles Eldridge, quoted in Sunderland, *Darwin's Enigma: Fossils and Other Problems*, 4th edition, Master Books, Santee, CA, 1988, p. 78.
[188]Luther D. Sunderland, "The Fossil Record—Reptile to Man," *Darwin's Enigma: Ebbing the Tide of Naturalism*, Master Books, Inc., Green Forest, AR, 1988, pp. 94-95.

David M. Raup, PhD, curator of geology at the Field Museum of Natural History in Chicago and past president of the Paleontological Society, also cites the horse example as an illustration of the nonexistence of transitional fossils in the entire fossil record. He observed that "instead of finding the gradual unfolding of life, what geologists of Darwin's time and geologists of the present day actually find is a highly uneven or jerky record; that is, species appear in the sequence very suddenly, show little or no change during their existence in the record, then abruptly go out of the record."[189]

Bird evolution

The geologic column also bears witness to the lack of evolution for flight:

> Robin J. Wootton: We would naturally expect the gradual evolution of flight to leave some evidence in the fossil record. But when fossil insects first appear in the geologic column, flying is fully developed.[190]

> Ariel A. Roth, PhD: The flying pterosaurs, birds, and bats also show up suddenly as fully functional flying organisms. The anatomical changes needed to develop flight, including transformations in bone, musculature, feathers, respiration, and nervous system, would take a long time, and the organisms undergoing such changes would surely leave some fossil record of intermediate stages. The feather of the bird supposedly evolved from the scales of some ancestral reptile. Anyone who has examined feathers under a microscope realizes that they are intricate and highly specialized structures. Would not the extended process of creating all these parts from reptile scales by undirected evolution, including the unsuccessful lines of development, have made some record in the rocks?[191]

No evolution?

> Canadian entomologist William R. Thompson, DSc, PhD: The general tendency to eliminate, by means of unverifiable speculations, the limits of the categories nature presents to us, is the inheritance of biology from the *Origin of Species*. To establish the continuity required by the theory, historical arguments are invoked, even though historical evidence is lacking. Thus are engendered those fragile towers of hypotheses based on hypotheses, where fact and fiction intermingle in an inextricable confusion.[192]

Dr. Thompson, editor of *The Canadian Entomologist* and author of some 150 scientific papers, was chosen to write the Introduction to the new edition of Darwin's *On the Origin of Species*, published in the Darwinian Centennial Year as part of the Everyman's Library Series. He wrote in part:

> William R. Thompson, DSc, PhD: As we know, there is a great divergence of opinion among biologists, not only about the causes of evolution but even about the actual process. This divergence exists because the evidence is unsatisfactory and does not permit any certain conclusion. It is therefore right and proper to draw the attention of the non-scientific public to the disagreements about evolution. . . . This situation, where men rally to the defense of a doctrine they are unable to define scientifically, much less demonstrate with scientific rigor, attempting to maintain its credit with the public by the suppression of criticism and the elimination of difficulties, is abnormal and undesirable in science.[193]

---

[189] David M. Raup, PhD, "Conflicts between Darwin and Paleontology," *Field Museum of Natural History Bulletin*, Vol. 50, 1979, pp. 22-29.
[190] Robin J. Wootton, "Flight: arthropods," *Paleobiology: A Synthesis*, In: Briggs and Crowther, Blackwell Scientific Publications, Oxford and London, 1990, pp. 72-75.
[191] Ariel A. Roth, PhD, "What Fossils Say About Evolution," *Origins: Linking Science and Scripture*, Review and Herald Publishing Association, 1998, p. 185.
[192] W. R. Thompson, introduction to *The Origin of Species*, (reprint, New York: Dutton, Everyman's Library, 1956, quoted in Henry M. Morris, PhD, and John D. Morris, PhD, *Science and Creation*, Master Books, Inc., Green Forest, AR, 1996, p. 29.
[193] W. R. Thompson, "Introduction" to *The Origin of Species* by Charles Darwin, Everyman's Library, E. P. Dutton & Company, NY, 1956; Thompson's Introduction has also been reprinted in the *Journal of the American Scientific Affiliation*, Vol. 12, March, 1960, pp. 2-9.

Chapter 8
Does Evolution Invest Chance with Intelligence?

One of the major reasons I can't believe in evolution (descent with modification, universal common ancestry, change over time, from mutations and natural selection) is because *evolution appears to invest chance with intelligence.*

Although no evolutionists would say that chance is intelligent, they would claim that natural selection acts like an intelligent screening process to filter the variations that were produced by chance. Therefore, according to my understanding of the theory of evolution, if we needed to see in three dimensions and hear in stereo, evolution's intelligent screening process gave us two eyes for binocular vision and two ears for stereo sound. It also gave us nothing that would harm us.

Even the British naturalist Darwin himself states such:

> Charles Darwin: Natural selection will never produce in a being anything injurious to itself, for natural selection acts solely by and for the good of each.[194]

One might ask, Why? From where did this one-way process originate? And is it even true? If evolution were correct, chance must not only have been a really intelligent screening device, but also, according to Darwin, would be caring. If true, chance's intelligent screening process gave us:

- Five senses in order to successfully navigate our environment.
- A way to make tears that would keep our eyes lubricated, washed, and sterilized.
- Eyelids and eyelashes and a blinking reaction in order to protect our eyes from foreign objects.
- The ability for our pupils to dilate to allow more light to enter our eyes in a dark environment.
- Muscles attached to our eye's lenses in order to focus that light on our retinas to continuously produce clear images.
- The anatomy and physiology to see and interpret what we see, in color.
- Skin growths on the sides of our heads to direct and focus sound waves onto the tympanic membranes of our ears, not only to amplify sound, but also to help us determine which direction it comes from.
- A memory in order to learn from our mistakes.
- One-way valves in the extremities of our circulatory system to keep blood from backing up and thus forcing it to move only in one direction.
- A very intricate, multi-step (more than 12) blood-clotting mechanism which keeps us from bleeding to death from minor injuries.
- A very complex immune system to make it possible for us to live past our first infection.
- Hair growth on certain parts of our bodies in order to insulate the skin, absorb sweat, filter air, provide a lubricant to avoid skin rash, and to enhance beauty.
- An optimistic and determined outlook so that we wouldn't give up when we fell down as we learned to walk.
- The ability to learn to walk and keep our balance.
- An incredibly complex system for sexual reproduction.
- Strategically located and unique sphincter muscles that are always tight even while we sleep, that would have to be consciously relaxed in order to eliminate waste.
- A code of ethics and a concept of morality to solve problems and live in harmony with our fellow man.
- Consciousness; the ability to appreciate beauty,

---
[194]Charles Darwin, "Difficulties on Theory," *On the Origin of Species by Means of Natural Selection or the Preservation of Favoured Races in the Struggle for Life*, John Murray, Albemarle Street, London, 1859, p. 201.

create art, and love one another.
- A limbic system to foster motivation and emotional behaviors.

This screening intelligence also knew that we would need to grow teeth in order to prepare food for digestion. Somehow chance allowed us to grow a tongue with brain-controlled muscles to enable speech and to help us keep our food between our teeth. Notice sometime what your tongue does while you eat.

It also gave us a much-needed lubricant called saliva to help move food from our mouths into our stomachs. It then gave us a series of muscles that were somehow designed to work together to keep our food moving through our digestive systems in a one-way action known as peristalsis.

Somehow, chance's intelligent screening device gave priority to those individuals whose hearts and lungs increased blood and air movement when they work hard or exercise. It also was kind enough to create sweat glands that would excrete moisture which by evaporation would help cool our bodies upon exertion.

And this mechanism gave us more brainpower than we would ever use and skull bones to protect our brains. These facts created a dilemma that still has not been resolved:

> Richard Milner: Alfred Russel Wallace, Charles Darwin's "junior partner" in discovering natural selection, had a disturbing problem: He did not believe the theory could account for the evolution of the human brain.... In *[On the] Origin of Species* (1859), Darwin had concluded that natural selection makes an animal only as perfect as it needs to be for survival in its environment. But it struck Wallace that the human brain seemed to be a much better piece of equipment than our ancestors really needed.
>
> After all, he reasoned, humans living as simple tribal hunter-gatherers would not need much more intelligence than gorillas. If all they had to do was gather plants and eggs and kill a few small creatures for a living, why develop a brain capable not only of speech, but also of composing symphonies and doing higher mathematics?... Nevertheless, Wallace's problem remains unsolved; the emergence of the human mind is still a mystery.[195]

Obviously, this intelligent screening device is benevolent.

It might be said that, initially, evolution depends on Chance (the absence of planning) in a void filled with Nothing, and that Everything has come, by chance, from Nothing.

And it might be said that creation depends on a Creator, a Planner, a Master Designer Who left nothing to Chance and looked at the void filled with Nothing, was lonely, and said, "Let there be light—and a growing universe, and recently our earth—for people who would all love as He loves (though His dream is not yet reality).

Serious doubt about chance has existed in the scientific community for over 50 years. In 1966, The Wistar Institute Symposium, "Mathematical Challenges to the Darwinian Interpretation of Evolution," brought together mathematicians and biologists of impeccable academic credentials, including British biologist Sir Peter B. Medawar, 1960 winner of the Nobel Prize in physiology. In his introductory address, Sir Peter acknowledged the existence of a widespread feeling of skepticism over the role of chance in evolution with these words: "Something is missing from orthodox theory."[196]

> Michael J. Denton, MD, PhD: It is the sheer universality of perfection, the fact that everywhere we look, to whatever depth we look, we find the elegance and ingenuity of an absolutely transcending quality, which so mitigates against the idea of chance. Is it really credible that random processes could have constructed a reality, the smallest element of which—a functional protein or gene—is complex beyond our own creative capacities, a reality which is the very antithesis of chance, which excels in every sense anything produced by the intelligence of man?[197]

Mathematicians agree that any requisite number beyond 1 in 10^{50} [1 followed by 50 zeros] has, statistically, a zero probability of occurrence.[198] And what did popular American astronomer

---

[195] Richard Milner, "Wallace's Problem, Evolution of Human Brain," *Encyclopedia of Evolution*, Facts on File, New York, NY, 1990, p. 457.
[196] Sir Peter B. Medawar, Remarks by chair in *Mathematical Challenges to the Darwinian Interpretation of Evolution*, Wistar Institute Symposium Monograph, Vol. 5, p. xi, 1966.
[197] Michael Denton, PhD, "The Puzzle of Perfection," *Evolution: a Theory in Crisis*, Adler & Adler, NY, 1986, p. 342.
[198] I. L. Cohen, *Darwin Was Wrong—A Study in Probabilities*, New Research Publications, Greenvale, NY, 1984, p. 205.

Carl Sagan, PhD, have to say about chance? Sagan estimated that the chance of life evolving on any given single planet, like the earth, is one chance in $1 \times 10^{2,000,000,000}$ [that's 1 with two billion zeros after it]. This figure is so large that it would take 6,000 books of 300 pages each just to write that number.[199] Put these facts together and one has to conclude that life cannot evolve from non-life.

Jean Rostand: Transformism (evolution) is a fairy tale for adults.[200]

---

[199] Bert Thompson & Brad Harrub, "15 Answers to John Rennie and *Scientific American's Nonsense,*" Montgomery, Alabama: Apologetics Press, September 2002, p. 31.

[200] Jean Rostand, a famous French biologist and member of the Academy of Sciences of the French Academy, *Age Nouveau*, February 1959, p. 12.

Chapter 9
Refreshing Acknowledgments

Some scientists claim there is no evidence for creation. Yet Yale professor Benjamin Silliman, the first president of the Association of American Geologists did just that:

> With the Bible in my hands, and the world before me, I think I perceive a perfect harmony between science and revealed religion. . . . It cannot be doubted that there is a perfect harmony between the works and the word of God.[201]

It's OK to ask questions. A statement by a Yale Professor of Paleontology motivates some profound questions:

> Richard S. Lull, PhD: Since Darwin's day, evolution has been more and more generally accepted, until now in the minds of informed, thinking men there is no doubt that it is the only logical way whereby the creation can be interpreted and understood. We are not so sure, however, as to the *modus operandi* [manner of operating], but we may rest assured that the great process has been in accordance with great natural laws, some of which are as yet unknown, perhaps unknowable.[202]

What? Commenting on this statement, respected chemist and entomologist Anthony Standen, executive editor of the 22-volume second edition of the *Kirk-Othmer Encyclopedia of Chemical Technology*, believed that scientists and their theories should be subjected to the same scrutiny and criticism that other professionals face:

> Anthony Standen: And so biologists continue to "rest assured." But one may be tempted to ask, if some of the great natural laws are as yet unknown, how do we know that they are there? And if some are perhaps unknowable, how do we know that they are "logical?"[203]

To illustrate the changing face of this controversy, I quote a physician from Australia:

> Bernard Brandstater, MBBS: Some data from science are truly impressive. The specialties are vast: cosmology, cell chemistry and DNA, geology, radiometric dating, genetics, and much more. These sciences have provided the main arguments for evolutionary theory. In the past they have produced data we must explain within our Creation story. But let me say plainly what you may not realize, and you may doubt: *These arguments, once threatening to faith, are now crumbling.* And that's not just my private judgment. Highly respected scientists are now questioning the foundations of secular [no Creator] cosmogony [the branch of science that deals with the origin of the universe, especially the solar system] and evolution. *Earlier theories are proving untenable*, and they are searching for new ones.[204]

A scientific awakening materializes

> Ariel Roth, PhD: The legendary British philosopher Antony Flew has written nearly two dozen books on philosophy, has been a champion icon for atheists for decades, and has been called the world's most influential philosophical atheist. However, he has recently found some of the evidence from science quite compelling and has changed his view from atheism to believing that some kind of God must be involved to explain what science is discovering. In his own words, he "had to go where the evidence leads." He points out that "the most impressive arguments for God's existence are those that are supported by recent scientific discoveries. . . . It now seems to me that the findings of more than fifty years of DNA research have provided material for a new and enormously powerful argument to design."[205]

---

[201]Benjamin Silliman, quoted by George P. Fisher, *The Life of Benjamin Silliman*, Vol. 2, Charles Scribner, NY, 1866, p. 148.
[202]Richard Swann Lull, PhD, *Organic Evolution*, The Macmillan Company, NY, 1948, p. 15.
[203]Anthony Standen, *Science Is a Sacred Cow*, E. P. Dutton & Company, NY, 1950, p. 106; quoted by Howard A. Peth, "Is Evolution Scientific?" *Blind Faith: Evolution Exposed*, Amazing Facts, Inc., Frederick, MD, Howard A. Peth, 1990, p. 7.

[204]Bernard Brandstater, MBBS, "The Centrality of Creation in Adventist Belief," paper presented to Adventist Theological Society Spring Symposium, Southwestern Adventist University, Keene, TX, April 17, 2010, p. 3, (emphasis in the original).
[205]Ariel Roth, PhD, "How Did Life Get Started?" *Science Discovers God: Seven Convincing Lines of Evidence for His Existence*, Autumn House Publishing, a division of Review and Herald Publishing, Hagerstown, MD, 2008, p. 91; quoting Antony Flew, Gary R. Habermas, "My Pilgrimage from Atheism to Theism: an Exclusive Interview with Former British Atheist Professor," *Philosophia Christi, Journal of the Evangelical Philosophical Society*, Orlando,

Craig J. Hazen, PhD: Antony Flew was a world-class atheist philosopher, with whom I had many "conversations" (via his books) when I was doing my D.Phil. at Oxford on the relation between science and theology, 1961-1964. I remember him as one of the clearest writers on various issues. . . . He wrote clearly, gracefully, and honestly.

The history of the relation between Flew and his interviewer, Gary Habermas, is a classic example of why we should befriend and respect those with whom we disagree, not berate, insult, demean, or ignore them because they are "not one of us."[206]

Here's another example of a highly respected scientist who changed his well-known worldview. Allan Rex Sandage, PhD, a one-time protégé of legendary astronomer Edwin Hubble, is known and respected as the greatest observational cosmologist in the world. He has worked at Mount Wilson and Mount Palomar observatories. The *New York Times* dubbed him the "Grand Old Man of Cosmology."[207]

In a 1985 Dallas, Texas, conference on science and religion, in a discussion about the origin of the universe, the panel was divided among scientists who believed in God and those who did not. The two groups were seated on separate sides of the stage. Everyone expected Sandage to sit with the doubters. But he didn't. He told the rapt audience that science had taken us to the first event, but it can't take us further to the First Cause. The sudden emergence of matter, space, time and energy pointed to the need for some kind of transcendence.[208]

> Dr. Sandage: It was my science that drove me to the conclusion that the world is much more complicated than can be explained by science. It is only through the supernatural that I can understand the mystery of existence.[209]

As Sandage would later write, "Many scientists are now driven to faith by their very work."[210] At the age of 50, Sandage became a Christian. When asked the famous question regarding whether it's possible to be a scientist and a Christian, Sandage replied, "Yes. The world is too complicated in all its parts and interconnections to be due to chance alone. I am convinced that the existence of life with all its order in each of its organisms is simply too well put together."[211]

Another scientist at the Dallas conference, Dean H. Kenyon, PhD, a biophysicist from San Francisco State University, repudiated the conclusions of his own book,[212] which seemed to be one of the most promising explanations for the mystery of how the first living cell could somehow self-assemble from nonliving matter. He declared that he had become critical of all naturalistic theories of origins because of the immense molecular complexity of the cell and the information-bearing properties of DNA. Kenyon now believed that the best evidence pointed toward a designer of life.[213]

A recent 'Bombshell' from a Cornell University professor

In order for evolution to "work," there must be a mechanism for change. John C. Sanford, PhD, a Cornell University geneticist with more than 30 patents and 80 scientific publications has published *Genetic Entropy and the Mystery of the Genome*, a 2008 book that shows how population genetics rules out random mutations and Darwin's natural selection as viable mechanisms for speciation and evolution. He concluded that the genome is deteriorating, not improving.[214]

Dr. Sanford states that the neo-Darwinian theory—that mutation plus natural selection can explain all aspects of life—is taught universally,

---

Florida, December 9, 2004, pp. 2, 5; www.biola.edu/antonyflew/flew-interview.pdf. *See also:* Antony Flew, Roy Abraham Varghese, There is a God: how the world's most notorious atheist changed his mind, HarperOne, NY, October 23, 2007.
[206]Craig J. Hazen, PhD, Professor of Comparative Religion, Biola University, "Atheist Becomes Theist," *Philosophia Christi, Journal of the Evangelical Philosophical Society*, Biola University, La Mirada, California, Winter 2004.
[207]John Noble Wilfod, "Sizing Up the Cosmos: An Astronomer's Quest," *New York Times*, March 12, 1991.
[208]Lee Strobel, "Where Science Meets Faith," *The Case for a Creator*, Zondervan, Grand Rapids, MI, 2004, pp. 69-70.
[209]Allan Rex Sandage, PhD, quoted by Sharon Begley, "Science Finds God," *Newsweek*, July 20, 1998, p. 46.
[210]Allan Rex Sandage, PhD, "A Scientist Reflects on Religious Belief," available at: www.leaderu.com/truth/1truth15.html, January 7, 2003.
[211]Allan Rex Sandage, PhD, "A Scientist Reflects on Religious Belief, *The Religion of Scientists*, Part 2: Great Scientists Who Believe," SDA Global.
[212]Dean H. Kenyon, PhD, and Gary Steinman, *Biochemical Predestination*, New York: McGraw Hill, 1969.
[213]Lee Strobel, "Where Science Meets Faith," *The Case for a Creator*, Zondervan, Grand Rapids, MI, 2004, pp. 70-71.
[214]John C. Sanford, PhD, *Genetic Entropy and the Mystery of the Genome*, FMS Publishing, Waterloo, NY, 2008.

and is almost universally accepted. "It is the constantly mouthed mantra, repeated endlessly on every college campus." Late in his career he states that he did something "that would seem unthinkable for a Cornell professor." He began to question the neo-Darwinian theory. To his amazement, he gradually realized that "the great and unassailable fortress" which had been built up around it is really a house of cards—extremely vulnerable and indefensible. "Its apparent invincibility derives largely from bluster, smoke, and mirrors. . . . What I eventually experienced was a complete overthrow of my previous understanding. . . . Furthermore, I began to see that this deep-seated faith in natural selection [causing evolution] is typically coupled with a degree of ideological commitment which can only be described as religious. I started to realize . . . that I might be offending the religion of a great number of people!"[215]

Dr. Sanford became convinced that the theory was most definitely wrong, and that it could be shown to be wrong to any reasonable and open-minded individual. "This realization was both exhilarating and frightening. I realized that I had a moral obligation to openly challenge this most sacred of cows." Dr. Sanford became convinced that it "is insidious on the highest level, having a catastrophic impact on countless human lives." He states that every form of objective analysis he performed convinced him that it is clearly false. "So now, regardless of the consequences," he wrote, "I have to say it out loud [reflecting on a Danish fairy tale by Hans Christian Andersen]: *the Emperor has no clothes!*"[216]

The mechanism for change

If you are going to develop complex systems or structures beyond the results of simple mutations, is there some mechanism that would cause new information to be added to our DNA? Two completely different mechanisms now promoted by evolution as mechanisms for change include natural selection and *beneficial mutations.* We will address a third, punctuated equilibria, in a later chapter. Natural selection is supposed to select a good mutation over any existing genes or other detrimental mutations that code for a new function. Then this good mutation must be inherited by offspring. However, it must be emphasized that natural selection and mutations are not enough. New genetic information must be added.[217]

> Speaking from his years as a Cornell University geneticist Dr. Sanford explains: The notion of gene duplication as a way to "evolve new information" has become very firmly entrenched within the evolutionary community. I believe this is partly because, "It must be true! How else could evolution have happened?" I also believe that when a mantra is mouthed often enough, it takes on the appearance of unassailable truth. But careful analysis of what information really is, and how it arises, combined with a healthy dose of common sense, should reveal to us that random duplications are consistently bad. It is my personal opinion that "evolution through random duplications" is for the most part a widely-held philosophical assumption, rather than a scientifically-defensible observation. I believe that while it sounds quite sophisticated and respectable, it does not withstand honest and critical assessment.[218]

The question is, can natural selection and survival of the fittest cause one kind to become a new kind? No. Can it add new information? No. And the reason is simple. Natural selection works ONLY with existing information. Let me reiterate, we have more than *400 different kinds* of dogs. But they are all dogs. We have different ethnic groups of people. But they are all humans. Natural Selection never adds anything new, and in fact, it usually causes a *loss* of information.[219]

> Elmer Noble, Glenn Nobel, Gerhard Schad, and Austin MacInnes—four PhDs: Natural selection can act only on those biological properties that already exist; it cannot create properties in order to meet adaptational needs.[220]

---

[215] John C. Sanford, PhD, "Prologue," *Genetic Entropy and the Mystery of the Genome,* FMS Publishing, Waterloo, NY, 2008, pp. vi-vii.
[216] John C. Sanford, PhD, "Prologue," *Genetic Entropy and the Mystery of the Genome,* FMS Publishing, Waterloo, NY, 2008, pp. vi-vii, (emphasis in the original).
[217] *The Fossil Record,* DVD, Answers in Genesis-USA, P. O. Box 510, Hebron, KY, 2004.
[218] John C. Sanford, PhD, "Appendix 4," *Genetic Entropy and the Mystery of the Genome,* FMS Publishing, Waterloo, New York, 2008, pp. vi-vii., p. 197, (emphasis in the original).
[219] *The Origin of Humans,* DVD, Answers in Genesis-USA, P.O. Box 510, Hebron, KY, 2004.
[220] Elmer Noble, PhD, Glenn Nobel, PhD, Gerhard Schad, PhD, Austin MacInnes, PhD, *Parasitology: The Biology of Animal Parasites,*

Werner Gitt, PhD: There is no known law of nature, no known process and no known sequence of events which can cause information to originate by itself in matter.[221]

In his Harvard University-published book, Ernst Mayr, PhD, notes: The fact that the [evolutionary] theory is now so universally accepted is not in itself proof of its correctness. . . . The basic theory is in many instances hardly more than a postulate [supposition assumed without proof].[222]

6th edition, Lea & Febiger, Philadelphia, London, 1989, p. 516.
[221] Werner Gitt, PhD, (Director of the German Federal Institute of Physics and Technology) "Information in Living Organisms," *In the Beginning was Information: A Scientist Explains the Incredible Design in Nature,* Master Books, a Division of New Leaf Publishing Group, Green Forest, AR, 2005, p. 106.
[222] Ernst Mayr, *Animal Species and Evolution,* Harvard University Press, Cambridge, MA, 1963, pp. 7-8.

II
Science and Life

Chapter 10
What Is Science?

Webster defines science as: 1. knowledge possessed or attained through study or practice. 2. accumulated and accepted knowledge that has been systematized and formulated with reference to the discovery of general truths or the operation of general laws.[223]

And the *American Heritage Dictionary* defines science as: the observation, identification, description, experimental investigation, and theoretical explanation of phenomena.

Linus Pauling, PhD, a two-time winner of the Nobel Prize once stated, "Science is the search for truth."[224]

> National Academy of Sciences: Science is a particular way of knowing about the world. In science, explanations are limited to those based on observation and experiments that can be substantiated by other scientists. Explanations that cannot be based on empirical evidence are not part of science.[225]

And what is a theory?

> National Academy of Sciences: In scientific terms, "theory" does not mean "guess" or "hunch" as it does in everyday usage. Scientific theories are explanations of natural phenomena built up logically from testable observations and hypotheses.[226]

How does science determine scientific fact, and how does that impact creation science?

> Clyde L. Webster, Jr., PhD: The strength of science exists in its ability to make observations of the physical world and predict future events. Herein also lies one of the greatest limitations of science. If the observations are limited only to incomplete evidence of past events and it is impossible to conduct appropriate experimentation in order to test hypotheses concerning the past, the strength of the scientific method has been neutralized and one is left with little more than speculation....
>
> The conclusions drawn from research, especially in the untestable [i.e., unfalsifiable] areas, are heavily influenced by the assumptions used to conduct the studies. Logically it is reasonable to maintain that the conclusions drawn and the theories proposed from such research are also influenced by one's personal philosophy and convictions.[227]

> Gareth J. Nelson and Norman Platnick: If our knowledge is limited to those things we can observe directly, the task of studying the history of life is indeed fraught with insurmountable difficulties, since we can hardly observe directly the past history of present-day organisms.[228]

> Harold G. Coffin, PhD: Many scientists and laypeople assume that evolution theory is scientific but creation theory is not. That the facts support only evolution and it alone is open to study by the scientific method of observation and experimentation....
>
> When someone first proposes a theory (better called a hypothesis), it often is poorly supported by the facts. But as research progresses, a hypothesis either becomes better documented and more widely accepted (changes to a theory), or [else] it continues to lack substantiating evidence, and another hypothesis eventually replaces it. Has the foundation for evolution theory become more firmly cemented by data obtained in the more than a century since Darwin? The answer is both Yes and No, depending on the definition of evolution. Yes, biology has thoroughly documented minor changes—microevolution is a "fact." No,

---

[223] *Webster's Third New International Dictionary of the English Language,* Unabridged, G. & C. Merriam Co, United States of America, 1976, p. 2032.
[224] Linus Pauling, PhD, "A Proposal: Research for Peace," *No More War,* Dodd, Mead & Co., NY, 1958, p. 209.
[225] "Introduction," *Science and Creationism: A View From the National Academy of Sciences,* National Academy of Sciences, Washington, DC, 1999, p. 1.
[226] "Is Evolution a Fact or Theory," *Science and Creationism: A View From the National Academy of Sciences,* National Academy of Sciences, Washington, DC, 1999, p. 28.
[227] Clyde L. Webster, Jr., PhD, "Nature of Science," *A Scientist's Perspective on Creation and the Flood,* Geoscience Research Institute, Loma Linda, CA, 1995, p. 26.
[228] Gareth J. Nelson and Norman Platnick, "Comparative Biology: Space, Time, and Form," *Systematics and Biogeography, Cladistics and Vicariance,* Columbia University Press, NY, 1981, p. 7.

we have not substantiated major changes despite much experimentation in laboratories and search for fossil evidence—macroevolution is still a theory. Individuals who should know that only minor changes can be proved and who fail to distinguish between micro- and macroevolution have published many careless claims.[229]

Science is supposed to be knowledge that is attained through systematized observation, study, or practice. The scientific method formulates a hypothesis, then tests the hypothesis by the collection of data through observation and if possible experimentation. Hopefully, the end result turns the hypothesis into scientific theory. If the theory cannot be substantiated, it is considered false.

Even back in 1960, G. A. Kerkut, Professor of Physiology and Biochemistry at the University of Southampton, England, wrote the book *Implications of Evolution,* to expose the weaknesses and fallacies in the evidence used to support the theory of evolution. Although an evolutionist, Kerkut was honest enough as a scientist to admit: "The evidence that supports it is not sufficiently strong to allow us to consider it anything more than a working hypothesis."[230]

Dr. John T. Bonner, one of the nation's leading biologists and a professor at Princeton University, wrote a review of Kerkut's book that was as startling as the book itself:

> John T. Bonner, PhD: This is a book with a disturbing message; it points to some unseemly cracks in the foundations. One is disturbed because what is said gives us the uneasy feeling that we knew it for a long time deep down but were never willing to admit it even to ourselves. . . . The particular truth is simply that we have no reliable evidence as to the evolutionary sequence of invertebrate phyla [organisms without vertebrae]. We have all been telling our students for years not to accept any statement on its face value but to examine the evidence, and, therefore, it is rather a shock to discover that we have failed to follow our own sound advice.[231]

I want to emphasize here and now that I still have lots of questions. I don't pretend to know all the answers. But neither does anyone else.

When I was a teenager, I felt like I had all the answers. As I grew older I started realizing just how little I knew. Questions: Am I searching for truth? How much do I really know? Half of everything? One percent? Neither did Einstein. Not even a millionth of one percent.

Considering that a thoracic surgeon goes to school for 29 years before he completes his education, and considering all the specialties and subspecialties in fields like medicine, surgery, law, engineering, astronomy, paleontology, and hundreds of other professions, we realize just how little we really understand about the known world.

And what if God and the Supernatural are just in the percentage of knowledge the atheist and naturalist do not yet know?

> Alex Williams and John Hartnett, PhD: At this point it is worth reflecting on a global trend that is emerging in all areas of science. In cosmology we find that the more that astronomers look into space, the more they find—layer upon layer of complexity. The more that mathematicians examine their equations, the more mathematics they discover. The deeper that particle physicists delve into the structure of the atom, the more structure they find. In biology, the more that the living cell is studied, the more they are finding—layer upon layer of complexity. The more that ecologists study the web of life on earth, the more interactions, the more dependencies, and the more connectedness they find. The more the human person is studied, the more complex we find the interactions to be between mind and body and between cells and organs within the body.
>
> No one is reaching a point where [he or she] can say, "We now know it all." We believe that this points to the unfathomable depths of the Creator. A materialist universe, on the other hand, would presumably have some kind of baseline of matter from which all else is built—but this is not what we are finding.[232]

Science acknowledges ignorance.

---

[229]Harold G. Coffin, PhD, "Summary—Is Creation a Viable Theory of Origins?" *Origin By Design,* Review and Herald Publishing Association, Washington, DC; Hagerstown, MD, 1983, p. 442.
[230]G. A. Kerkut, *Implications of Evolution,* Pergamon Press, NY, 1960, p. 157.
[231]John T. Bonner, Review of Kerkut's book, *American Scientist,* June 1961, p. 240; Quoted by Howard A. Peth, "Is Evolution Scientific?" *Blind Faith: Evolution Exposed,* Amazing Facts, Inc., Frederick, MD, 1990, p. 11.
[232]Alex Williams and John Hartnett, PhD, "Tools for Explaining the Universe," *Dismantling the Big Bang: God's Universe Rediscovered,* Master Books, Inc., Green Forest, AR, p. 105.

Paul Hilpman, PhD: I can tell you this, and any scientist will tell you, that we haven't the foggiest idea of where matter and energy came from. We can just say through the conservation and principles of the laws of thermodynamics that nothing is created or destroyed. But where it came from initially, we haven't the slightest idea.... I'm not at all going to say that there aren't geologists or scientists who want to prove that evolution is right. And they will only report the evidence that supports them and won't admit to problems.[233]

It would be difficult to better illustrate what science ought to be than the contents of an amazing 443-page book, *The Encyclopedia of Ignorance*. Some of the world's leading scientists leapt at the opportunity to contribute chapters on what they did not know about their specialties. As the editor stated, "The more eminent they were, the more ready to run to us with their ignorance."[234] Is it any surprise that this remarkable encyclopedia devotes more space to the problems of evolution—especially the problems evolutionists have not solved concerning how humans might have evolved—and the mysteries of astronomy, than to any other topics?

In a refreshing acknowledgment, to illustrate the fact that most scientists are honest, the following quotation is taken from a university textbook under the subheading, "Mathematical Challenges to the Evolution Theory:"

William D. Stansfield, PhD: It would be quite wrong for the student of evolution to obtain the impression from reading this or any other text on evolution that adequate answers have now been found to all the problems in this field. Nothing could be further from the truth! Many of the questions that troubled Darwin are still being raised today. Are the processes of random variation and natural selection really sufficient to account for the diversity of life that we see around us? One of the most widely publicized recent debates of this age-old question took place . . . at the Wistar Institute of Anatomy and Biology in Philadelphia between a group of mathematicians and biologists. The mathematicians charged that if natural selection had to choose from the astronomically large number of available alternative systems by means of the mechanisms described in current evolution theory, the chances of producing a creature like ourselves is virtually zero.[235]

---

[233]Paul Hilpman, PhD, quoted in DVD Debate No. 5, University of Missouri, "Does geology support creation, or evolution?" Creation Science Evangelism, 29 Cummings Road, Pensacola, FL, 1995.
[234]Ronald Duncan, "Editorial Preface," (ed.), Miranda Weston-Smith (ed.), *The Encyclopedia of Ignorance*, A Wallaby Book, Pocket Books, NY, 1977, p. ix.

---

[235]William D. Stansfield, PhD, "Man," *The Science of Evolution*, MacMillan Publishing Co., Inc., New York; Collier Macmillan Publishers, London, 1977, p. 571.

Chapter 11
Why Is Creation an Important Issue?

The preponderance of scientific evidence

Jonathan D. Sarfati, PhD: Many people have the belief that "science" has proven the earth to be billions of years old, and that every living thing descended via evolutionary processes from a single cell, which itself is the result of a chance combination of chemicals. However, science deals with repeatable observations in the present, while evolution/long-age ideas are based on assumptions from outside science about the unobservable past. Facts do not speak for themselves—they must be interpreted according to a framework. It is not a case of religion/creation/subjectivity versus science/evolution/objectivity. Rather, it is the biases of the religions of Christianity and of humanism interpreting the same facts in diametrically opposite ways.[236]

Harold G. Coffin, PhD: Since both creation theory and evolution theory lie outside the realm of science, we cannot make a decision on the basis of which one is science and which one is not. We must determine which theory the total range of available evidences at hand *best supports* and which *comes closest* to the method of operation and results we have learned to expect of science.[237]

What is some of this evidence? One of the strongest arguments in favor of creationism is biogenesis—the fact that life comes only from life and that spontaneous generation of life is totally impossible.

The evolutionary theory of uniformitarianism (that change has always been uniformly gradual over billions of years) is no longer believed, at least as strongly as once accepted. This can be attributed to the Ice Age(s), and the giant asteroid that hit the edge of the Yucatan Peninsula (which some evolutionists believe killed the dinosaurs 65 million years ago).[238] So, throughout this book you will see numerous scientific examples which raise reasonable doubt about the theory of uniformitarianism.

We will look at the possibility of a global Flood in the days of Noah covering the tops of the highest mountains and ask where did all the water come from and where did it go? One point will emerge: evidence in favor of one worldview will be evidence against other worldviews. Then we will acknowledge that even though nobody has all the answers, and even though everyone has to exercise some faith, the preponderance of evidence for creationism points toward a worldview with a hopeful future and ample scientific support.

How do creationism and evolutionism relate to science?

Christianity and Darwinism are incompatible. Evolutionist Sir Arthur Keith agrees, but adds a significant twist when he says, "The conclusion I have come to is this: the law of Christ is incompatible with the law of evolution. . . . The two laws are at war with each other."[239]

Many evolutionists claim that their worldview is standing on a solid foundation of scientific data and that creationism is simply a religion based on biblical myths. They recoil at the suggestion that their worldview also is like a religion, complete with dogma, doctrine, evangelism, and indoctrination. Let's put this in perspective. Creationists and evolutionists have faith either that God or no God created the universe and earth and life—and both can find evidence they believe strongly supports these faiths. It takes a very open mind to weigh the evidences. Are most students today being taught the faith-based

---

[236] Edited by John Ashton, PhD, Jonathan D. Sarfati, PhD, "Jonathan D. Sarfati," *In Six Days: Why Fifty Scientists Choose to Believe in Creation*, Master Books, Inc., Green Forest, AR 72638, 2000, p. 75.
[237] Harold G. Coffin, PhD, "Summary—Is Creation a Viable Theory of Origins?" *Origin By Design*, Review and Herald Publishing Association, Washington, DC; Hagerstown, MD, 1983, p. 430, (emphasis added).
[238] Evolution, "What killed the dinosaurs?" *Evolution Library for Teachers and Students*, PBS Website, May 22, 2015.
[239] Sir Arthur Keith, "Human Life: Its Purpose or Ultimate End," *Evolution & Ethics*, G. P. Putnam's Sons, New York, 1946, 1947, p. 15. Keith wrote the forward to Darwin's book for the 1959 100-year anniversary reprint.

religion called evolution? Here are some evolutionists who agree:

> Michael Ruse, PhD: Evolution is promoted by its practitioners as more than mere science. Evolution is promulgated as an ideology, a secular religion—a full-fledged alternative to Christianity with meaning and morality. . . . Evolution is a religion. I am an ardent evolutionist, . . . but I must admit that this is one complaint . . . [where] the literalists are absolutely right. This was true of evolution in the beginning, and it is true of evolution still today.[240]

> Louis Trenchard More, PhD: The more one studies paleontology, the more certain one becomes that evolution is based on faith alone, exactly the same sort of faith which it is necessary to have when one encounters the great mysteries of religion.[241]

> G. H. Harper: For some time, it has seemed to me that our current methods of teaching Darwinism are suspiciously similar to indoctrination. . . . The teacher of Darwin's theory . . . undoubtedly is concerned to put across the conclusion that natural selection causes evolution, while he cannot be concerned to any great extent with real evidence, because there isn't any.[242]

> Eminent biologist and zoologist Leonard Harrison Matthews, ScD, FRS: The fact of evolution is the backbone of biology, and biology is thus in the peculiar position of being a science founded on an unproved theory—is this a science or a faith?[243]

Thomas Henry Huxley was more responsible for the *acceptance* of evolution than Darwin himself. As the foremost champion of evolution, he became known as "Darwin's Bulldog." He "became a brilliant press agent and enthusiastic salesman for the theory—debating, defending, and promoting it with untiring voice and pen."[244] Yet in 1896, Huxley made an honest confession:

> Thomas Henry Huxley: To say, therefore, in the admitted absence of evidence, that I have any belief as to the mode in which the existing forms of life have originated would be using words in a wrong sense. . . . I have no right to call my opinion anything but an act of philosophical faith.[245]

> Physical anthropologist David Pilbeam, PhD: I know that, at least in paleoanthropology, data are still so sparse that theory heavily influences interpretations. Theories have, in the past, clearly reflected our current ideologies instead of the actual data.[246]

A creationist responds.

> Harold G. Coffin, PhD: It is obvious . . . that evolution, as generally understood, is not a fact. Statements of the "fact" of evolution strike one as whistling in the dark. Much faith—more than required of a creationist—is needed by the believer in evolution, unless he blindly accepts what he is told by others.

> With all the time, technology, and materials available, man has not been able to produce a perpetual motion machine. In fact, the United States patent office has officially decided that such a machine is impossible and will not consider requests for patenting any such contraption. But mechanistic evolution requires that organisms build themselves out of nothing and then continually improve themselves as they maintain their existence![247]

---

[240]Michael Ruse, PhD, professor of physiology and zoology at the University of Guelph, www.omniology.com/*How Evolution Became Religion.*
[241]Louis Trenchard More, *The Dogma of Evolution*, Princeton University Press, New Jersey, 1925; Quoted by Howard A. Peth, "Is Evolution Scientific?" *Blind Faith: Evolution Exposed*, Amazing Facts, Inc., Frederick, MD, Howard A. Peth, 1990, p. 14.
[242]G. H. Harper, "Darwinism and Indoctrination," *School Science Review*, Vol. 59, December 1977, pp. 258, 265.
[243]Leonard Harrison Matthews, ScD, FRS, *Introduction to Darwin's On The Origin of Species*, J. M. Dent & Sons Ltd., London, 1971, p. xi.
[244]Howard Peth, "Is Evolution Scientific?" *Blind Faith: Evolution Exposed*, Amazing Facts, Inc., Fredrick, MD, 1990, p. 15.
[245]Thomas Henry Huxley, *Discourses Biological and Geological*, 1896, pp. 256, 257; quoted by Howard A. Peth, "Is Evolution Scientific?" *Blind Faith: Evolution Exposed*, Amazing Facts, Inc., Frederick, MD, Howard A. Peth, 1990, p. 15.
[246]David Pilbeam, PhD, "Rearranging Our Family Tree," *Human Nature*, June 1978, p. 45.
[247]Harold G. Coffin, PhD: "Science and Faith," *Creation—Accident or Design*, Review and Herald Publishing Association, Washington, DC, 1969, pp. 462-463.

Chapter 12
How Did Life Begin?

There is no generally accepted theory on how life started from non-life. Boys and girls are just taught that it happened.

National Academy of Sciences: For those who are studying the origin of life, the question is no longer whether life could have originated by chemical processes involving nonbiological components. The question instead has become which of many pathways might have been followed to produce the first cells. . . .

The study of the origin of life is a very active research area in which important progress is being made, although the consensus among scientists is that none of the current hypotheses has thus far been confirmed. The history of science shows that seemingly intractable problems like this one may become amenable to solutions later, as a result of advances in theory, instrumentation, or the discovery of new facts.[248]

Is this former statement realistic? Or is it just wishful thinking?

Ariel A. Roth, PhD: Without doubt scientific discoveries over the past decades have not been kind to evolution. Probably the most severe challenge evolution faces is the question of the origin of life itself.[249]

Stephen C. Meyer, PhD: The revolution in the field of molecular biology has revealed so great a complexity and specificity of design in even the simplest cells and cellular components as to defy materialistic explanation. Even scientists known for a staunch commitment to materialistic philosophy now concede that materialistic science in no way suffices to explain the origin of life.

As origin of life biochemist Klaus Dose has said, more than 30 years of experimentation on the origin of life in the fields of chemical and molecular evolution have led to a better reception of the immensity of the problem of the origin of life on earth rather than to its solution. At present all discussions on principle theories and experiments in the field either end in stalemate or confession of ignorance.[250]

Leonard R. Brand, PhD: Today's advances in biochemistry are not reducing the problems for abiogenesis [the spontaneous generation of life]. The more information accumulates on the nature of life, the more it indicates that natural law does not have the answer to the origin of life.[251]

David L. Cowles, PhD, and L. James Gibson, PhD: The first postulated step in abiogenesis is the production of simple organic molecules (for example, amino acids) from inorganic materials. Although these molecules have been synthesized, the conditions required are not plausible on an early earth. The next step is polymerization—the linking of the small molecules together. While a few natural conditions allowing polymerization have been found, none help form the precise, complicated sequences characteristic of molecules in living cells. The gap between *what random polymerization processes can be shown to produce and the simplest living cell is enormous.*

Another feature characterizing living things is the ability to reproduce detailed copies of themselves, which in turn are also able to reproduce. This highly complex process involves a whole suite of different molecules, all interacting with one another in a precisely directed way.

However, the entire complicated system of molecules is required in order for the cell to be able to copy itself. If any part of the chain of interacting molecules is missing, the entire process fails and the cell cannot function or reproduce itself. This fact has long been recognized as a formidable

---

[248]"The Origin of the Universe, Earth, and Life," *Science and Creationism: A View From the National Academy of Sciences*, National Academy of Sciences, Washington, DC, 1999, p. 3, (emphasis added).
[249]Ariel A. Roth, PhD, "Is Science in Trouble?" *Origins: Linking Science and Scripture*, Review and Herald Publishing Association, Hagerstown, MD, 1998, p. 323.
[250]Stephen C. Meyer, PhD, *The Explanatory Power of Design*, in William A. Dembski, ed. *Mere Creation: Science, Faith and Intelligent Design*, Downers Grove, IL: Inter-Varsity Press, 1998, p. 118, see also: Klaus Dose, "The Origin of Life: More Questions Than Answers," *Interdisciplinary Science Review*, 1998, p. 13.
[251]Leonard R. Brand, PhD, "The Origin of Life," *Faith, Reason, and Earth History*, Andrews University Press, Berrien Springs, MI, 1997, p. 109.

challenge for the evolutionary theory of the origin of life.[252]

In reading Darwin's book, *On the Origin of Species*, I was surprised to see references to creation in a number of places. It seems as though at one point he may have believed that life was created in a simple form and evolved to its present complexity over countless eons of time:

> Charles Darwin: I should infer from analogy that probably all the organic beings which have ever lived on this earth have descended from some one primordial form, into which life was first *breathed*.
>
> ... The whole history of the world, as at present known, although of a length quite incomprehensible by us, will hereafter be recognized as a mere fragment of time, compared with the ages which have elapsed since the first creature, the progenitor of innumerable extinct and living descendants, was *created*.[253]

Nevertheless, many evolutionists today do not believe in a Creator who spoke or "breathed" life into the first primordial creature or first human, and that leads to their belief in the unscientific theory of the spontaneous generation of life.

According to the University of Jena Trial Records, Ernst Haeckel, MD, claimed also that spontaneous generation must be true, not because it had been proven in the laboratory, but because otherwise it would be necessary to believe in a creator.[254]

So, what about the spontaneous generation of life?

> *Glencoe Biology:* For early scientists, the idea that mud produced fish and that rotting meat produced flies were reasonable explanations for what people observed. After all, maggots seemed to simply materialize on meat and then change into flies. These and other observations led scientists to believe in the idea of *spontaneous generation*, a process by which life was thought to be produced from nonliving matter.
>
> In 1668, an Italian physician, Francesco Redi, designed a controlled experiment to test the idea of spontaneous generation.... Although Redi ... disproved spontaneous generation of larger organisms, microorganisms were so numerous and widespread that it was believed that they arose spontaneously from a vital force in the air.
>
> In the mid-1800s, a French scientist, Louis Pasteur, decided to test this idea of a vital force in air. To disprove spontaneous generation once and for all, Pasteur realized that he would have to set up an experiment in which only air, and not microorganisms, was allowed to come in contact with a nutrient broth....
>
> Pasteur showed that no microorganisms would arise in the broth, even in the presence of air. With his experiment, Pasteur claimed to have "driven partisans of the doctrine of spontaneous generation into the corner." *Biogenesis*, the idea that living organisms come only from other living organisms, became a cornerstone of biology.[255]

Spontaneous generation, the emergence of life from nonliving matter, has never been observed. This fact has been so consistent that it is has led to the Law of Biogenesis. Materialist philosophy, the basis for evolution, conflicts with this law by claiming that life came from nonliving matter through natural processes.[256] And yet, leading evolutionists are forced to accept some form of spontaneous generation *no matter how illogical and unscientific the conclusion:*

> George Wald, PhD: The reasonable view was to believe in spontaneous generation; the only alternative, to believe in a single, primary act of supernatural creation. There is no third position.[257]
>
> ... One has only to contemplate the magnitude of this task to concede that the spontaneous generation of a living organism is impossible.

---

[252] David L. Cowles, PhD, L. James Gibson, PhD, "Does the Theory of Evolution Explain the Diversity of Life?" *Understanding Creation*, Pacific Press Publishing Association, Nampa, Idaho, 2011, p. 68, see also: S. C. Meyer, *Signature in the Cell: DNA and The Evidence for Intelligent Design*, Harper Collins Publishers, NY, 2009, (emphasis added).

[253] Charles Darwin, "Recapitulation and Conclusion," *On the Origin of Species By Means of Natural Selection, or the Preservation of Favoured Races in the Struggle for Life*, John Murray, Albemarle Street, London, 1859, pp. 484, 488, (emphasis added).

[254] Records from University of Jena trial in 1875. Dr. Edward Blick, Norman, Oklahoma.

[255] "The History of Life," *Glencoe Biology, The Dynamics of Life*, Glencoe/McGraw Hill, New York, NY; Columbus, OH; Mission Hills, California; Peoria, Illinois, 1995, p. 410-411, (emphasis in the original).

[256] L. James Gibson, PhD, Director, Geoscience Research Institute, note to author, July 23, 2013; Walt Brown, Jr., PhD, "Life Sciences," *In the Beginning: Compelling Evidence for Creation and the Flood*, Center for Scientific Creation, Phoenix, AZ, 2001, p. 5.

[257] George Wald, PhD, "The Origin of Life," *Scientific American*, Vol. 190, August 1954, p. 46.

Yet here we are—as a result, I believe, of spontaneous generation.[258]

Scientists who believe in creation respond to the concept:

> Clyde L. Webster Jr., PhD: One of the most serious problems with evolution is its inability to account for the spontaneous generation of even the simplest form of life. From a biochemical point of view the odds against the spontaneous generation of a simple protein are so immense that spontaneous generation of more complex material is virtually impossible. Science is also unable to account for the spontaneous generation of the *complex interrelated systems* necessary for life.[259]

> Ariel A. Roth, PhD: The problem of origins has become even more complicated with the discovery of "programmed" systems such as the genetic code, elaborate gene control systems, and the correcting systems for DNA replication. To the best of our knowledge, such complex kinds of programs do not arise spontaneously—they appear to represent intelligent design such as we would expect from a Creator.[260]

Andrew Knoll, PhD, professor of biology at Harvard University was asked, "How does life form?" He responded, "The short answer is we don't really know how life originated on this planet. There have been a variety of experiments that tell us some possible roads, but we remain in substantial ignorance."[261]

> National Academy of Sciences: Will we ever be able to identify the path of chemical evolution that succeeded in initiating life on earth? Scientists are designing experiments and speculating about how early earth could have provided a hospitable site for the segregation of molecules in units that might have been the first living systems. The recent speculation includes the possibility that the first living cells might have arisen on Mars, seeding earth via the many meteorites that are known to travel from Mars to our planet.[262]

Notice the words "speculating," "could have," "might have (twice)," "speculation," and "the possibility." Is that science? It is fine for a scientist to entertain possibilities and to speculate at the growing edge of his research—his questions may suggest a new hypothesis to examine through observation and experimentation. However, since we were not "there" to observe and experiment about the origin of our universe, and when there is no real, hard evidence to support the speculation, what is offered as speculation, over time, with repetition, can become accepted as fact.

Eighteen years later, the following quotation from *PNAS (Proceedings of the National Academy of Sciences of the United States of America)* reflects this move from speaking of models as "speculating" to wording them now as if they are facts. Here is a model of how life had to form in warm little ponds undergoing wet/dry cycles on earth, with RNA coming from outer space via a meteorite:

> Ben K.D. Pearce, Ralph E. Pudritz, Dmitry A. Semenov, and Thomas K. Kenning: Before the origin of simple cellular life, the building blocks of RNA (nucleotides) had to form and polymerize in favorable environments on early Earth. At this time, meteorites and interplanetary dust particles delivered organics such as nucleobases (the characteristic molecules of nucleotides) to warm little ponds whose wet-dry cycles promoted rapid polymerization.[263]

No longer meteorites "could have" or "might have" delivered RNA building blocks, now they "delivered" them. This shift may be because the problem of finding a satisfying explanation of the origin of life and its macromolecules has become so acute that secular scientists have turned to *space* as a source of macromolecules and genetic information, thus quietly conceding that life could never have self-originated on earth.

Evolutionist Richard Dawkins, PhD, also suggested that life came from outer space. But how did it get started *there?* As you read the following comment, ask yourself, is this science or is it just more conjecture?

---

[258] George Wald, PhD, "The Origin of Life," *Scientific American*, Vol. 191, August 1954, pp. 44-53.
[259] Clyde L. Webster, Jr., PhD, "Origins," *A Scientist's Perspective on Creation and the Flood*, Geoscience Research Institute, Loma Linda, CA, 1995, p. 10, (emphasis added).
[260] Ariel A. Roth, PhD, "A Few Final Words," *Origins: Linking Science and Scripture*, Review and Herald Publishing Association, Hagerstown, MD, 1998, p. 356.
[261] http://www.time.com/time/magazine/article/0,9171,979365-2,00.html, (emphasis added).
[262] "The Origin of the Universe, Earth, and Life," *Science and Creationism: A View From the National Academy of Sciences*, National Academy of Sciences, Washington, DC, 1999, pp. 6-7.
[263] Origin of the RNA world: The fate of nucleobases in warm little ponds. Ben K.D. Pearce, Ralph E. Pudritz, Dmitry A. Semenov, and Thomas K. Kenning. PNAS October 24, 2017. 114 (43) 11327-11332; https://doi.org/10.1073/pnas.1710339114

Richard Dawkins, PhD: It could come about in the following way: it could be that, at some earlier time somewhere in the universe a civilization evolved probably by some kind of Darwinian means to a very high level of technology and designed a form of life that they seeded onto perhaps this planet . . . and that designer could well be a higher intelligence from elsewhere in the universe.[264]

Notice, Dawkins acknowledged a possible designer and a higher intelligence. Have scientists ever produced life in the laboratory? Absolutely not. Nobody, including noted members of the sophisticated scientific community of the 21st Century, has even come close to creating life. And the more we learn about what has been taught to millions of children about how life *might* have started by naturalistic means, the more impossible it has become.

To illustrate the *tentative* nature of science education regarding the beginning of life by the process of "abiogenesis" [living creatures came from non-living matter], in just three pages of a classic textbook I found the following honest acknowledgments and forewarnings: "How these elements present in the atmosphere *could have formed* simple organic compounds important to life is a challenging scientific puzzle. Some scientists have performed experiments that suggest *that amino acids might have formed* chains spontaneously in the early atmosphere. . . . Amino acids *could have combined* and formed complex proteins. . . . Such results *suggest* many ways that vital organic molecules *might have formed* on the young earth. . . . Similar mechanisms *might have led* to the formation of carbohydrates, lipids, and nucleic acids." And interspersed were the following caveats: "probably descended," "Oparin suggested," "must have occurred," "could have formed," "might have formed," "Oparin hypothesized," "would have resulted," "Oparin assumed," "might have been concentrated," "might have developed," and "might have eventually crossed over the border into the living world."[265]

In an article in *Reader's Digest* published shortly before his death, William Jennings Bryan, former Secretary of State and the most widely known creationist of his generation, stated that Darwin used the phrase "we may well suppose" 800 times.[266]

On the evening of January 17, 2012, I was amazed to see what appeared to be a scientific history of the universe on Direct TV that had so many "waffle words," I wrote some of them down. They included: "if," "possibly," "perhaps," "we imagine," "we think," "the most likely," "it could be supposed," "the hope," "one possibility is," " the notion that," "how potential life forms might," "could take," "could have," "what looks like," "beginning to entertain the notion," "we think," "that may have started," "microbial life might," "it looks like probably," "one possibility is," "you might find," "speculative," "likely has," and "might sound far fetched."

Here would be a good place to reiterate what science is:

> National Academy of Sciences: Science is a particular way of knowing about the world. In science, explanations are limited to those based on observation and experiments that can be substantiated by other scientists. Explanations that cannot be based on empirical evidence are not part of science.[267]

As cited in an earlier, brief historical review, Stanley Miller attempted to simulate the earth's early conditions as hypothesized by Oparin, Urey, and other scientists. Miller's experiment produced some of the chemicals of life.[268] He took four gasses (methane, ammonia, water vapor, and hydrogen), ran them through some tubes and into a spark chamber supposed to simulate lightning. Because a red goo containing some amino acids formed in the bottom of the apparatus, he concluded that he had created the beginning elements of life from non-living elements. But these spark-discharge experiments produced more

---

[264]Richard Dawkins, PhD, "Expelled: No Intelligence Allowed," Independent Film directed by Nathan Frankowski and hosted by Ben Stein, 2008.
[265]"The Appearance of Life on Earth," *Modern Biology*, Holt, Rinehart and Winston, Austin, New York, San Diego, Chicago, Toronto, Montreal, 1989, pp. 209-211, (emphasis added).
[266]William Jennings Bryan, *Reader's Digest*, August 1925 (Vol. 4, No. 400), see also: "Mr. Bryan on Evolution," *Impact*, No. 213, Institute for Creation Research, El Cajon, CA, March 1991, p. iii.
[267]"Introduction," *Science and Creationism: A View From the National Academy of Sciences*, National Academy of Sciences, Washington, DC, 1999, p. 1.
[268]*Heath Biology*, 1991, D.C. Heath and Company, Lexington, MA, p. 250.

different kinds of amino acids that do not occur in living organisms than the 20 that do.[269]

The famous Miller-Urey 1953 experiment attempted to create a "chemical soup" similar to what was then thought to be on earth 3.5 billion years ago. Stanley Miller, then a graduate student at the University of Chicago, and Harold Urey, his faculty advisor, believed—and the popular media accepted—that their experiment's "lightning" created **organic** amino acids in their **inorganic** "primordial soup"—thus providing "evidence" that nonlife would be able to create life.[270]

However, the scientific community dismissed the experimental results as not "good science." In 1962, Urey himself wrote, "All of us who study the origin of life find that the more we look into it, the more we feel it is too complex to have evolved anywhere. We all believe as an article of faith that life evolved from dead matter on this planet. It is just that its complexity is so great, it is hard for us to imagine that it did."[271]

And Urey's former student Miller said in a 2007 paper: "The origin of life remains one of humankind's last great unanswered questions, as well as one of the most experimentally challenging research areas. . . . Despite recent progress in the field, a single definitive description of the events leading up to the origin of life on Earth some 3.5 billion years ago remains elusive."[272]

In an interview with Jonathan Wells, PhD, author of *Icons of Evolution*, Lee Strobel asked what scientists would get if they repeated the Miller experiment today with an accurate atmosphere? "Embalming fluid," said Dr. Wells.[273]

Even in 2011, a "study of twenty-two high school textbooks found that nineteen discuss the Miller-Urey experiment as a possible explanation of the origin of life."[274]

And what does Dr. Wells think of the fact that the Miller simulation is featured in current textbooks, often with pictures? "It's misleading. It's wrong to even give the impression that science has empirically shown how life could have originated. . . . It's becoming clearer and clearer to me that this is materialistic philosophy masquerading as empirical science. *The attitude is that life had to have developed this way because there is no other materialistic explanation.*"[275]

What Miller did not include in his "primitive atmosphere" was oxygen, because life could not evolve in the presence of oxygen. He knew that oxygen would quickly oxidize whatever the chemicals tried to combine.

But the earth has always had oxygen. Oxygen is found in the lowest rocks [of the geologic column].[276]

> Erich Dimroth and Michael M. Kimberley in *Canadian Journal of Science:* In general, we find no evidence in the sedimentary distribution of carbon, sulfur, uranium, or iron, that an oxygen-free atmosphere has existed at any time during the span of geological history recorded in well preserved sedimentary rocks.[277]

Even old-earth geologists agree:

> Harry Clemmey and Nick Badham in *Geology:* It is suggested that from the time of the earliest dated rocks at 3.7 [billion years] ago, earth had an oxygenic atmosphere.[278]

As you can see from the above quotation, even though erosion rates would cut billions of years off the age of the earth, evolutionary scientists accept the ancient-age concept. Radiometric dating provides them with the "scientific evi-

---

[269] Miller S. L., Orgel L. E., *The Origins of Life on Earth*, Prentice Hall, Inc., Englewood Cliffs, NJ, 1974, pp. 87-88.
[270] See Douglas Ell, *Counting to God: a personal journey through science to belief,* The Miller-Urey Experiment, from Science, Religion, and Philosophy section of http://www.attitudemedia.com, 2014.
[271] Harold C. Urey, quoted in *Christian Science Monitor,* January 4, 1962, p. 4.
[272] Stanley L. Miller and H. James Cleaves, "Prebiotic Chemistry on the Primitive Earth," in *Systems Biology: Volume 1, Genomics*, ed. Isidore and Gregory Stephanopoulos, Oxford University Press, New York, 2007, p. 3.
[273] Lee Strobel, "Doubts About Darwinism," *The Case for A Creator,* Zondervan, Grand Rapids, MI, 2004, pp. 37-41, (emphasis added).
[274] Casey Luskin, "Not Making the Grade: An Evaluation of 22 Recent Biology Textbooks and their Use of [the] *Selection Icons of Evolution,*" September 26, 2011, http://www.evolutionnews.org/discoveryInstitute_Textbook Review, pdf.
[275] Lee Strobel, "Doubts About Darwinism," *The Case for A Creator,* Zondervan, Grand Rapids, MI, 2004, pp. 37-41, (emphasis added).
[276] Michael Denton, PhD, "The Enigma of Life's Origin," *Evolution: a Theory in Crisis,* Burnett Books, Limited, London, 1985, p. 261.
[277] Erich Dimroth and Michael M. Kimberley, "Precambrian Atmospheric Oxygen: Evidence in the Sedimentary Distributions of Carbon, Sulfur, Uranium, and Iron," *Canadian Journal of Science,* Vol. 13, No. 9, September 1976, p. 1161.
[278] Harry Clemmey, Nick Badham, "Oxygen in the Precambrian Atmosphere: An Evaluation of the Geological Evidence," *Geology,* Vol. 10, March 1982, p. 141.

dence" they depend on because it predicts some expected dates in the geologic column, though they vary from young to billions of years. Younger layers are predicted to be only half a million years old, and for the older layers, all dates are extrapolated to be about equal to 540 million years. Radiometric dating has even led some creationists to believe in an old [perhaps Mars-like] earth. And it may be. There is nothing in Genesis 1 verses 1 and 2 to say otherwise. The rest of Genesis and of Scripture however point to a recent creation of *life* on earth. Later, we will see that assumptions aplenty throw a serious cloak of reasonable doubt around the subject of the origin of life on earth by purely materialistic/natural means.

The earth had more oxygen "In the beginning" than had been thought:

> Commonwealth Scientific and Industrial Research Organization: Primordial air may have been "breathable." The earth may have had an oxygen-rich atmosphere as long ago as three billion years and possibly even earlier, three leading geologists claimed.[279]

And there are other problems with thinking early earth had no oxygen (as Miller had presumed).

Most small proteins have 70 to 100 amino acids in precise order, all left-handed. Half of Miller's amino acids were right handed (backward).

> Carl Werner, MD: *Amino acids* produced in *Miller's* apparatus are both *right- and left-handed* amino acids, *but right-handed amino acids are poisonous* to living organisms. Right-handed amino acids render proteins nonfunctional.

Miller's experiment produced only a few rudimentary amino acids, not the full complement of 20 amino acids that are used by living organisms today. Other scientists have subsequently designed laboratory experiments and have produced the remaining amino acids, but the same problems still limit the ability of amino acids to form spontaneously, *without* investigator interference.[280]

> Ariel A. Roth, PhD: An incredibly huge chasm looms between the simple disorganized molecules of the much-acclaimed Miller type of experiment and the intricate structure of a living cell, including its multitude of controlled operating systems. Unfortunately, biology textbooks rarely note this fact.[281]

> Michael J. Denton, MD, PhD: The really significant finding that comes to light from comparing the proteins' amino acid sequences is that it is impossible to arrange them in any sort of evolutionary series. . . . There is little doubt that if this molecular evidence had been available a century ago it would have been seized upon with devastating effect by the opponents of evolution theory like Agassiz and Owen, and the idea of organic evolution might never have been accepted.[282]

> Stephen C. Meyer, PhD: Geological and geochemical evidence suggests that prebiotic atmospheric conditions were hostile, not friendly to the production of amino acids and other essential building blocks of life.[283]

Miller relied heavily on the atmospheric theories of his doctoral adviser, Nobel laureate Harold C. Urey, PhD. Miller chose a mixture of hydrogen, methane, ammonia, and water vapor, which was consistent with what many scientists thought back then. But most scientists don't believe that anymore. Philip H. Abelson, PhD, a geophysicist with the Carnegie Institution, wrote, "What is the evidence for a primitive methane-ammonia atmosphere on earth? The answer is that there is no evidence for it, but much against it."[284]

> Stephen C. Meyer, PhD: Miller's amino acids reacted very quickly with the other chemicals in the chamber, resulting in a brown sludge that's not life-friendly at all. . . . Even if amino acids existed in the theoretical prebiotic soup, they would

---

[279]CSIRO Australia, "Primordial Air May Have Been Breathable," *ScienceDaily*, Commonwealth Scientific and Industrial Research Organization, January 9, 2002.
[280]Carl Werner, MD, "Criticisms of the Stanley Miller Experiment," *Evolution: the Grand Experiment: the Quest for an Answer*, New Leaf Press, Green Forest, AR, 2007, p. 207 (emphasis added).
[281]Ariel A. Roth, PhD, "How Did Life Get Started?" *Science Discovers God: Seven Convincing Lines of Evidence for His Existence*, Autumn House Publishing, a division of Review and Herald Publishing, Hagerstown, MD, 2008, p. 85.
[282]Michael J. Denton, PhD, "A Biochemical Echo of Typology," *Evolution: A Theory in Crisis*, Adler & Adler, Bethesda, MD, 1986, pp. 289-291.
[283]Stephen C. Meyer, PhD, *The Explanatory Power of Design*, in William A. Dembski, ed. *Mere Creation: Science, Faith and Intelligent Design*, Downers Grove, IL: Inter-Varsity Press, 1998, p. 118.
284 Philip H. Abelson, "Chemical Events on the Primitive Earth," *Proceedings of the National Academy of Sciences USA 55*, 1966, pp. 1365-1372.

have readily reacted with other chemicals. This would have been another tremendous barrier to the formation of life. The way that origin-of-life scientists have dealt with this in their experiments has been to remove these other chemicals in the hope that further reactions could take the experiment in a life-friendly direction.

So instead of simulating a natural process, they interfered in order to get the outcome they wanted. *And that is intelligent design.*[285]

In 1995, John Cohen in the journal *Science* said that "experts now dismiss Miller's experiment because 'the early atmosphere looked nothing like the Miller-Urey simulation.'"[286]

Brooks J. Shaw, in *Origin and Development of Living Systems:* Another question involves the lack of evidence in the earth's rocks for the assumed "primordial soup" in which all the molecules supposedly formed. If at one time in the distant past there existed an ocean rich with organic molecules in which life might by chance arise, the rocks [the geologic column] do not show any sign of it. Rocks rich in organic matter are conspicuously absent in the deeper layers representing the time during which life supposedly evolved.[287]

British atheist, Sir Fred Hoyle: The notion that not only the bipolymers but the operating programme of a living cell could be arrived at by chance in a primordial organic soup here on the earth is evidently nonsense of a high order.[288]

Ernst Mayr, PhD: In the last 75 years [published in 2001] an extensive literature dealing with this problem has developed and some six or seven competing theories for the origin of life have been proposed. Although no fully satisfactory theory has yet emerged, the problem no longer seems as formidable as at the beginning of the twentieth century.

In spite of all the theoretical advances that have been made toward solving the problem of the origin of life, the cold fact remains that no one has so far succeeded in creating life in a laboratory. . . . Many more years of experimentation will likely pass before a laboratory succeeds in actually producing life. *However, the production of life cannot be too difficult, because it happened on earth* apparently as soon as conditions had become suitable for life, *around 3.8 billion years ago.*[289]

The assumption "3.8 billion years ago" is rooted in the materialistic presupposition that there can be no Creator or Designer. Does Mayer's conclusion grow from his assumption?

Mayr says, "The production of life cannot be too difficult." So . . .

Warren L. Johns, JD: Why not confront the ultimate challenge to human intelligence and create life in a lab from non-living matter? . . . If life from unintelligent non-life can result theoretically from an accidental whim of nature, then why couldn't human intelligence be recruited to design and create a simple, living cell from inert matter?[290]

This is such a problem to those who believe that life evolved that they are continually seeking a solution. From time to time someone claims it has been solved, but only after checking the details does one find that the problem remains. In Germany, in 1994, a doctoral candidate, Guido Zadel, claimed he had solved the problem of mixed left- and right-handed amino acids. Supposedly, a strong magnetic field will bias a reaction toward either the left-handed or right-handed form. Origin-of-life researchers were excited. Zadel's doctorate was awarded. At least 20 groups then tried to duplicate the results (repeating the original experiment is *real* science), but always unsuccessfully. Later, Zadel admitted he manipulated his data.[291]

Furthermore, hundreds of amino acids must combine to make proteins, yet they unbind in water faster than they bind.

George B. Johnson and Peter H. Raven in *Biol-*

---

[285]Stephen C. Meyer, PhD, quoted in Lee Strobel, "The Evidence of Biological Information," *The Case for A Creator*, Zondervan, Grand Rapids, MI, 2004, p. 228, (emphasis added).
[286]John Cohen, "Novel Center Seeks to Add Spark to Origins of Life," *Science* 270, 1995, pp. 1925-1926.
[287]Brooks J., Shaw G, 1973, *Origin and Development of Living Systems*, Academic Press, London and New York, p. 359; C.B. Thaxton, W. L. Bradley, R. L. Olsen, 1984, *The Mystery of Life's Origin: Reassessing Current Theories*, Philosophical Library, NY, p. 65.
[288]Sir Fred Hoyle, "The Big Bang in Astronomy," *New Scientist*, New Science Publications, London, November 19, 1981, p. 527.
[289]Ernst Mayr, PhD, "The Rise of the Living World," *What Evolution Is*, Basic Books, New York, NY, 2001, pp. 42-43 (emphasis added).
[290]Warren L. Johns, JD, "Superstitious Nonsense: Life from Spontaneous Generation?" *Genesis File*, www.GenesisFile.com, Lightning Source, LaVergne, TN, 2010, p. 11.
[291]Daniel Clery and David Bradley, "Underhanded 'Breakthrough' Revealed," *Science*, Vol. 265, July 1, 1994, p. 21.

ogy—*Principles and Explorations:* Scientists have not been able to cause amino acids dissolved in water to join together to form proteins. The energy-requiring chemical reactions that join amino acids are reversible and do not occur spontaneously in water.[292]

From 2006 to 2012, Harvard University sponsored an "Origins of Life Initiative." Begun in 2005 as the "Origins of Life in the Universe Initiative,"[293] then as an "interfaculty initiative in 2007."[294] "In March of 2009, *The Origins of Life Initiative* brought together a number of distinguished scientists to discuss how life could have begun. . . . The theme [became], 'We just don't know. . . .'"[295]

. . . As of mid-2012, before the Harvard initiative took down its list of research papers, its website had included no new biology papers in three years.

Perhaps frustrated by their inability to penetrate the molecular biology obstacles to the origin of life, the *Harvard Origins of Life Initiative* has changed course toward *the search for Earthlike planets*. The Initiative redesigned its website to focus on astronomy and away from the intractable problems of the origin of life.[296]

. . . Or maybe to search for life on other planets.

In 2013, Dimitar Sasselov, professor of anatomy and director of the *Origins of Life Initiative* at Harvard, said that the question of how life started here on earth "is one of the big unsolved questions humanity has always asked."[297] And yet, said the article, for various reasons, it has been difficult to answer.

The problems are complex enough to inspire many thousands of research ventures. In 2016, on the BBC website about the origin of life, an 88-page article citing recent research in the concluding remarks said: "Maybe life began in a volcanic pond like this one [pictured on the website] in Yellowstone National Park, US."[298]

Like their forerunners, today's researchers maintain high hopes and unwavering faith that ongoing research will reveal the origin of life.

---

[292]George B. Johnson, Peter H. Raven, "Evaluating the Spontaneous Origin Hypothesis," *Biology: Principles & Explorations*, Holt, Reinhardt and Winston, Austin, New York, Orlando, Atlanta, San Francisco, Boston, Dallas, Toronto, London, 1996, p. 235.
[293]www.thecrimson.com/article/9/12/harvard-out-to-uncover-lifes-origin/
[294]Overview, *Origins of Life Initiative*, 2007, Harvard University, p. 1. origins.harvard.edu
[295]Douglas Ell, "The Origin of Life" (chapter 10), *Counting to God: A Personal Journey Through Science to Belief*, Attitude Media, 2014, (emphasis added). Free on Kindle app players: https://www.attitudemedia.com/pdf/?file=CountingtoGodBook.

[296]Douglas Ell, "The Origin of Life" (chapter 10), *Counting to God: A Personal Journey Through Science to Belief*, Attitude Media, 2014, (emphasis added). Free on Kindle app players.https://www.attitudemedia.com/pdf/?file=CountingtoGodBook
[297]https://harvardmagazine.com/2013/09/life-s-beginnings
[298]www.bbc.com/earth/story/20161026-the-secret-of-how-life-on-earth-began

Chapter 13
Irreducible Complexity

For a living organism to function, all of its different functioning parts had to exist from its beginning. They could not have evolved separately and somehow come together to "work."

> Sean D. Pitman, MD: Just like the chicken and the egg paradox, it seems like the function of the most simple living cell is dependent upon all its parts being there in the proper order simultaneously. Some have referred to such systems as "irreducibly complex" in that if any one part is removed, the higher "emergent" function of the collective system vanishes. This apparent irreducibility of the living cell is found in the fact that DNA makes the proteins that make the DNA. Without either one of them, the other cannot be made or maintained. Since these molecules are the very basics of all life, it seems rather difficult to imagine a more primitive life form to evolve from. No one has been able to adequately propose what such a life form would have looked like or how it would have functioned. Certainly, no such life form or pre-life form has been discovered. Even viruses and the like are dependent upon the existence of pre-established living cells to carry out their replication. They simply do not replicate by themselves. How then could the first cell have evolved from the non-living soup of the "primitive" prebiotic oceans?[299]

> Michael J. Denton, MD, PhD: Rocks of great antiquity have been examined over the past two decades and in none of them has any trace of abiotically produced organic compounds been found.... Considering the way the prebiotic soup is referred to in so many discussions of the origin of life as an already established reality, it comes as something of a shock to realize that there is absolutely no positive evidence for its existence.[300]

What is "irreducible complexity?" The bacterial flagellum, a micromachine, is just one illustration of irreducible complexity detailed in the book *Darwin's Black Box: The Biochemical Challenge to Evolution*, authored by Michael J. Behe, PhD. ("Black box" is a term scientists use to describe any system, machine, device, or theoretical construct that they find interesting but inexplicable. They know what it can do but not how it is constructed nor how it works.) In 1996, the award-winning best-selling Dr. Behe made an overwhelming case against Darwin's theory on the biochemical level.[301]

Who is Dr. Behe, and how is he qualified to address the subject? Behe earned a degree in chemistry with honors from Drexel University and a doctorate in biochemistry at the University of Pennsylvania. He conducted post-doctoral research at the University of Pennsylvania and at the National Institutes of Health. He also served on the Molecular Biochemistry Review Panel of the Division of Molecular and Cellular Biosciences at the National Science Foundation.

He has authored 40 articles for peer-reviewed scientific journals such as *DNA Sequence*, the *Journal of Molecular Biology, Nucleic Acids Research, Biopolymers, Proceedings of the National Academy of Sciences USA, Biophysics*, and *Biochemistry*. He has lectured at the Mayo Clinic and dozens of schools, including Yale, Carnegie-Melon, University of Aberdeen, Temple, Colgate, Notre Dame, and Princeton. He is a member of the American Society for Biochemistry and Molecular Biology, the Society for Molecular Biology and Evolution, and other professional organizations.[302]

Dr. Behe outlines his premise:

> In Darwin's day, scientists could see the cell under a microscope, but it looked like a little blob of Jello, with a dark spot as the nucleus. The cell could do interesting things—it could divide, it could move around—but they didn't know how it did anything.... Most scientists speculated that

---

[299] Sean D. Pitman, MD, "The Chicken or the Egg, DNA or Protein?" *True.Origin Archive*, January 2007.
[300] Michael J. Denton, PhD, "The Enigma of Life's Origin," *Evolution: A Theory in Crisis*, Adler & Adler, Bethesda, MD, 1986, p. 261.
[301] Lee Strobel, "The Evidence of Biochemistry," *The Case for a Creator*, Zondervan, Grand Rapids, MI, 2004, pp. 195-196.
[302] Lee Strobel, "The Evidence of Biochemistry," *The Case for a Creator*, Zondervan, Grand Rapids, MI, 2004, pp. 195-196.

the deeper they delved into the cell, the more simplicity they would find. But the opposite happened. . . . We've learned the cell is horrendously complicated, and that it's actually run by micromachines of the right shape, the right strength, and the right interactions. The existence of these machines challenges a test that Darwin himself provided.[303]

Charles Darwin: If it could be demonstrated that any complex organ [or cell run by "micromachines"] existed, which could not possibly have been formed by numerous, successive, slight modifications, my theory would absolutely break down.[304]

Dr. Behe: You see, a system or device is irreducibly complex if it has a number of different components that all work together to accomplish the task of the system, and if you were to remove one of the components, the system would no longer function. An irreducibly complex system is highly unlikely to be built piece by piece through Darwinian processes, because [all the parts of] the system [have] to be fully present in order for it to function. . . . As we discover more and more of these irreducibly complex biological systems, we can be more and more confident that we've met Darwin's criterion of failure.[305]

James Perloff, BSN: Blood clotting swings into action when we get a cut. A clot may look simple to the naked eye. However, through a microscope, it is a very complex process involving more than a dozen steps. A person with hemophilia is missing just one clotting factor and is at high risk for bleeding. Someone missing several components would have no chance for survival at all. To paraphrase Dr. Behe very simply, if blood clotting had evolved step-by-step over eons, creatures would have bled to death before it was ever perfected—and its incremental stages never passed on to subsequent generations. The system is irreducibly complex.[306]

Behe, himself, expounded on his position:

Dr. Behe: Life is actually based on molecular machines. They haul cargo from one place in the cell to another; they turn cellular switches on and off; they act as pulleys and cables; electrical machines let current flow through nerves; manufacturing machines build other machines; solar-powered machines capture the energy from light and store it in chemicals. Molecular machinery lets cells move, reproduce, and process food. In fact, every part of the cell's function is controlled by complex, highly calibrated machines. . . . Remember, the audacious claim of Darwinian evolution is that it can put together complex systems with no intelligence at all.[307]

Behe's description of complex molecular machines is confirmed by others—scientists, as well as those who penned the Judeo-Christian Scriptures. One of the Psalms David sang to His Creator was Psalm 139, verse 14 (here in the *Living Bible* version):

Thank you for making me so wonderfully complex. It is amazing to think about. Your workmanship is marvelous—and how well I know it.

Illustra Media: In the process known as transcription, a molecular machine first unwinds a section of the DNA helix to expose the genetic instructions needed to assemble a specific protein molecule. Another machine then copies these instructions to form a molecule known as messenger RNA. When transcription is complete, the slender RNA strand carries the genetic information . . . out of the cell nucleus. The messenger RNA strand is directed to a two-part molecular factory called a ribosome. . . . Inside the ribosome, a molecular assembly line builds a specifically sequenced chain of amino acids. These amino acids are transported from other parts of the cell and then link into chains often hundreds of units long. Their sequential arrangement determines the type of protein manufactured. When the chain is finished, it is moved from the ribosome to a barrel-shaped machine that helps fold it into the precise shape critical to its function. After the chain is folded into a protein, it is then released and shepherded by another molecular machine to the exact location where it is needed.[308]

---

[303] Michael J. Behe, PhD, quoted by Lee Strobel, "The Evidence of Biochemistry," *The Case for a Creator*, Zondervan, Grand Rapids, MI, 2004, pp. 196, 197, (emphasis added).
[304] Charles Darwin, "Difficulties on Theory," *On the Origin of Species by Means of Natural Selection or the Preservation of Favoured Races in the Struggle for Life*, John Murray, Albemarle Street, London, 1859, p. 189, (bracketed copy added).
[305] Michael J. Behe, PhD, quoted by Lee Strobel, "The Evidence of Biochemistry," *The Case for a Creator*, Zondervan, Grand Rapids, MI, 2004, pp. 197-198, (bracketed copy added).
[306] James Perloff, BSN, "Evidence Against the Theory of Evolution," *The Case Against Darwin*, Refuge Books, Burlington, MA, 2002, pp. 36-37, see also: Michael Behe, *Darwin's Black Box, The Biochemical Challenge To Evolution*, The Free Press, NY, 1996, pp. 77-97.
[307] Michael J. Behe, PhD, quoted by Lee Strobel, "The Evidence of Biochemistry," *The Case for a Creator*, Zondervan, Grand Rapids, MI, 2004, pp. 198-199.
[308] *Unlocking the Mystery of Life: The Scientific Case for Intelligent Design*,

Irreducible Complexity

Dr. Behe described his best example of irreducible complexity—the bacterial flagellum:

> It was discovered in 1973 that the flagellum performs like a rotary propeller. Only bacteria have them. . . . Just picture an outboard motor on a boat and you get a pretty good idea of how the flagellum functions, only the flagellum is far more incredible. The flagellum's propeller is long and whiplike, made out of a protein called flagellin. This is attached to a driveshaft by hook protein, which acts as a universal joint, allowing the propeller and driveshaft to rotate freely. Several types of proteins act as bushing material to allow the drive shaft to penetrate the bacterial wall and attach to the rotary motor.
>
> The flagellum's propeller can start spinning at ten thousand revolutions per minute. . . . The propeller can stop spinning within a quarter turn and instantly start spinning the other way at ten thousand rpms. [My Toyota Camry red-lines at 6,500 rpms.] Howard Berg of Harvard University called it the *most efficient motor in the universe.* It's way beyond anything we can make, especially when you consider its size. . . . The motor itself would be maybe 1/100,000 of an inch.
>
> [The bacterial flagellum] has sensory systems that . . . tell it when to turn on and when to turn off, so that it guides it to food, light, or whatever it's seeking. . . . Genetic studies have shown that between thirty and thirty-five proteins are needed to create a functional flagellum. I haven't even begun to describe all of its complexities; we don't even know the roles of all its proteins. But at a minimum you need at least three parts—a paddle, a rotor, and a motor—that are made up of various proteins. Eliminate one of those parts and you don't get a flagellum that only spins at five thousand rpms; you get a flagellum that simply doesn't work at all. So, it's irreducibly complex—and a huge stumbling block to Darwinian theory.[309]

Bruce Alberts, PhD, president of the National Academy of Sciences, also commented on the complexity of these molecular machines:

> We have always underestimated the cell. . . . The entire cell can be viewed as a factory that contains an elaborate network of interlocking assembly lines, each of which is composed of a set of large protein machines. . . . Why do we call them machines? Precisely because, like machines invented by humans to deal efficiently with the macroscopic world, these protein assemblies contain highly coordinated moving parts.[310]

Mark Whorton, PhD, and Hill Roberts, MA: In the early twentieth century, Henry Ford revolutionized manufacturing by introducing the assembly line. However, it turns out that biology beat him to the punch, for each cell houses a miniature assembly line. The design specifications for each sub-system in the cell is copied from the DNA; supply workers then read this blueprint and couriers fetch the right parts and bring them to the assembly line; assembly workers systematically add the pieces together according to the directions until the assembly is complete; quality control workers inspect the completed product, and finally a delivery worker transports the finished product to its destination. This is not merely an analogy, this is exactly the process by which the cell operates. DNA provides the design specifications for the cellular components; molecules known as messenger RNA copy a specific segment of the assembly instructions; transfer RNA then translates the information coded in DNA into the specific sequence of amino acids; and finally, ribosomal RNA assembles the amino acids into the proper sequence to form the specified protein molecule.[311]

Add intelligent design

Lee Strobel, an award-winning investigative reporter for the *Chicago Tribune*, once believed that creationism was "an archaic belief system hurtling toward oblivion," that the evidence for evolution was ironclad, that miracles were impossible, and that science was on the path to ultimately explaining everything in the universe. To him, "science represented the empirical, the trustworthy, the hard facts, the experimentally proven." He tended to dismiss everything else as being "narrow opinion, conjecture, superstition—and mindless faith." Evolution had taught him there was no need for God if evolution's abiogenesis were true—that *"living organisms could emerge by themselves out of the primordial soup and then develop naturally over the eons*

---

produced by Illustra Media, available at www.illustramedia.com.
[309] Michael J. Behe, PhD, quoted by Lee Strobel, "The Evidence of Biochemistry," *The Case for a Creator,* Zondervan, Grand Rapids, MI, 2004, pp. 204-206, (bracketed supplied).
[310] Bruce Alberts, PhD, "The Cell As a Collection of Protein Machines," *Cell* 92, February 8, 1998.
[311] Mark Whorton, PhD, Hill Roberts, MA, "Impossibilities in Chemistry: The Origin of Life," *Understanding Creation, a Biblical and Scientific Overview,* a Holman Reference Book, B & H Publishing Group, 127 Ninth Avenue, North Nashville, TN, p. 324.

into more and more complex creatures. . . ."[312]

He cross-examined authorities in various scientific disciplines about the most current findings in their fields. He sought doctorate-level professors who have unquestioned expertise, and who refuse to limit themselves only to the politically correct world of naturalism or materialism. He wanted the freedom to pursue *all* possibilities. As an investigative reporter he was trained not only to ask questions, but to go wherever the answers would take him.[313]

Stephen C. Meyer, PhD, earned his doctoral degree from the prestigious Cambridge University in England where he analyzed the scientific and methodological issues in origin-of-life biology. In a period of more than two decades, Meyer became one of the most knowledgeable and compelling voices in the burgeoning Intelligent Design movement.

In an interview with Lee Strobel he stated, "I believe that the testimony of science supports theism [the God hypothesis]. While there will always be points of tension or unresolved conflict, the major developments in science in the past five decades have been running in a strongly theistic direction. Science, done right, points toward God. . . . The fact that most scientists now believe that energy, matter, space, and time had a beginning is profoundly anti-materialistic. . . . In short, naturalism is on hard times in cosmology; the deeper you get into it, the harder it is to get rid of the God hypothesis."[314]

> Dr. Meyer: The problem with irreducibly complex systems is that they perform no function until all the parts are present and working together in close coordination with one another. So, natural selection cannot help you build such systems; it can only preserve them once they've been built. And it's virtually impossible for evolution to take such a huge leap by mere chance to create the whole system at once.[315]

Books written by Dr. Stephen C. Meyer include *Darwin's Doubt* and *Signature in the Cell*.

An example of irreducible complexity is the bombardier beetle:

> Andrew McIntosh, PhD: This creature requires an explosive mixture (hydrogen peroxide and hydroquinone), a combustion chamber to contain the chemicals, exhaust nozzles to eject the mixture into which two catalysts are also injected (the enzymes catalase and peroxidase)—all this at the right moment to make the violent reaction to take place as the mixture leaves the back end. The bombardier beetle manages it all with ease, along with the capacity to send four or five bombs in succession into the face of a predator, controlled by muscles and directed by a reflexive nervous system. This is combustion theory and practice *par excellence!*
>
> All the above requirements would have to be in place at the same "evolutionary moment!" There is no way any "intermediate" [missing link] could survive because of the risk of either (1) blowing him/herself to smithereens (because he has the combustible mixture and the catalyst, but no exhaust system), or (2) slowly eroding his/her insides by having a combustible mixture, all the necessary exhaust tubes, but no catalyst, or (3) being eaten by predators despite trying to blow them away with catalysts through a fine exhaust system, but no combustible mixture! For the creature to function, everything must all be in place together.[316]

---

[312] Lee Strobel, "The Images of Evolution," *The Case for a Creator*, Zondervan, Grand Rapids, MI, 2004, pp. 16-19, (emphasis added).

[313] Lee Strobel, "The Images of Evolution," *The Case for a Creator*, Zondervan, Grand Rapids, MI, 2004, pp. 28-29.

[314] Stephen C. Meyer, PhD, quoted by Lee Strobel, "Where Science Meets Faith," *The Case for a Creator*, Zondervan, Grand Rapids, MI, 2004, pp. 71-72, 77.

[315] Stephen C. Meyer, PhD, quoted by Lee Strobel, "Where Science Meets Faith," *The Case for a Creator*, Zondervan, Grand Rapids, MI, 2004, p. 79.

[316] Edited by John Ashton, PhD, Andrew McIntosh, PhD, "Andrew McIntosh, PhD," *In Six Days: Why Fifty Scientists Choose to Believe in Creation*, Master Books, Inc., Green Forest, AR 72638, 2000, pp. 168-169.

Chapter 14
Are There Questions? Who's Asking?

An increasing number of books criticizing materialism and evolutionary theory have been published; many by individuals who either believe in evolution or who at least do not believe in creation:

- Michael J. Behe, PhD, *Darwin's Black Box: The Biochemical Challenge to Evolution*
- Francis H. C. Crick, FRS—*Life Itself: Its Origin and Nature*
- Michael J. Denton, MD, PhD—*Evolution: A Theory in Crisis*
- Soren Lovtrup, Darwinism—*The Refutation of a Myth*
- Mark Ridley, *Problems of Evolution*
- Robert Shapiro, PhD, *Origins: A Skeptic's Guide to the Creation of Life on Earth*

Here are two quotations from one of the above books:

> Francis Crick: An honest man, armed with all the knowledge available to us now, could only state that in some sense, the origin of life appears at the moment to be almost a miracle, so many are the conditions which would have had to have been satisfied to get it going.[317] . . . Every time I write a paper on the origin of life I swear I will never write another one, because there is too much speculation running after too few facts.[318]

Richard Hutton, Executive Producer of the controversial PBS TV series, "Evolution," was asked, "What are some of the larger questions still unanswered by evolutionary theory?"

He replied, "The origin of life. There is no consensus at all here—lots of theories, little science. That's one of the reasons we didn't cover it in the series. The evidence wasn't very good."[319]

Francis Collins, MD, PhD, director of the National Human Genome Research Institute, who had much to do with the mapping of the human genetic pattern (our three billion base DNA formula), believes that "a higher power must also play some role in what we are and what we become."[320]

> Warren L. Johns, JD: Finite minds, capable of creating computers, have never successfully designed and built a living cell from non-living, inorganic matter, much less shaped a strand of information-packed DNA.[321]

It seemed fitting that when scientists announced they had finally mapped the three billion codes of the *human genome*—a project that filled the equivalent of 75,490 pages of the *New York Times*—references to the Divine abounded. Geneticist Francis S. Collins, MD, PhD, head of the Human Genome Project, said DNA was "our own instruction book, previously known only to God."[322]

Stephen C. Meyer, PhD, a proponent of creation, expounds on the relevance of information and its relationship to DNA:

> Dr. Meyer: I'm not saying intelligent design makes sense simply because other theories fail. Instead, I'm making an inference to the best explanation, which is how scientists reason in historical matters. Based on the evidence, the scientist assesses each hypothesis on the basis of its ability to explain the evidence at hand. Typically, the key criterion is whether the explanation has "causal power," which is the ability to produce the effect in question.
>
> In this case, the effect in question is informa-

---

[317] Francis Crick, *Life Itself: Its Origin and Nature*, Simon and Schuster, NY, 1981, p. 88.
[318] Francis Crick, *Life Itself: Its Origin and Nature*, Simon and Schuster, NY, 1981, p. 153.
[319] Richard Hutton, "Evolution: The Series," WashingtonPost.com, Live Online, Wednesday, September 28, 2001.
[320] Francis S. Collins, MD, PhD; L. Weiss, K. Hudson, 2001, "Heredity and Humanity," *The New Republic*, 224(26):27-29.
[321] Warren L. Johns, JD, "Superstitious Nonsense, Life from Spontaneous Generation?" *Genesis File*, www.GenesisFile.com, Lightning Source, LaVergne, TN, 2010, p. 16.
[322] Quoted in Larry Witham, "The Tree of Life," *By Design: Science and the Search for God*, Encounter Books, San Francisco, 2003, p. 172; Lee Strobel, "The Evidence of Biological Information," *The Case for a Creator*, Zondervan, Grand Rapids, MI, 2004, p. 221.

tion. We've seen that neither chance nor chance combined with natural selection, nor self-organizational processes have causal power to produce information. But we do know of one entity that does have the required causal power to produce information, and that's intelligence. We're not referring to that entity on the basis of what we don't know, but on the basis of what we do know. That's not an argument from ignorance.

The coding regions of DNA have *exactly* the same relevant properties as a computer code or language.... Whenever you find a sequential arrangement that's complex and corresponds to an independent pattern or functional requirement, this kind of information is *always* the product of intelligence. Books, computer codes, and DNA all have these two properties. We know books and computer codes are designed by intelligence, and the presence of this type of information in DNA also implies an intelligent source.

Scientists in many fields recognize this connection between information and intelligence. When archaeologists discovered the *Rosetta Stone* [a black stone Egyptian decree in Ancient Egyptian hieroglyphs and demotic (simplified) scripts, and in Ancient Greek], they didn't think its inscriptions were the product of random chance or self-organizational processes. Obviously, the sequential arrangements of symbols were conveying information, and it was a reasonable assumption that intelligence created it. The same principle is true for DNA.[323]

Furthermore, the concept of natural selection can explain the ability to adapt to the environment, but it cannot explain origin or complexity, such as the anatomy and physiology of the eye or the ear, together providing us with the ability to keep our balance. Intelligent design explains complexity. Natural selection *selects*. It does not *create* new kinds.

> John C. Sanford, PhD, also a proponent of creation: What is the mystery of the genome? Its very existence is its mystery. Information and complexity which surpass human understanding are programmed into a space smaller than an invisible speck of dust. Mutation/selection cannot even begin to explain this. It should be very clear that our genome could not have arisen spontaneously. The only reasonable alternative to a spontaneous genome is a designed genome. Isn't that an awesome mystery—one worthy of our contemplation?[324]

After being exposed to this creation/evolution controversy all my life, and after studying it intensely, I am totally bewildered by the fact that macroevolution is considered to be a scientific and a preferred theory of origins.

Just look at the human body, for example. Each one of our organs plays a vital role in our continued health and well-being. Each of these organs had to come into existence simultaneously and in complete maturity in order to perform its role effectively. And each organ had to be connected to the circulatory system (including arteries, veins, and the small capillaries which connect the two) in order to be nourished and oxygenated and waste removed. This concept leads to numerous questions. How did all of these organs become supported with oxygen and nourishment unless they were connected to the circulatory system, digestive system, and respiratory system? How would all of these organs be supported without a skeletal system? Would the brain be protected without a bony skull cap? How does the digestive system know how to tell the brain that hunger can be satisfied by food intake? How does every cell know how to take what it needs from a carrot or potato or glass of milk and turn it into what that muscle cell or nerve cell needs? How do the lungs know how to exchange carbon dioxide for oxygen? How does the heart know how to beat faster upon exertion?

The heart of an 85-year-old man has beaten approximately 3.5 billion times. According to Dr. Irwin A. Moon of the Moody Institute of Science, "A pump fashioned of the finest steel by the

---

[323] Stephen C. Meyer, PhD, quoted by Lee Strobel, "The Evidence of Biological Information," *The Case for a Creator*, Zondervan, Grand Rapids, MI, 2004, pp. 237-238, (emphasis in the original).

[324] John C. Sanford, PhD, "Is the Downward Curve Real?" *Genetic Entropy & The Mystery of the Genome*, FMS Publications, Waterloo, NY, 2008, p. 154.

most skilled craftsman could not begin to match the endurance of the human heart."[325]

How did tiny bones in the ear and the nervous system connect to the brain? How did the eye develop an aperture to control the amount of light entering the eye, an optic nerve connection to the brain, and the ability to focus? And how did the brain develop the ability to interpret this input as meaningful sounds and vision?

These are honest questions. And I don't believe I'm the only one who asks them.

---

[325] Dr. Irwin A. Moon, *Red River of Life*, Moody Science Classics, A Moody Institute of Science Presentation, Chicago, IL, Reviewed by the American Scientific Affiliation, 1957, 1968, 1998.

Chapter 15
Where Did Humanity Come From?

In instructional materials about science most states legally require factual accuracy, stressing the importance of observation. The State of California and the National Academy of Sciences explain:

> California State Board of Education: Students should never be told that "many scientists think this or that." Science is not decided by vote, but by evidence. Nor should students be told that "scientists believe." Science is not a matter of belief; rather, it is a matter of evidence that can be subjected to the tests of observation and objective reasoning. . . . Show students that nothing in science is decided just because someone important says it is so or because that is the way it was always done [tradition].[326]

> National Academy of Sciences: Scientific investigators seek to understand natural phenomena by observation and experimentation. Scientific interpretations of facts and explanations that account for them therefore must be testable by observation and experimentation.[327]

Yet look at the facts. Public school students are indoctrinated with only one worldview: evolution. It appears that many teachers and textbooks use outdated or false information. If the evolution of humans from an ape-like ancestor were true there should be two documentable evidences: the fossil record, and a mechanism for change.

What do the textbooks say?

> *Holt Biology—Visualizing Life:* Look closely at your hand. You have five flexible fingers. Animals with five flexible fingers are called primates. Monkeys, apes and humans are examples of primates. . . . Primates most likely evolved from small, insect-eating rodent-like mammals that lived about 60 million years ago.[328]

> *Biology:* But all researchers agree on certain basic facts. We know, for example, that humans evolved from ancestors we share with other living primates such as chimpanzees and apes.[329]

Java Man

If we look at the evidence we can determine quite plainly that not "all researchers agree on certain basic facts." Let's look at some of the "facts." Java Man (*Pithecanthropus erectus*), discovered in 1891 by Dr. Eugene Dubois, was reconstructed from an apelike scull cap and a human-like thighbone found a year later and 40 feet away. These two artifacts were put together and claimed to be a 500,000-year-old ape-man.[330]

Lee Strobel acknowledges being deceived

> Lee Strobel: *The World Book Encyclopedia's* two-page spread highlighted a parade of prehistoric men. Second in line was a lifelike bust of Java man from the American Museum of Natural History, accompanied by an outline showing his profile. With a sloping forehead, heavy brow, jutting jaw, receding chin and bemused expression, he was exactly what a blend of ape and man should look like. For me, studying his face and looking into his eyes helped cement the reality of human evolution.
>
> I was blithely ignorant, however, of the full Java man story. . . . The lifelike depiction of Java man, which had so gripped me when I was young, was little more than speculation fueled by evolutionary expectations of what he should have looked like if Darwinism were true.[331]

Rudolph Virchow, MD (1821–1902), regarded as the father of modern pathology, and one who knew and understood human anatomy, responded at the time: "In my opinion this creature was an animal, a giant gibbon, in fact. The thigh bone

---

[326]California State Board of Education, "Science and Pseudoscience," quoted by Phillip E. Johnson, JD, "Darwinist Education," *Darwin on Trial*, Regnery Gateway, Washington, DC, 1991, p. 143.
[327]Conclusion," *Science and Creationism: A View From the National Academy of Sciences*, National Academy of Sciences, Washington, DC, 1999, p. 25.
[328]George B. Johnson, "Human Evolution," *Holt Biology: Visualizing Life*, Holt, Rinehart, and Winston, Inc., Austin, New York, Orlando, Chicago, 1994, p. 221, (emphasis in the original).
[329]Kenneth R. Miller, PhD, Joseph Levine, PhD, "Humans," *Biology*, Prentice-Hall, Inc., Upper Saddle River, NJ, 2000, p. 756.
[330]*The Origin of Humans*, DVD, Answers in Genesis-USA, PO Box 510, Hebron, KY, 2004.
[331]Lee Strobel, "Doubts About Darwinism," *The Case for a Creator*, Zondervan, Grand Rapids, MI, 2004, p. 61.

has not the slightest connection with the skull." Virchow's expertise was ignored and still is.[332]

Unfortunately, Dubois refused to allow other scientists to see the bones.

Time magazine, as recently as 1994, treated Java man as a legitimate evolutionary ancestor.[333]

Piltdown Man

The *Piltdown Man* was so believable that a *New York Times* headline stated, "DARWIN THEORY PROVED TRUE."[334] Henry Fairfield Osborn, "head of the American Museum of Natural History," proclaimed Piltdown man to be "a discovery of transcendent importance to the prehistory of man."[335] For 40 years Piltdown Man appeared in textbooks and encyclopedias as evidence of evolution. It was a hoax.

Until 1953, almost every evolutionary scientist in the world accepted the "fact" of Piltdown Man. Then, two British scientists managed to gain access to the bones, which had been locked away in the British Museum for decades. As they examined them they discovered that they were not ancient relics at all, but relatively modern bones (both jaw and skull) that had been stained with bichromate to make them look ancient. When they examined the teeth, they found that they belonged to an orangutan and were covered with suspicious looking file marks.

> *New Scientist:* In 1953 Piltdown man was shown to be a fake, but for nearly 20 years the specimen was a crucial piece of palaeoanthropological evidence against which every other fossil had to be compared; and the three gentlemen most responsible for promoting it to that exalted status—[Arthur Smith] Woodward [keeper of geology at the British Museum of Natural History in London], Arthur Keith, and Grafton Elliot-Smith—were subsequently knighted for their contributions to science.[336]

An attorney evaluates the evidence:

> David C. Read, JD: The hoaxer, whomever he was, must have had extensive knowledge of anatomy, chemistry, paleontology, geology, and anthropology. He had to know to file down the ape teeth so that they would look more human and fit with the human upper jaw. He had to know to break off the symphysis and condyle so no one could articulate [join] the jawbone with the skull and see that they did not match. He had to know what chemicals would give the fossils a patina [stain] of age. He had to know how to seed the gravel pit with index fossils, so that his "missing link" would be dated as expected. He had to know paleolithic stone tools and how to make them or where to find them. Obviously, the hoaxer was either a professional paleontologist or a very well-trained amateur.
>
> It is perhaps more interesting to analyze the hoaxer's motive than to try to guess his identity. If the hoaxer had been a creationist bent on embarrassing the Darwinists, he would have exposed the hoax within a couple of years after the scientific establishment had committed to Piltdown man's authenticity. But he did not. The hoax remained undisclosed for forty years and was never exposed by the hoaxer. We can only conclude that the hoaxer had no intention of embarrassing Darwinian science.[337]

Nebraska Man

The Nebraska Man was discovered in Nebraska in 1922. It was declared to be a one-million-year-old intermediate link (even though dating mechanisms would not exist until decades later). The scientific community at the University of Nebraska reconstructed the entire Nebraska Man, his wife and family, and their environment.[338] Professor Henry Fairfield Osborn, a confirmed evolutionist and paleontologist who directed the American Museum of Natural History in the 1920s, announced the Nebraska Man.

> Founder and President of the Creation Resource Foundation Dennis R. Petersen: "Dental experts at the American Museum of Natural History studied the fossil tooth carefully and concluded it

---

[332] *The Origin of Humans*, DVD, Answers in Genesis-USA, PO Box 510, Hebron, KY, 2004.
[333] Michael D. Lemonic, "How Man Began," *Time*, March 14, 1994, pp. 80-87.
[334] "Darwin Theory Proved True," *The New York Times*, December 22, 1912, p. C1.
[335] Henry Fairfield Osborn, as reported by Stephen Jay Gould, PhD, "Piltdown Revisited, *The Panda's Thumb*, W.W. Norton & Company, Inc., NY, 1980, p. 120.
[336] John Reader, "Whatever happened to Zinjanthropus?" *New Scientist*, New Science Publications, London, March 26, 1981, pp. 802-805.
[337] David C. Read, JD, "Ancient Man—The Darwinian View," *Dinosaur*, Clarion Call Books, Keene, TX, 2009, pp. 361-362.
[338] Daniel A. Biddle, PhD, David A. Bisbee, and Jerry Bergman, PhD, "Debunking Human Evolution Taught in Public Schools, Junior/Senior High Edition, *A Guidebook for Christian Students, Parents, and Pastors*, Genesis Apologetics, 2016, p. 93, genesisapologetics.com.

I visited a museum in North Dakota, and found another example of a reconstruction from almost nothing. Read the caption on the right and notice all the conclusions made.

was from a species closer to man than ape. These evolutionistic experts delighted in the first North American discovery of a missing link. . . .

An Englishman [Osborn] who was involved in the Piltdown discovery a few years before persuaded the widely read *Illustrated London News* to publish an artist's rendering of Nebraska Man and his mate. A full two-page spread was drawn and distributed worldwide, showing a naked pair of stupid looking ape-people. The club swinging male no doubt impressed a whole generation with the idea that human ancestors were far inferior to the biblical Adam and Eve.[339]

Henry Fairfield Osborn: It is hard to believe that a single water-worn tooth . . . can signalize the arrival of the anthropoid primates in North America. . . . We have been eagerly anticipating some discovery of this kind, but were not prepared for such convincing evidence.[340]

You read that correctly. The evidence was "the fossil tooth"—"a single water-worn tooth." All of the displays and art designed to illustrate evolution were reconstructed from one tooth—which turned out to be the tooth of an extinct pig.[341]

Neanderthal Man

The *Neanderthal Man,* an old man with arthritis, is still used as an example of evolution. The first Neanderthal skeleton was found near Dusseldorf, Germany, in the Neander Valley in 1856. His larger brain capacity (the average reported Neanderthal brain capacity was 1,450 cc) was a problem for evolutionists who believe that the more we have evolved, the larger our brain capacity. Evolution sees early man and his brain as smaller than later (more evolved) man (whose brain capacity is now said to be 1,365 cc). So, how can a Neanderthal be "early" man if his

---

[339]Dennis R. Petersen, "The Final End of All Supposed 'Missing Links,'" *Unlocking the Mysteries of Creation, the Explorer's Guide to the Awesome Works of God*, Bridge-Logos Publishers, Alachua, FL, 2002, p. 137.
[340]Henry Fairfield Osborn, "Hespropithecus, the First Anthropoid Primate Found in America," *American Museum Novitates*, No. 37, April, 1922, pp. 1-5.
[341]Gregory, W. K. (1927). "Hesperopithecus apparently not an ape nor a man," *Science*, December 16, 1927, pp. 579-581, see also: "Nebraska Man," Wikipedia, *The Free Encyclopedia*.

brain is *larger?* Maybe there really is no "evolution" implied in Neanderthal Man. After many years several anatomists evaluated the bones and discovered that they had been misconstructed. When they reconstructed the bones correctly, Neanderthal Man looked somewhat like modern man.[342] The Neanderthal Man is one of the best known of all human remains. In spite of the many illustrations showing him as stooped, it is now known that his posture was erect:

> David C. Read, JD: [Marcellin] Boule was world renowned, one of the top men in his field. But he was a Darwinist. He believed that men have evolved from apes, and when he examined ancient human remains, he expected to see ape-like qualities. When he studied the remains of the Old man of La Chapelle-aux-Saints [the first Neanderthal Man], Darwinian expectations blinded him to the possibility that what he was seeing was the result of disease and did not reflect the normal Neanderthal condition. His reconstruction of Neanderthal was more the result of his Darwinian beliefs than anything in the bones themselves. . . . And Boule's interpretation of Neanderthal became part of the body of scientific knowledge, as reflected in museum displays and textbooks, for half a century.[343]

> Harold W. Clark, MS: It seems to be an obsession with anthropologists to make their restorations of prehistoric man look wild and beastly. Much of the resemblance between their restorations and the apes is due to their vivid imaginations.[344]

A physical anthropologist, provides perspective:

> David Phillips: [Compared with today's humans] "Neanderthal anatomy differences are extremely minor and can be for the most part explained as a result of a genetically isolated people that lived a rigorous life in a harsh, cold climate.[345]

There is new DNA evidence that modern humans contain Neanderthal DNA, so the two groups *interbred,* essentially making them the same species, as defined by dictionaries, such as the *American Heritage Dictionary:*

> Species—a group of closely related organisms that are very similar to each other and are usually capable of interbreeding and producing fertile offspring. (The assumed very old age for Neanderthals rests on evolution's presuppositions.)

Peking Man

Peking Man was reconstructed just from pieces of skull, found in Peking, China, in the 1920s. All evidence was lost in World War II.

Lucy

Lucy and other *australopithecine [Australopithecus afarensis]* fossils are found mostly in South Africa. Lucy was discovered in 1974 by Donald C. Johanson, PhD. He found about 40 percent of the fossil, claimed that it was 3.5 million years old, and that it was bipedal (that it walked upright on two legs, comfortably, for long periods of time, like human beings).

The creature was pictured in a 1998 textbook with human feet[346] in spite of the fact that *no feet were found with Lucy.* Although Lucy had two legs, two arms, a face, and no tail, she bore *no similarity in appearance to humans.* Her long arms, jaws, and upper leg bone are similar to chimpanzees. Her legs and brain capacity were very ape-like. She was small (43 inches) and had hands similar to pygmy chimpanzees.

Anatomists Jack Stern, PhD, and Randall Susman, PhD, reported their 1983 study of the *Australopithecus afarensis* in the *American Journal of Physical Anthropology* and described the anatomy of Lucy's species as having hands and *feet* that were long and curved, *typical* of a tree-dwelling *ape*.[347]

In 1993, Paris Anthropologist Christine Tardieu reported, "Its locking mechanism [which allows humans to stand up comfortably for long periods of time] was not developed."[348] In 1994, the *Journal of Human Evolution* reported a biochemical

---

[342] *The Origin of Humans,* DVD, Answers in Genesis-USA, PO Box 510, Hebron, KY, 2004.
[343] David C. Read, JD, "Ancient Man—The Darwinian View," *Dinosaurs,* Clarion Call Books, Keene, TX, 2009, p. 367, [brackets added].
[344] Harold W. Clark, "Cave Men and Stone Ages," *Fossils, Flood, and Fire,* Outdoor Pictures, 1968, p. 210.
[345] David Phillips, "Neanderthals Are Still Human," *Impact #220,* Institute for Creation Research, El Cajon CA, May, 2000, see also: D. Phillips, "Neanderthals Are Still Human!" *Acts & Facts,* 29 (5).
[346] William K. Purves, Gordon H. Orians, H. Craig Heller, David Sadava, "The Origin of Life on Earth," *Life: the Science of Biology,* Sinauer Associates, Inc., Sunderland, MA, 1998, p. 687.
[347] Richard Milton, "Down from the Trees," *Shattering the Myths of Darwinism,* Park Street Press, Rochester, VT, 1997, p. 207.
[348] "Evolutionists In Love With Lucy," *Way of Life Literature,* wayoflife.org., August 16, 2016; *see also,* "Lying Evolutionary Art, Lucy, the Cute Little Ape-Woman," *Way of Life Literature,* wayoflife.org.; see also, *New Scientist,* January 20, 1983, p. 173.

study of the hip and thigh indicated that the *australopithecine* "walk differed significantly from that of humans, involving a sort of waddling gait."[349]

Archaeologists unearthed Lucy in Ethiopia in 1974, two weeks before grant money expired. Lucy was a three-foot tall, tree-climbing ape. Remember, only about 40 percent of the fossil, claimed to be bipedal, was found. When the four or five thoroughly crushed skull bones were put together, they were made to look half human/half ape. "Lucy's knee," now known to be the Hadar Knee, was found 200 feet lower and more than a mile away, yet *National Geographic* called it "Lucy" five times.[350]

An 880-page, 1989 textbook portrays eight figures under the heading "Hominid Evolution," beginning with *A. afarensis* (Lucy) and ending with Cro-Magnon, all illustrated with human feet.[351]

A 1992 textbook, *Life: The Science of Biology*, has an artist's conception of Lucy portrayed as having human feet. In a 2001 edition of the same textbook, two other *australopithecines* also are portrayed with human feet.[352]

It turns out that feet are a major problem for evolutionists. In 1985, *National Geographic* illustrated "40,000 Years of Bipedalism" with nine subjects running from left to right, illustrating an ape-like being evolving into man. All nine, starting with *Australopithecus afarensis* (Lucy) and ending with Homo sapiens (modern) are illustrated with human feet.[353]

In 1994, another artist's conception in *Time* magazine depicts four million years of bipedalism with eight subjects running from left to right, illustrating an ape-like being evolving into a man. All eight *(Australopithecus afarensis, Australopithecus africanus, Paranthropus boisei, Paranthropus robustus, Homo habilis, Homo erectus, Homo neanderthalensis,* and *Homo sapiens)*, are illustrated with human feet.[354]

The St. Louis Zoo created a statue of Lucy also portraying her with human feet. Every other *australopithecine* that has been found has curled toes. Professor David Menton of Washington University said the statue is "a complete misrepresentation. And I believe they know it is a misrepresentation."[355]

In response, Bruce L. Carr, the zoo's director of education said [in 1996], "Zoo officials have no plans to knuckle under. We cannot be updating every exhibit based on every new piece of evidence. [Lucy was discovered 22 years earlier in 1974]. We look at the overall exhibit and the impression it creates. We think the overall impression this exhibit creates is correct."[356]

Notice the following scientific corrections:

E. Stokstad: I walked over to the cabinet, pulled out Lucy, and shazam! She had the morphology that was classic for knuckle walkers.[357]

Richmond and Strait: Regardless of the status of Lucy's knee, new evidence has come forth that Lucy has the morphology (appearance) of a knuckle-walker.[358]

Charles E. Oxnard, PhD, Professor of anatomy and leading expert on *australopithecine* fossils: The various *australopithecines* are, indeed, more different from both African Apes and humans in most features than these latter are from each other.[359] . . . The *australopithecines* known over the last several decades . . . are now irrevocably removed from a place in the evolution of human bipedalism [could walk upright]. . . . All this should make us wonder about the usual presentation of human evolution in introductory textbooks.[360]

How old are apes and the "more recent"

---

[349]Cristine Berge, "How did *Australopithicus* walk? A biomechanical study of the hip and thigh of *Australopithicus afarensis,*" *Journal of Human Evolution*, Vol. 26, No. 4, 1994, p. 271.
[350]"The Primate that Walks," *National Geographic*, November 1985, p. 593.
[351]"Hominid Evolution," *Modern Biology*, Holt, Rinehart and Winston, Austin, New York, San Diego, Chicago, Toronto, Montreal, 1989, pp. 258-259.
[352]Purves, Orians, and Heller, *Life: The Science of Biology*, 1992, p. 604; William K. Purves, David Sadava, Gordon H. Orians, H. Craig Heller, "Deuterostomate Animals," Life: the Science of Biology, Sinauer Associates, Inc., Sunderland, MA, 2001, p. 598.
[353]"40,000 Years of Bipedalism," *National Geographic*, November 1984, pp. 574-577.

[354]Michael D. Lemonick, "How Man Began," *Time*, March 14, 1994, p. 82.
[355]David Menton, *St. Louis Post Dispatch*, July 22, 1996, p. 1.
[356]Bruce L. Carr, *St. Louis Post Dispatch*, July 22, 1996, p. 1.
[357]E. Stokstad, "Hominid Ancestors May Have Knuckle Walked," *Science*, March 24, 2000, 287(5461): 2131-2132.
[358]Richmond and Strait, "Evidence that Humans Evolved from Knuckle-Walking Ancestor," *Nature*, 2000.
[359]Charles E. Oxnard, PhD, *Fossils, Teeth and Sex: New Perspectives on Human Evolution*, University of Washington Press, Seattle and London, 1987, p. 227.
[360]Charles E. Oxnard, PhD, *The Order of Man: A Biomathematical Anatomy of the Primates*, 1984, p. 332.

humans? The footprints of apes are actually handprints. (To see pictures, do a Google search for "close-up of foot of an ape"; it shows clearly that the foot of an ape looks like a human hand.) That is, the feet on their lower limbs have a thumb and can grasp like a hand. They can grasp small objects between their toes, and manipulate them as if the feet were hands. This "hand-like" foot-with-opposable-thumb on the apes says from evolution's perspective that such animals are not "yet" human.

However, there are new discoveries of "modern human" footprints (meaning there is no grasping thumb at a right angle to the toes ["fingers"], no "opposable thumb" on the feet of these "now humans"; these "modern human" footprints have been found on the island of Crete and dated to something like 3.5 million years—yet, from the evolutionary perspective, that is *earlier* than ANY "modernish hominin" could have had modern feet.

Contrary to the standard evolutionary perspective, *"Australopithecus afarensis* . . . is ape and not transitional morphology between so-called 'ancient upright walking apes' and modern humans. Lucy was the nickname for an incomplete Ethiopian skeleton found by the American paleoanthropologist Donald Johanson in 1974. . . . [Lacking some] pelvis bones—Lucy soon became the benchmark fossil for the species *Australopithecus afarensis*.[361]

J. R. Miller continues and here quotes Rupe: "Of the 400 bone fragments found after Lucy, and classified as *Afarensis*, only 50 are related to the foot and all of them indicate Lucy had apelike feet. . . . [Lucy] is a tree-dwelling creature . . . who mostly walked on all fours—not human" (Rupe, 101-104).[362]

Ardipithecus ramidus

Ardipithecus ramidus is one of the newest "hominids" to appear in the textbooks. It was discovered in Aramis, Ethiopia, in 1994, and declared to be 4.4 million years old—almost a million years older than Lucy. *Time* magazine reported, "Bones from the Ethiopian desert prove that human ancestors walked on earth 4.4 million years ago."[363]

On the same day, *Newsweek* reported, *"Ramidus* confirms once and for all that the common ancestor lived just a little more than 4.4 million years ago,"[364] and that it was found in 4.4 million-year-old sediment.[365]

What was actually found? Fossils were collected from the surface at seventeen different areas over a one-mile range. Most of eight teeth were damaged. Arm bones were fragmented. Parts of the base of the skull were found 500 meters [1,640 feet] away.

Questions: Does all this material belong to the same creature? And does all this material belong to the same species?

Where were these discoveries made? *Time* magazine, in an article entitled, "One Less Missing Link," claimed the fossils "were enclosed in sedimentary rock that was neatly sandwiched between layers of *volcanic ash.*"[366]

Newsweek claimed the fossils were "locked in 4.4 million-year-old sediment."[367]

However, the fossil discoverers stated: "All hominid specimens were surface finds."[368]

And what about the teeth of *ramidus?*

> From *Nature:* The dm [deciduous molar] 1 has been critically important in studies of *Australopithecus* since the discovery of the genus 70 years ago. . . . The Aramis dm 1 [tooth of *ramidus]* is morphologically far closer to that of a chimpanzee than to any known hominid.[369]

This was reported in *Nature,* one of the world's most prestigious science journals. See if you can

---

[361] J. R. Miller in Miller's *The Descent of Man Recorded in the Bones*, in "the case for Lucy as Ape: Part 5 of 6," July 9, 2018 (on-line post), <at morethancake.org>.

[362] J. R. Miller, quoting Rupe, in The Descent of *Man Recorded in the Bones,* in "The Case for Lucy as Ape: part 5 of 6," July 9, 2018 (online post), at <morethancake.org> See Christopher Rupe, and John Sanford, PhD, 2017. *Contested Bones.* London Docklands: FMS Publications.

[363] Michael Lemonick, "One Less Missing Link," *Time,* October 3, 1994, p. 69.

[364] Sharon Begley, "Out of Africa, A Missing Link," *Newsweek,* October 3, 1994, p. 56.

[365] Sharon Begley, "Out of Africa, A Missing Link," *Newsweek,* October 3, 1994, p. 57.

[366] Michael D. Lemonick, "One Less Missing Link," *Time,* October 3, 1994, p. 68, (emphasis added).

[367] Sharon Begley, "Out of Africa, a Missing Link," *Newsweek,* October 3, 1994, p. 56.

[368] Bernard Wood, "The Oldest Hominid Yet," *Nature,* Vol. 371, September 22, 1994, p. 280.

[369] T. D. White, G. Suwa, and B. Asfaw, *"Australopithecus ramidus,* a new species of early hominid from Aramis, Ethiopia," *Nature,* Vol. 371, September 22, 1994, pp. 306-311.

find this evidence reported in any textbooks.

Authors in the journal *Nature* made additional statements which did not compare *ramidus* with man, but with the chimpanzee. Describing the cranial fossil, they said it "is morphologically far closer to that of a chimpanzee than to any known hominid." And they described the arm bones as "a mosaic of [characteristics] usually attributed to hominids and/or great apes."[370]

In order to be a credible theory, all the evidence must be examined.

> Former atheist James Perloff: Most textbooks avoid showing *comprehensive* tables of the discovered human fossils—doing so exposes the contradictions.[371]

> Phillip E. Johnson, JD: The fossils provide much more discouragement than support for Darwinism when they are examined objectively, but objective examination has rarely been the object of Darwinist paleontology. The Darwinist approach has consistently been to find some supporting fossil evidence, claim it as proof for "evolution," and then ignore all the difficulties.[372]

In 1968, Harold W. Clark noted that, after more than 30 years, the scientific findings supported the contention of Austin H. Clark, an American zoologist and author of more than 600 publications:[373]

> Dr. Austin H. Clark: There is a sharp, clean-cut, and very marked difference between man and the apes. Every bone in the body of a man is at once distinguishable from the corresponding bone in the body of any of the apes....
>
> Man is not an ape, and in spite of the similarity between them there is not the slightest evidence that man is descended from an ape.[374]

Another scientist agrees.

> Henry M. Morris, PhD: As far as the actual fossil evidence is concerned, man has always been man, and the ape has always been an ape.
>
> There are no intermediate or transitional forms leading up to man, any more than there were transitional forms between any of the other basic kinds of animals in the fossil records.[375]

Donkeys and dolphins

The *Orce Man*, promoted as the oldest example of man in Eurasia, was later identified as a four-month-old donkey-skull fragment.[376]

A dolphin's rib had been labeled "human collarbone" and had been promoted as evidence for evolution.[377]

What conclusion do science journals come to?

> Henry Gee in the prestigious journal *Nature*: Fossil evidence of human evolutionary history is fragmentary and open to various interpretations. Fossil evidence of chimpanzee evolution is absent altogether.[378]

> John Reader in *New Scientist*: Judged by the amount of evidence upon which it is based, the study of fossil man (palaeoanthropology) hardly deserves to be more than a sub-discipline of paleontology or anthropology.... The specimens themselves are often so fragmentary and inconclusive, that more can be said about what is missing than about what is present.... Darwin's work inspired the notion that fossils linking modern man and extinct ancestors would provide the most convincing proof of human evolution. Preconceptions have led evidence by the nose in the study of fossil man.[379]

In 1999, Henry Gee, the chief science writer for the journal *Nature*, candidly acknowledged, "The intervals of time that separate fossils are so huge that we cannot say anything definite about their possible connections to ancestry and descent."[380] He claimed that *all the fossil evidence*

---

[370] T. D. White, G. Suwa, and B. Asfaw, *"Australopithecus ramidus, a new species of early hominid from Aramis, Ethiopia,"* Nature, Vol. 371, September 22, 1994, pp. 306-311.
[371] James Perloff, "The Reigning World Chimp," *Tornado in a Junkyard: The Relentless Myth of Darwinism*, Refuge Books, Arlington, MA, 1999, p. 106.
[372] Phillip E. Johnson, JD, "The Vertebrate Sequence," *Darwin on Trial*, Regnery Gateway, Washington, DC, 1991, p. 84.
[373] Harold W. Clark, "Cave Men and Stone Ages," *Fossils, Flood, and Fire*, Outdoor Pictures, 1968, p. 219.
[374] Austin H. Clark, *The New Evolution: Zoogenesis*, Williams and Wilkin, Baltimore, MD, 1930, p. 224.
[375] Henry M. Morris, PhD, "Apes or Men?" *Scientific Creationism*, CLP Publishers, San Diego, CA, 1974, p.178.
[376] "Ass Taken for Man," *Daily Telegraph*, May 14, 1984.
[377] Ian Anderson, "Humanoid collarbone exposed as dolphin's rib," *New Scientist*, New Science Publications, London, April 28, 1983, p. 199.
[378] Henry Gee, "Return to the planet of the apes," *Nature*, Vol. 412, July 12, 2001, p. 131.
[379] John Reader, "Whatever happened to Zinjanthropous?" *New Scientist*, New Science Publications, London, March 26, 1981, p. 802.
[380] Henry Gee, "Nothing Besides Remains," *In Search of Deep Time: Beyond the Fossil Record to a New History of Life*, the Free Press, NY. 1999, p. 23.

for human evolution of several thousand generations of living creatures can be fitted into a small box.[381] Consequently, he concluded, that the conventional picture of human evolution is "a completely human invention created after the fact, shaped to accord with human prejudices. . . . Each fossil represents an isolated point, with no knowable connection to any other given fossil, and all float around in an overwhelming sea of gaps."[382] Then he said quite bluntly: "To take a line of fossils and claim that they represent a lineage is not a scientific hypothesis that can be tested, but an assertion that carries the same validity as a bedtime story—amusing, perhaps even instructive, but not scientific."[383]

> James Perloff: *Paleoanthropologists* use cranial capacity (skull size) to judge the evolutionary status of our supposed ancestors, but even in modern humans, adult cranial capacity ranges from 700 to 2200 cubic centimeters,[384] and has no bearing on intelligence. People's bone structure greatly varies, based on heredity, age, sex, health, and climate. Some are big-boned, some small-boned. There are Sumo wrestlers and pygmies. Doubtless, our ancient forebears were also diverse in their looks. How, then, can one assign a fossil bone to a distinct place in human history?[385]

Giuseppe Sermonti, PhD, in *Genetics:* Many schools proclaim as a matter without any doubt that man has derived from the African apes. . . . This is a falsehood which any honest scientist should protest against. It is not balanced teaching.

That which science has never demonstrated should be erased from any textbook and from our minds and remembered only as a joke in bad taste. One should also teach people how many hoaxes have been plotted to support the theory of the simian (ape) origins of man.[386]

Implications

One scientist makes a profoundly revealing admission:

> Stephen Grocott, PhD, from the University of Western Australia: Creation and evolution are actually both outside the realms of science. . . . Neither "process" is currently observable, testable, or repeatable. Please note that when speaking of evolution, I'm talking of the appearance of new (not rearranged) genetic information leading to greater and greater complexity of genetic information. I'm also talking about the appearance [origin] of life starting from inanimate chemicals. . . .
>
> Evolution needs increasing complexity, increasing information. We don't see it occurring today and no one was there to observe it in the past. Evolutionists counter by saying that it is too slow to observe. Even if this were true, it still means that evolution is not scientific because it is not observable or testable. . . . If one believes in evolution, then one has to account for the origin of life—the very first step. Without this, the whole subject of evolution hangs on nothing.[387]

---

[381] Henry Gee, "Are We Not Men," *In Search of Deep Time: Beyond the Fossil Record to a New History of Life*, the Free Press, NY, 1999, p. 202.
[382] Henry Gee, "Nothing Besides Remains," *In Search of Deep Time: Beyond the Fossil Record to a New History of Life*, the Free Press, NY. 1999, p. 32.
[383] Henry Gee, "Darwin and His Precursors," *In Search of Deep Time: Beyond the Fossil Record to a New History of Life*, the Free Press, NY, 1999, pp. 116-117.
[384] Stephen Molnar, *Races, Types, and Ethnic Groups: The Problem of Human Variation*, Prentice-Hall, Englewood Cliffs, NJ, 1975, p. 57.
[385] James Perloff, "The Reigning World Chimp," *Tornado in a Junkyard, The Relentless Myth of Darwinism*, Refuge Books, Arlington, MA, 1999, p. 103.
[386] Giuseppe Sermonti, PhD, "Not from the Apes," *Creation ex nihilo*, June-August 1993, p. 13.
[387] Edited by John Ashton, PhD, Stephen Grocott, PhD, "Stephen Grocott, PhD," *In Six Days: Why Fifty Scientists Choose to Believe in*

Paul Lemoine, once President of the Geological Society of France, Director of the Natural History Museum in Paris, and chief editor of the *Encyclopedie Francaise:* Evolution removes all morality. Evolution is a kind of dogma which its own priests no longer believe, but which they uphold for their people. It is necessary to have the courage to state this if only so that men of a future generation may orient their research into a different direction.[388]

Malcolm Muggeridge: I myself am convinced that the theory of evolution, especially the extent to which it has been applied, will be one of the great jokes in the history books of the future. Posterity will marvel that so flimsy and dubious an hypothesis could be accepted with the incredible credulity that it has.[389]

On the other hand King David once wrote, "I will praise thee; for I am fearfully and wonderfully made" (Psalm 139:14).

Creation, Master Books, Inc., Green Forest, AR, 2000, pp. 147-148.
[388]Paul Lemoine, director of the National Museum of Natural History, *Encyclopedie Francaise,* Vol. 5, 1937.

[389]Malcolm Muggeridge, Pascal Lectures, University of Waterloo, Ontario, Canada.

III

Transitional Fossils—
Intermediates

Chapter 16
Missing Links

How should we relate to the subject of *missing links?* We read about them from time to time in well-known science journals, but what is the real evidence?

We can't say that all evolutionists would acknowledge the scarcity of transitional fossils. However, in 2007, Occidental College geologist Donald Prothero, PhD, asserted, "Creationists claim there are no transitional fossils, aka missing links. Biologists and paleontologists, among others, know this claim is false . . . but the fossil record—which is far from complete—is full of them nonetheless."[390]

However, in his 2001 book, *What Evolution Is,* Ernst W. Mayr, PhD, Professor Emeritus in the Museum of Comparative Zoology at Harvard University, and hailed as the Darwin of the 20th century, acknowledged the obvious:

> Ernst Mayr, PhD: Given the fact of evolution, one would expect the fossils to document gradual steady change from ancestral forms to the descendants. But this is not what the paleontologist finds. Instead, he or she finds gaps in just about every phyletic series. New types often appear quite suddenly, and their immediate ancestors are absent in the earlier geological strata.[391] All species are separated from each other . . . ; intermediates between species are not observed.[392]

If the fossil record were already full of transitional fossils (no longer "missing" links) in 2007, then the following 2009 scenario is puzzling:

> Creationist Ray Comfort: Scientists have unveiled a 47-million-year-old fossilized skeleton of a monkey hailed as the missing link in human evolution. This 95 percent-complete "lemur monkey"—dubbed Ida—is described by experts as the "eighth wonder of the world." The search for a direct connection between humans and the rest of the animal kingdom has taken 200 years—but it was presented to the world today at a special news conference in New York. . . . Researchers say proof of this transitional species *finally* confirms Charles Darwin's theory of evolution. . . . [Does this appear to be big news?] Sir David Attenborough said Darwin "would have been thrilled" to have seen the fossil—and says it tells us who we are and where we came from. This is the one that connects us directly with them. . . . Now people can say "okay, we are primates, show us the link." . . . The link they would have said *up to now* is missing—well it's no longer missing.[393]

So, is the fossil record "full of transitions," or is it not? Is Ida proof of Darwinism? CBS News had a different interpretation: "So while we don't know exactly what Ida means to human origins, she's proof we are endlessly fascinated by where we came from."[394]

The *Wall Street Journal* addressed the real issue: "The discovery has little bearing on a separate paleontological debate centering on the identity of a common ancestor of chimps and humans, which could have lived about six million years ago and still hasn't been found."[395]

If evolution were true, the fossil record should document millions of "missing" links.

> Henry M. Morris, PhD, and Gary E. Parker, EdD: Why should we be able to classify plants and animals into types or species at all? In a fascinating editorial feature in *Natural History,* Stephen Gould writes that biologists have been quite successful in dividing up the living world into distinct and discrete species. . . . "But," says Gould, "how could the existence of distinct species be justified by a theory [evolution] that proclaimed ceaseless change as the most fundamental fact of nature?" For an evolutionist, why should there be species at all? If all life forms have been produced by

---

[390] Donald Prothero, *Evolution: What the Fossils Say and Why it Matters,* Columbia University Press, 2007; http://www.livescience.com/animals/090211-transitional-fossils.html.
[391] Ernst Mayr, PhD, "What is the Evidence for Evolution?" *What Evolution Is,* Basic Books, New York, NY, 2001, p. 14.
[392] Ernst Mayer, *The Growth of Biological Thought: Diversity, Evolution and Inheritance,* The Belknap Press of Harvard University Press, Cambridge, MA: 1982, p. 584.
[393] Quoted by Ray Comfort in "The Missing Link Finally Found," *Nothing Created Everything,* A WND Books book, [web site] WorldNetDaily, Los Angeles, CA, 2009, pp. 127-128, (emphasis added).
[394] CBS News, May 19, 2009.
[395] Gautam Naik, "Fossil Discovery is Heralded," *The Wall Street Journal,* May 15, 2009.

gradual expansion through selected mutations from a small beginning gene pool, organisms really should just grade into one another without distinct boundaries.[396]

Robert L. Carroll, PhD: Fossils would be expected to show a continuous progression of slightly different forms linking all species and all major groups with one another in a nearly unbroken spectrum. In fact, most well-preserved fossils are as readily classified in a relatively small number of major groups as are living species.[397]

The relatively few groups then and now? . . . with major gaps? . . . so many missing links? Why not one continuous group, then; and now?

Howard A. Peth: The very possibility of a classification is strong evidence against evolution. For example, organisms are neatly categorized in terms of species, genus, family, order, class, phylum, and kingdom. *But the ease with which we can "pigeonhole" basic kinds—with clear-cut gaps between—does not indicate an evolutionary relationship. Just the opposite is indicated,* for if all organisms arose by slow descent from a common ancestor, there should be a continuous blend from one kind into another. It should be impossible to tell where one species stops and another begins, so that any system of classification would be quite impossible.[398]

David B. Kitts, PhD, at the University of Oklahoma agrees:

Dr. Kitts: Despite the bright promise that *paleontology provides a means of seeing evolution,* it has presented some nasty difficulties for evolutionists, the most notorious of which is the presence of gaps in the fossil record. *Evolution requires intermediate forms between species, and paleontology does not provide them.*[399]

From Darwin's own *Origin* and well into the 21st Century, these intermediate forms are still missing.

Ariel A. Roth, PhD: A lot of evolutionists don't seem to understand the real problem of the fossil record. They point to isolated suggestions of intermediate parts or forms. Unfortunately, that is not what they must have in order to demonstrate that evolution actually occurred. By now we have identified many millions of fossils, comprising more than 250,000 different species. The more we find, the surer it appears that the lack of intermediates is a real fact.

If evolution had actually taken place, as organisms tried to evolve through billions of years, with the occasional successes and the many failures expected, we should find a solid continuity of intermediates, not just a few questionable exceptions.[400]

Michael J. Denton, MD, PhD: The universal experience of paleontology . . . [is that] while the rocks have continually yielded new and exciting and even bizarre forms of life, . . . what they have never yielded is any of Darwin's myriads of transitional forms. Despite the tremendous increase in geological activity in every corner of the globe and despite the discovery of many strange and hitherto unknown forms, the infinitude of connecting links has still not been discovered and the fossil record is about as discontinuous as it was when Darwin was writing the *Origin*. The intermediates have remained as elusive as ever and their absence remains, a century later, one of the most striking characteristics of the fossil record.[401]

David M. Raup, PhD, of the University of Chicago–Field Museum of Natural History and others add their perspective:

Dr. Raup: In the years after Darwin, his advocates hoped to find predictable progressions. In general, these have not been found—yet the optimism has died hard, and some pure fantasy has crept into textbooks.[402]

Gareth Nelson also questions the much-discussed fossil record:

Gareth J. Nelson, Professorial Fellow of the University of Melbourne, retired from the American Museum of Natural History in New York City: The idea that one can go to the fossil record and

---

[396]Henry M. Morris, PhD, and Gary E. Parker, EdD, "Darwin and the Nature of Biologic Change," *What Is Creation Science?* Master Books, El Cajon, CA, 1987, pp. 121-122.
[397]Robert L. Carroll, PhD, *Patterns and Processes of Vertebrate Evolution,* Cambridge University Press, Cambridge, MA, 1997, pp. 8-9.
[398]Howard A. Peth, "Classification and Comparative Anatomy" *Blind Faith: Evolution Exposed,* Amazing Facts, Inc., Frederick, MD, 1990, p. 56.
[399]David B. Kitts, PhD, (School of Geology and Geophysics), University of Oklahoma, "Paleontology and Evolutionary Theory," *Evolution,* Vol. 28, September 1974, p. 467, (emphasis added).

[400]Ariel A. Roth, PhD, "Can a Scientist Dare to Believe in God?" *Science Discovers God: Seven Convincing Lines of Evidence for His Existence,* Autumn House Publishing, a division of Review and Herald Publishing, Hagerstown, MD, 2008, p. 153, (emphasis in the original).
[401]Michael J. Denton, PhD, "The Fossil Record," *Evolution: A Theory in Crisis,* Adler & Adler, Bethesda, MD, 1986, p. 162.
[402]David M. Raup, PhD, (University of Chicago-Field Museum of Natural History), "Evolution and the Fossil Record," *Science,* Vol. 213, July 17, 1981, p. 289.

expect to empirically recover an ancestor-descendant sequence, be it of species, genera, families, or whatever, has been and continues to be a pernicious illusion.

The phrase "the fossil record" sounds impressive and authoritative. As used by some persons it becomes, as intended, intimidating, taking on the aura of esoteric truth as expounded by an elite class of specialists. But what is it, really, this fossil record? Only data in search of interpretation. All claims to the contrary that I know, and I know of several, are so much superstition.[403]

James Perloff: If evolutionary theory is true, the geologic record should reveal the innumerable transitional forms Darwin spoke of. We shouldn't find just a handful of questionable fossils, but billions of intermediates validating his theory. Instead, the fossil record shows animals complete—not in developmental stages—the very first time they are seen. This is just what we would *expect* if animals were created, instead of evolved.[404]

Mark Czarnecki: A major problem in proving the theory has been the fossil record; the imprints of vanished species preserved in the earth's geological formations. This record has never revealed traces of Darwin's hypothetical intermediate variants—instead species appear and disappear abruptly, and this anomaly has fueled the creationist argument that each species was created by God.[405]

In his chapter on "Difficulties of the Theory," Charles Darwin himself acknowledged the problem: " . . . why if species have descended from other species by sensibly fine gradations, do we not everywhere see innumerable transitional forms? Why is not all nature in confusion instead of species being, as we shall see, well defined?"[406]

> David C. Read, JD: Darwin noted that paleontologists have thoroughly explored only Europe; he argued that more missing links might be found in other parts of the world. That argument cannot be made today. By some estimates, 99 percent of all fossil collecting has been done since 1859 [the year Darwin's book was first published], and the lack of transitional forms is as acute as ever.[407]

> Stephen Jay Gould, PhD: All paleontologists know that the fossil record contains precious little in the way of intermediate forms; transitions between major groups are characteristically abrupt.[408]

As you can see, the lack of missing links is exposed not only by creationists, but also by evolutionists, such as Dr. Niles Eldredge, an invertebrate paleontologist at the American Museum of Natural History, speaking in a published interview.

> Niles Eldredge, PhD: A smooth transition from one form of life to another which is implied in the theory is . . . not borne out by the facts. The search for "missing links" between various living creatures, like humans and apes, is probably fruitless . . . because they probably never existed as distinct transitional types. . . . No one has yet found any evidence of transitional creatures. This oddity has been attributed to gaps in the fossil record which gradualists expected to fill when rock strata of the proper age had been found. *In the last decade, however, geologists have found rock layers of all divisions of the last 500 million years and no transitional forms were contained in them.* If it is not the fossil record which is incomplete then it must be the theory.[409]

While some say transitional fossils are missing because there were very few and none were preserved, the last two authors, Gould and Eldredge, believe transitional fossils are missing because relatively rapid evolutionary jumps occurred over these gaps, a theory they called "punctuated equilibria."[410] Many geneticists are shocked by the proposal and wonder why they would propose something so contradictory to genetics.[411]

---

[403] Gareth J. Nelson, Google "Gareth J. Nelson quotes," 2012.
[404] James Perloff, "Evidence Against the Theory of Evolution," *The Case Against Darwin*, Refuge Books, Burlington, MA, 2002, pp. 44-45 (emphasis added).
[405] Mark Czarnecki, "The Revival of the Creationist Crusade," *Maclean's*, January 19, 1981, p. 56.
[406] Charles Darwin, Difficulties of the Theory," *On the Origin of Species by Means of Natural Selection or the Preservation of Favoured Races in the Struggle for Life*, John Murray, Albemarle Street, London, 1859, p. 171.
[407] David C. Read, JD, "Does the Fossil Record Support Darwinism?" *Dinosaurs*, Clarion Call Books, Keene, TX, 2009, p. 268.
[408] Stephen Jay Gould, PhD, "Return of the Hopeful Monster," *The Panda's Thumb*, W.W. Norton & Company, NY, 1982, p. 189; "The Return of Hopeful Monsters," Natural History, Vol. 86, 1977, p. 22.
[409] Niles Eldredge, PhD, "Missing, Believed Nonexistent," *Manchester Guardian (The Washington Post Weekly)*, Vol. 119, No. 22, November 26, 1978, p. 1, (emphasis added).
[410] Stephen Jay Gould, PhD, "Evolution's Erratic Pace," *Natural History*, American Museum of Natural History, Vol. 86, No. 5, May 1977, pp. 13-15
[411] Walt Brown, Jr., PhD, "The Scientific Case for Creation," *In The Beginning: Compelling Evidence for Creation and the Flood*, Center for Scientific Creation, Phoenix, AZ, 2001, p. 53

Certain evolutionists became so frustrated from not being able to find missing links that they came up with the idea of "punctuated equilibrium," which hypothesizes that the *intermediate stages of evolution happened quickly*. Dr. Richard B. Goldschmidt of Yale University supports O. H. Schindewolf who stated that, "The first bird hatched from a reptilian egg."[412]

Gary E. Parker, EdD: The reptile-bird example was the focus of an enthusiastic exchange between creationist Duane Gish and [evolutionist] Stephen Gould in *Discover* magazine for May and July, 1981. *Discover* had asked Dr. Richard Bliss and me to participate in a "magazine debate" with Stephen Gould and Carl Sagan, but Gould and/or Sagan declined. The magazine published instead an anti-creationist article by Gould and followed it with a letter by creationist Gish in which he mentioned that some evolutionists believe the first bird hatched out of a reptile egg. Gould responded that Gish was misrepresenting evolution and that anyone who believed that ought to be laughed off the intellectual stage. Gish responded by proving that the reptile-bird "jump" had indeed been used as a specific example quoted approvingly by Goldschmidt, whose ideas Gould is resurrecting.

And that's not the end of the story yet. There is a [1958] children's book called *The Wonderful Egg*[413] that achieved a specific recommendation from the prestigious American Association for the Advancement of Science, the American Council on Education, and the Association for Childhood Education International. The book starts innocently enough with a mother dinosaur laying an egg. After asking, "Did a mother dinosaur lay that egg to hatch into a baby dinosaur?" the book answers "no" to *Brontosaurus, Stegosaurus,* and a list of other dinosaurs. Then comes the dramatic conclusion. The egg laid by the mother dinosaur wasn't a dinosaur egg at all: "It was a wonderful new kind of egg." And what did this dinosaur egg hatch into? "It hatched into a baby bird, the first baby bird in the whole world. And the baby bird grew up to be a beautiful bird with feathers. The first beautiful bird that ever sang a song high in the tree tops . . . of long, long ago."[414]

Out of curiosity, I bought this 44-page picture book from an Internet seller on Amazon. com. On the back of the front and back covers a stamp says it once belonged to a school library in Camarillo, California. The fact that it is a children's book is obvious from the simple wording. The opening paragraph tells us when this "wonderful egg" was laid: "Long, long ago, more than one hundred million years ago." The book pictures a white egg the size, shape, and appearance of a chicken egg, sitting in what it identifies as "a blue-green mossy nest." The following pages ask if the egg was laid by one of 12 dinosaurs including the *Brontosaurus* (with the wrong head), *Triceratops, Stegosaurus, Pteranodon, Ornithomimus, Elasmosaurus, Tyrannosaurus, Trachodon, Corythosaurus, Parasaurolophus, Tylosaurus,* and *Ankylosaurus.* These are all big dinosaurs ranging from 10 to 70 feet in length that laid leathery eggs. In addition to illustrating this new creature as a song bird, the book illustrates what it looked like, and then identifies it as a chicken-size *Archaeopteryx*.[415]

Although a children's book from 1958 doesn't necessarily need a scientific rebuttal, it is this very idea that is the center of "punctuated equilibrium." A discerning parent should carefully consider such books before purchase and ask questions—in this case questions like:

- How does anybody know that the *Archaeopteryx* was a song bird or even a bird?
- Were any dinosaur nests made from blue-green moss?
- Which of these dinosaurs built such a little nest?
- If this dinosaur fed its chick, with what?
- And how did this big dinosaur transfer food into the baby bird's little beak?
- How did the dinosaur incubate the egg?
- Did it sit on this "wonderful egg"?
- Did another dinosaur of the same kind lay another "wonderful," non-leathery egg at the same time that hatched into an opposite-sex *Archaeopteryx?*
- How could the first *Archaeopteryx* hatched from

---

[412]Richard B. Goldschmidt, MD, PhD, "Macroevolution," *The Material Basis of Evolution,* Yale University Press, New Haven, London, 1940, 1982, p. 395; O. H. Schindewolf, *Palaeontologie, Entwicklungslehre und Genetik,* Berlin, Borntrager, 1936, p. 108.
[413]Dahlov Ipcar, *The Wonderful Egg,* Doubleday, Garden City, NY, 1958.
[414]Gary E. Parker, EdD, "The Fossil Evidence," What is Creation Science? Master Books, El Cajon, CA, 1987, pp. 147-148.
[415]Dahlov Ipcar, *The Wonderful Egg,* Doubleday & Company, Inc., Garden City, NY, 1958.

a dinosaur egg reproduce itself without mating with another *Archaeopteryx* of the opposite sex hatched from a similar dinosaur egg at about the same time?

A more serious question is, "Why do we see recognized scientists and recognized scientific organizations stretching credibility beyond comprehension by supporting such unscientific guesswork?" Here's why. The lack of transitions in the fossil record impaled evolutionists on the horns of a dilemma. What to do? They had to come up with something that "explained" the evidence, or lack thereof, without acknowledging the Creator. Punctuated equilibria to the rescue. A bird hatching from a dinosaur egg. A very creative and imaginative guess, considering the absence of evidence.

The British naturalist himself had already rejected this possibility:

> Charles Darwin said, ". . . natural selection can act only by taking advantage of slight successive variations; she can never take a leap, but must advance by the shortest and slowest steps."[416]
>
> Theodosius Dobzhansky: Systemic mutations [large mutations which transform one species at once into another] have never been observed, and it is extremely improbable that species are formed in so abrupt a manner.[417]

Think it through. Punctuated equilibria [termed "punk eek" by the irreverent] is an attempt to support a theory from *lack of evidence*. Lack of evidence, they say, showed that it happened. For those who only believe in something if it has scientific evidence I ask, How scientific is that? Actually, it was the fossil evidence that the new theory tried to explain. The evidence was stasis (a period of stability) and gaps.

> Stephen C. Meyer, PhD: The fossils of the *Cambrian explosion* absolutely cannot be explained by Darwinian theory or even by the concept of *punctuated equilibrium*, which was specifically formulated in an effort to explain away the embarrassing fossil record. When you look at the issue from the perspective of biological information, the best explanation is that an intelligence was responsible for this otherwise inexplicable phenomenon.[418]

Jonathan Wells, PhD, summarizes his position:

> My conclusion is that the case for Darwinian evolution is bankrupt. The evidence for Darwinism is not only grossly inadequate, it's systematically distorted. I'm convinced that sometime in the not too distant future—I don't know, maybe twenty or thirty years from now—people will look back in amazement and say, how could anyone have believed this? Darwinism is merely materialistic philosophy masquerading as science, and people are recognizing it for what it is.[419]

---

[416] Charles Darwin, "Difficulties on Theory," *On the Origin of Species of Species by Means of Natural Selection or the Preservation of Favoured Races in the Struggle for Life*, John Murray, Albemarle Street, London, 1859, p. 194.

[417] Theodosius Dobzhansky, *Genetics and the Origin of Species*, Second edition, Columbia University Press, NY, 1941, p. 80.

[418] Stephen C. Meyer, PhD, quoted by Lee Strobel, "The Evidence of Biological Information," *The Case for a Creator*, Zondervan, Grand Rapids, MI, 2004, pp. 238-239.

[419] Jonathan Wells, PhD, quoted in Lee Strobel, "Doubts About Darwinism," *The Case for a Creator*, Zondervan, Grand Rapids, MI, 2004, pp. 65-66.

Chapter 17
The Evidence Is Compelling

This chapter documents deeper reasons why the fossil record fails to yield the type of evidence that Darwinism predicted. We begin with a British science journalist:

> Richard Milton: The case for Darwinism would be made convincingly if someone were to produce a sequence of fossils from a sequence of adjacent strata . . . showing indisputable signs of gradual progressive change of the same basic stock, but above the species level (as opposed to subspecific variation [e.g., new kinds of dogs]). Ideally this should be demonstrated in a long sequence, ten or twenty or fifty successive fossil species, showing major generic evolution—but a short sequence would be enough.
>
> But this simple relationship is not what is shown in the sequence of the rocks. Nowhere in the world has anyone met this simple evidential criterion with a straightforward fossil sequence from successive strata. Yet there are so many billions of fossils available from so many thousands of strata that the failure to meet this modest demand is inexplicable if evolution has taken place in the way Darwin and his followers have envisaged. It ought to be relatively easy to assemble not merely a handful but hundreds of species arranged in lineal descent. Schoolchildren should be able to do this on an afternoon's nature study trip to the local quarry, but even the world's foremost paleontologists have failed to do so with the whole earth to choose from and the resources of the world's greatest universities at their disposal.[420]

The pattern, typical of the overwhelming majority of cases, is that life forms appear in the fossil record, they remain the same for "long periods of geologic time," and then they disappear from the record:[421]

> Stephen Jay Gould, PhD: The history of most fossil species includes two features particularly inconsistent with *gradualism:*
>
> 1. *Stasis.* Most species exhibit no directional change during their tenure on earth. They appear in the fossil record looking pretty much the same as when they disappear; morphological change [i.e., of form and structure] is usually limited and directionless.
>
> 2. *Sudden appearance.* In any local area, a species does not arise gradually by the steady transformation of its ancestors; it appears all at once and "fully formed."[422]

According to the September 7, February 2009 *Guardian*, in London the Natural History Museum's department of palaeontology houses one of the world's great fossil collections, made up of some 9 million specimens from all over the world. Some of these specimens were discovered by Charles Darwin himself, including vertebrate, invertebrate, and plant fossils.

The museum's palaeontology collection . . . makes it globally important.

When Luther D. Sunderland asked its senior paleontologist, Dr. Colin Patterson, why he did not show evolutionary transitions (missing links) in his book *Evolution*, he answered:

> Dr. Patterson: I fully agree with your comments on the lack of direct illustrations of evolutionary transitions in my book. If I knew of any, fossil or living, I would certainly have included them. You suggest that an artist should be asked to visualize such transformations, but where would he get the information from? I could not, honestly, provide it, and if I were to leave it to artistic license, would that not mislead the reader? . . . Yet [Dr. Stephen Jay] Gould and the American Museum [of Natural History] people are hard to contradict when they say that *there are no transitional fossils.* As a paleontologist myself, I am much occupied with the philosophical problems of identifying ancestral forms in the fossil record. You say that I should at least "show a photo of the fossil

---
[420] Richard Milton, "The Record of the Rocks," *Shattering the Myths of Darwinism*, Park Street Press, Rochester, VT, 1997, p. 110.
[421] David C. Read, JD, "Does the Fossil Record Support Darwinism?" *Dinosaurs*, Keene, TX, 2009, p. 270 [Read is a creationist.]
[422] Stephen Jay Gould, PhD, "Evolution's Erratic Pace," *Natural History*, American Museum of Natural History, Vol. 86, No. 5, May 1977, p. 13-15, see also: Stephen Jay Gould, PhD, "The Episodic Nature of Evolutionary Change," *The Panda's Thumb*, W.W. Norton & Company, NY, 1982, p. 182.

in which each type organism was derived." I will lay it on the line—there is not one such fossil for which one could make a watertight argument. . . . It is easy enough to make up stories of how one form gave rise to another, and to find reasons why the stages should be favored by natural selection. But such stories are not part of science, for there is no way of putting them to the test.[423]

On November 5, 1981, Dr. Patterson presented the keynote address to the American Museum of Natural History. Although not a creationist, his comments were astonishing:

> Dr. Patterson: One morning I woke up and something had happened in the night and it struck me that I had been working on this stuff for twenty years and there was not one thing I knew about it. That's quite a shock to learn that one can be so misled so long. Either there was something wrong with me or there was something wrong with evolutionary theory. Naturally, I know there is nothing wrong with me, so for the last few weeks I've tried putting a simple question to various people and groups of people.
>
> Question is: Can you tell me anything you know about evolution, any one thing that is true? I tried that question on the geology staff at the Field Museum of Natural History and the only answer I got was silence. I tried it on the members of the Evolutionary Morphology Seminar in the University of Chicago, a very prestigious body of evolutionists, and all I got there was silence for a long time and eventually one person said, "I do know one thing—it ought not to be taught in high school."[424]

Gary Bates: Although these are only excerpts from Patterson's very frank and startling lecture that evening (the full text is even more revealing), it is plain to see the doubts he was having. It also shows that creationist usage of such quotes by Patterson does not amount to "creationist foul play."

Dr. Patterson's penchant for openness did not do him any service with the pro-evolutionary scientific establishment, who often expressed anger and dismay at his comments when they could not make excuses for them. His experience and expertise as holder of one of the most prestigious scientific posts in the world did not grant him immunity from pressure for having dared to express doubts about the evolutionary worldview. It is a sad reminder that political and ideological correctness can be more important than any so-called "objective facts" in determining scientific acceptance of an idea.[425]

Vance Ferrell: Over the course of a year, Luther Sunderland interviewed the three leading paleontologists in charge of the largest fossil collections of the world: Dr. Colin Patterson at the British Museum of Natural History in London, Dr. David Raup at the Field Museum of Natural History in Chicago, and Dr. Niles Eldredge at the American Museum of Natural History in New York City. With their permission Sunderland made taped recordings of each interview. In charge of 50 percent of all the collected fossils of the world, each man was a lifetime expert in paleontology—and each one admitted that there were no transitional species![426]

Creationists vs. transition fossils

Creationists will critique transitional fossils in a variety of ways. They might claim that a transitional fossil is not proof of an evolutionary relationship since you can't prove that it is, in fact, an ancestor of any later organism. Without actually going back in time and watching the birth/hatching/etc. of each successive organism in an evolutionary chain, we can not "prove" that an evolutionary relationship exists.

> Evolutionist Niles Eldredge, PhD: Paleontologists ever since Darwin have been searching (largely in vain) for the sequences of insensibly graded series of fossils that would stand as examples of the sort of wholesale transformation of species that Darwin envisioned as the natural product of the evolutionary process. Few saw any reason to demur—though it is a startling fact that of the half dozen reviews of *On the Origin of Species* written by paleontologists that I have seen, all take Darwin to task for failing to recognize that most species remain recognizably themselves, virtually unchanged throughout

---

[423]Copy of letter, dated April 10, 1979, from Dr. Patterson to Sunderland; Luther Sunderland, *Darwin's Enigma*, 1988, p. 90, (emphasis added).
[424]Dr. Colin Patterson, "Evolutionism and Creationism," Transcript of Address at the American Museum of Natural History, New York, NY, November 5, 1981, p.1. Dr. Patterson was a senior paleontologist and editor of their journal at the British Museum of Natural History in London. He authored the book *Evolution*.
[425]Gary Bates, "That quote!—about missing transitional fossils: embarrassed evolutionists try to 'muddy the waters," Creation Ministries International, Creation.com, December 27, 2015.
[426]Vance Ferrell, a creationist, "The Latest Evolution Crisis," *The Evolution Handbook*, Evolution Facts, Altamont, TN, 2006, p. 916, (emphasis in the original), see also: Philip Johnson, JD, *Darwin on Trial*, Regnery Gateway, Washington, DC, 1991, p. 76.

their occurrence in the geological sediments of various ages.[427]

Evolutionist Stephen Jay Gould saw a big problem

> Stephen Jay Gould, PhD: The fossil record offered no support for gradual change. . . . New species almost always appeared suddenly in the fossil record with no intermediate links to ancestors in older rocks of the same region.[428]

> Then, Dr. Gould, in defending punctuated equilibrium: "The extreme rarity of transitional forms in the fossil record persists as a trade secret of paleontology. The evolutionary trees that adorn our textbooks have data only at the tips and nodes of their branches; the rest is inference, however reasonable, not the evidence of fossils. . . . Paleontologists have paid an exorbitant price for Darwin's argument. We fancy ourselves as the only true students of life's history, yet to preserve our favorite account of evolution by natural selection we view our data as so bad that we never see the very process we profess to study.[429]

The various trees of life are all based on faith in the theory of evolution. *Mary Leakey*, the famous British archaeologist and anthropologist said, "All those trees of life with their branches of our ancestors, that's a lot of nonsense."[430]

Michael J. Denton, MD, PhD, studied fish, amphibians, reptiles and mammals for evidence from molecular biology to support their evolutionary progression:

> Dr. Denton: Instead of revealing a multitude of transitional forms through which the evolution of the cell might have occurred, molecular biology has served only to emphasize the enormity of the gap. . . . No living system can be thought of as being primitive or ancestral with respect to any other system, nor is there the slightest empirical hint of an evolutionary sequence among all the incredibly diverse cells on earth. . . . The system of nature conforms fundamentally to a highly ordered hierarchic scheme from which all direct evidence for evolution is emphatically absent.[431]

> Tom S. Kemp, curator, Oxford University Museum: Gaps at a lower taxonomic level, species and genera, are practically universal in the fossil record of the mammal-like reptiles. In no single adequately documented case is it possible to trace a transition, species by species, from one genus to another.[432] . . . As is now well known, most fossil species appear instantaneously in the fossil record, persist for some millions of years virtually unchanged, only to disappear abruptly. . . .[433]

In 2009, the mainstream scientific community celebrated the 200th anniversary of Charles Darwin's birthday and the 150th anniversary of the publication of his book, *On The Origin of Species*. While acknowledging that there were no examples of transition from the fossil record that would illustrate and scientifically validate his hypothesis, Darwin thought given enough time these missing links would be found. But that hasn't happened. All of the extremely rare, so-called missing links thus far have turned out to be either one species or another closely related organism that reflect limited change, some of which are used in textbooks to this day to illustrate the theory of evolution.

> Michael Pitman, PhD: For the past, paleontologists are aware that the fossil record contains precious little in the way of intermediate forms. Most new species, genera and families, and nearly all categories above the level of families, appear in the record suddenly and are not led up to by known, gradual, completely continuous transitional sequences. A critical analysis of zoological groups, including fossil and recent species, shows that most of them "froze" in their present state and a very long time ago.[434]

A scientist who believes in creation adds his perspective:

---

[427]Niles Eldredge, PhD, "Progress in Evolution?" *New Scientist*, Vol. 110, New Science Publications, London, June 5, 1986, p. 55.
[428]Stephen Jay Gould, PhD, "Evolution's Erratic Pace," *Natural History*, American Museum of Natural History, Vol. 86, No. 5, May 1977, p. 14.
[429]Stephen Jay Gould, PhD, "Evolution's Erratic Pace," *Natural History*, American Museum of Natural History, Vol. 86, No. 5, May 1977, p. 13-15, see also: "The Episodic Nature of Evolutionary Change," *The Panda's Thumb*, W.W. Norton & Company, NY, 1982, pp. 181-182.
[430]Mary Leakey, Associated Press, December 10, 1996.

[431]Michael J. Denton, MD, PhD, "The Enigma of Life's Origin," and "The Biochemical Echo of Typology," *Evolution: A Theory in Crisis*, Adler & Adler, Bethesda, MD, 1986, pp. 249-250, 278.
[432]Tom S. Kemp, *Mammal-like Reptiles and the Origin of Mammals*, Academic Press, New York, NY, 1982, p. 319.
[433]Tom S. Kemp, "A Fresh Look at the Fossil Record," *New Scientist*, Vol. 108, New Science Publications, London, December 5, 1985, p. 66.
[434]Michael Pitman, PhD, "Sports [mutations], Survival and the Hone," *Adam and Evolution*, Rider & Company, London, Melbourne, Sydney, Auckland, Johannesburg, 1984, p. 69.

Clyde L. Webster Jr., PhD: The fossil record supports the model of *initially created-kinds subject to variation* as well as, or better than, it supports evolutionary models. The lowest multicellular fossils appearing in the geologic sequence are complex species! *The argued progression of simple to complex fossils in the geologic column can possibly be attributed to order of burial rather than to order of evolution.* This interpretation is strengthened when observation of the limited change and lack of new structures arising from genetic variation is taken into consideration![435]

Should evolution be taught in public schools? No. Allow its true believers to establish their own schools where tuition can fund the teaching of evolution. Public schools should limit their education to the facts of science, not to the interpretation or philosophical indoctrination of materialism. Anatomy, physiology, biophysics, pharmacology, pathology, even geology, biology, paleontology, and astronomy can be taught in public schools without reference to origins. Why should tax dollars of the general population, including people of various religions, be used to teach a subject that cannot be observed or subjected to experimentation? Teach the facts and allow the students to determine origins through their own research.

Through interlibrary loan I found a 570-page, 1981 biology textbook that devoted only 13 pages to Darwin, evolution, and fossils. A stamp in the front indicated that it once belonged to a Catholic school in Cincinnati, Ohio. And the evidence outlined only microevolution (variation within species). I scanned every page, and although I saw no references to the origin of life or how any complex structure or creature evolved from a common ancestor, the book was quite informative. The 13 pages provided only historical background on survival of the fittest, natural selection, and mutations, which most creationists would endorse. So, I found this book[436] to be a good example of how science can be taught to middle and high school students without major indoctrination of its readers into any materialist worldview.

Challenges in evolution education are not new. In 1974, evolutionist Barbara J. Stahl, PhD, published a book that exposed a major weakness in undergraduate textbooks on evolution:

Dr. Stahl: Anyone who has written or read a textbook becomes acutely aware of what cannot be included within its covers. The writer sets down an orderly summary of the knowledge in his field and of necessity omits mention of most of the problems with which workers have had to cope in coming to their conclusions. . . .

It is especially regrettable if undergraduate students of vertebrate evolution do not know of the problems that currently absorb embryologists, anatomists, and paleontologists, because the chance to wrestle with these problems—the fragmentary evidence and conflicting possibilities that exist—engenders more interest than does a catalog of conclusions.[437]

A science writer for *Time* magazine provided his perspective:

Michael D. Lemonick: Despite more than a century of digging, the fossil record remains maddeningly sparse. With so few clues, even a simple bone that doesn't fit into the picture can upset everything. Virtually every major discovery has put deep cracks in the conventional wisdom and forced scientists to concoct new theories, amid furious debate.[438]

Stephen C. Meyer, PhD, and Charles Reed address the controversy over the issue of scientific blind spots:

Dr. Meyer: The materialistic worldview has exercised dominance on intellectual life in Western culture for a hundred and fifty years. It has become the default worldview in science, philosophy, and academia in general. It's presupposed. Some people who dissent from it have experienced intense hostility and sometimes persecution. That could discourage others from exploring this area or speaking out favorably toward it. . . .

Finally, within the scientific culture there are belief systems that are philosophically very ques-

---

[435] Clyde L. Webster Jr., PhD, "Evolutionary Processes," *A Scientist's Perspective on Creation and the Flood*, Geoscience Research Institute, Loma Linda, CA, 1995, p. 14, (emphasis added).
[436] Albert Kaskel, Paul J. Hummer, Jr., Lucy Daniel, "The Changing World of Life," *Biology: An Everyday Experience*, Charles E. Merrill Publishing Co., Columbus, OH, 1981, pp. 494-507.
[437] Barbara J. Stahl, PhD, "Preface," *Vertebrate History: Problems in Evolution*, McGraw-Hill, NY, 1974, p. vii.
[438] Michael D. Lemonick, "How Man Began," *Time*, March 14, 1994, pp. 80-87.

tionable. For instance, many believe that science must only allow naturalistic explanations, which excludes from consideration the design hypothesis. Many scientists put blinders on, refusing to acknowledge that evidence, and a kind of "group think" develops. . . .

I don't think its right to invoke a self-serving rule that says only naturalistic explanations can be considered by science. Let's have a new period in the history of science where we have methodological rules that actually foster the unfettered seeking of truth. Scientists should be allowed to follow the evidence wherever it leads—even if it leads to the conclusion that makes some people uncomfortable.[439]

Charles A. Reed: We can easily get stuck in a rut, going round and round, reinforcing our preconceived notions by the happy process of talking only to those who agree with us and avoiding any dangerous new thoughts that might expose us to critical comment. Surely this is one of the greatest blocks to the creative process that should infuse the scholarly world. . . . Many conferences are made up only of "accepted" scholars who blandly sweep over the most fundamental questions and plunge on with their "accepted" lines of inquiry.[440]

Paul R. Ehrlich and L. C. Birch: Our theory of evolution . . . is thus "outside of empirical science." . . . No one can think of ways in which to test it. Ideas, either without basis or based on a few laboratory experiments carried out in extremely simplified systems have attained currency far beyond their validity. They have become part of an evolutionary dogma accepted by most of us as part of our training.[441]

Evolutionists reject creationism because they state correctly that it is a religion based on faith and miracles, and they don't believe in miracles. But creationism is also based on science, the study of the laws of nature, and what can be observed and measured in the natural world.

But time—millions of years—is evolution's "god." And, according to the famous Harvard professor and evolutionist, professor George Wald, PhD, it performs miracles:

Dr. Wald: I concede the spontaneous origin of life to be "impossible." It is impossible as we judge events in the scale of human experience. . . . The important point is that since the origin of life belongs in the category of at-least-once phenomena, time is on its side. However improbable we regard this event . . . given enough time it will almost certainly happen at least once. . . . Time is in fact the hero of the plot. . . . Given so much time, the "impossible" becomes possible, the possible probable, and the probable virtually certain. One has only to wait: time itself performs miracles.[442]

I'm curious to know how many evolutionary scientists would consider the last part of that statement to be scientific or even logical.

Michael J. Denton, MD, PhD: The complexity of the simplest known type of cell is so great that it is impossible to accept that such an object could have been thrown together suddenly by some kind of freakish, vastly improbable, event. Such an occurrence would be indistinguishable from a miracle.[443]

To account for the development of life, evolutionists need millions of years. Correction: evolutionists *desperately need* millions of years. Without millions of years, evolution would be impossible. And we already learned that erosion rates should have destroyed the geologic column many times, which rules out millions of years. According to the *Encyclopædia Britannica*, the continents should have been eroded away at least 200 times.

Correcting for the erosional effects of modern agriculture, etc., it seems that our continents could have been eroded away at least 100 times during the estimated billions of years of their existence.

---

[439]Stephen C. Meyer, PhD, quoted by Lee Strobel, "Where Science Meets Faith," *The Case for a Creator*, Zondervan, Grand Rapids, MI, 2004, pp. 84-86.
[440]Charles Reed, *Origins of Agriculture*, The Hague: Mouton, 1977, p. 89.
[441]Paul R. Ehrlich and L. C. Birch, "Evolutionary History and Population Biology," *Nature*, April 22, 1967, p. 352.
[442]George Wald, PhD, "The Origin of Life," *Scientific American*, Scientific American, Inc., NY, August 1954, pp. 47-48.
[443]Michael J. Denton, MD, PhD, "The Enigma of Life's Origin," *Evolution: A Theory in Crisis*, Adler & Adler, Bethesda, MD, 1986, p. 264.

Chapter 18
The Cambrian Explosion

There are almost no evolutionary ancestors in the lowest levels of the geologic column: the *Precambrian layer*, which is believed by some creationists to be earth's original surface. Instead, the next layer up, the Cambrian layer, has so many completely formed and mature fossil specimens that their existence has been identified as *"the Cambrian explosion."* What is the "old-earth" view of earth history from the evolutionary perspective?

> *Life Science:* The largest span of time in the Geologic Time Scale . . . is sometimes referred to simply as the *Precambrian* (pree **KAM** bree un). It covers the first 4 billion years of earth's history. Scientists know very little about the Precambrian because there are few fossils from these ancient times. . . . The Precambrian covers about 87 percent of earth's history.[444]

> Daniel I. Axelrod, PhD: One of the major unsolved problems of geology and evolution is the occurrence of diversified, multicellular marine invertebrates in Lower Cambrian rocks on all continents and their absence in rocks of greater age. . . . A simple way of putting it is that currently we have about 38 phyla of different groups of animals, but the total number of phyla discovered during the Cambrian including those in China, Canada, and elsewhere adds up to over 50 phyla. That means there are more phyla in the very, very beginning, where we found the first fossils of animal life, than exist now.[445]

> Paul K. Chien, PhD: Stephen Jay Gould, PhD, has referred to this as the *reverse cone of diversity*. The theory of evolution implies that things get more complex and get more and more diverse from one single origin. But the whole thing turns out to be reversed—we have more diverse groups in the very beginning, and in fact more and more of them die off over time, and we have less and less now.[446]

> Ariel A. Roth, PhD: *The Cambrian explosion is not just a case of all the major animal phyla appearing about the same place in the geologic column. It is also a situation of no ancestors to suggest how they might have evolved.* Paleontologists have thoroughly studied the rocks just below the Cambrian explosion, in which we would expect to find the intermediates. It has been a virtually futile search.[447]

> Evolutionist Stephen Jay Gould, PhD: Our more extensive labor has still failed to identify any creature that might serve as a plausible immediate ancestor for the Cambrian faunas [animals].[448] . . . The most famous such burst, the Cambrian explosion, marks the inception of modern multicellular life. Within just a few million years, nearly every major kind of animal anatomy appears in the fossil record for the first time. . . . The Precambrian record is now sufficiently good that the old rationale about undiscovered sequences of smoothly transitional forms will no longer wash.[449]

The Cambrian explosion can be predicted from the Creation Story. Although a university textbook hints at the concept, it suggests time might solve Darwin's problem.

> William D. Stansfield, PhD: During the Cambrian Period there suddenly appeared representatives of nearly all the major groups of animals (phyla) now recognized. It was as if a giant curtain had been lifted to reveal a world teeming with life in fantastic diversity. The Cambrian "curtain" has become the touchstone of the creation theory. Darwin was aware of the problem this created for evolutionists and it remains a problem today.

Evolutionists keep hoping that new discoveries will eventually fill in the missing pieces of the

---

[444]"Fossil Record," *Life Science*, Prentice-Hall, Inc., Upper Saddle River, NJ, 2002, pp. 161-162.
[445]Daniel I. Axelrod, PhD, "Early Cambrian Marine Fauna," *Science*, Vol. 128, July 4, 1958, p. 7.
[446]Paul Chien, (Chair, Biology Department, University of San Francisco), "Explosion of Life," www.origins.org/real/ri9701/chien.html, p. 2. Interviewed June 30, 1997.
[447]Ariel A. Roth, PhD, "What Fossils Say About Evolution," *Origins: Linking Science and Scripture*, Review and Herald Publishing Association, Hagerstown, MD, 1998, p. 184.
[448]Stephen Jay Gould, PhD, "A Short Way to Big Ends," *Natural History*, Vol. 95, January 1986, p. 18.
[449]Stephen Jay Gould, PhD, "An Asteroid to Die for," *Discover*, October 1989, p. 65.

fossil puzzle, but the chances of success may be less than those of finding the proverbial "needle in the haystack."[450]

Could I suggest a different interpretation of this scientific evidence: that these animals did not evolve, but were instead created? To reiterate, we've already seen that erosion rates around the world do not support millions of years. True believers in evolution acknowledge this predicament, but continue to support their worldview:

> Evolutionist Richard Dawkins, PhD: And we find many of them [Cambrian fossils] already in an advanced state of evolution, the very first time they appear. It is as though they were just planted there, without any evolutionary history. Needless to say, this appearance of sudden planting has delighted creationists.[451]

Richard B. Goldschmidt, MD, PhD, acknowledges that time has not solved the problem:

> When a new phylum, class, or order appears, there follows a quick, explosive (in terms of geological time) diversification so that practically all orders or families known appear suddenly and without any apparent transitions. . . . In spite of the immense amounts of the paleontological material and the existence of long series of intact stratigraphic sequences with perfect records for the lower categories, transitions between the higher categories are missing.[452]

Sometimes evolutionists suggest that fossils of intermediate organisms are missing because they were soft-bodied and would not be as easily preserved as organisms with hard parts.[453] Such an argument does not appear to be based on reality because many soft-bodied organisms are well-preserved as fossils. The Cambrian explosion presents one of the greatest problems for evolutionary theory. The Burgess Shale organisms in the Cambrian are mainly soft-bodied, and many are excellently preserved, yet the expected intermediates below the Cambrian explosion are missing.[454]

> Michael Pitman, PhD: Cambrian animals include sponges, corals, jellyfish, worms, mollusks, trilobites, crustaceans, in fact every one of the major invertebrate forms of life. Nor is the record entirely biased. Marvelous imprints of both soft-bodied and shelled animals occur in the Burgess shales of British Columbia. Complex organs such as intestines, stomachs, bristles and spines are found: eyes and feelers indicate the presence of nervous systems.[455]

I am reminded here also of the first words of Scripture, "In the beginning, God created . . . ."

> David C. Read, JD: Occasionally, even soft body parts are found fossilized or preserved in contour by the surrounding rock. *Fossilization of soft body parts requires extremely rapid burial in an environment with little oxygen—typically burial in thick silt under water.*[456]

> Ariel A. Roth, PhD: The abrupt appearance of major kinds of animals and plants looks more like creation by God rather than gradual evolutionary development. Evolution needs a lot of time to accommodate the virtually impossible events necessary for producing such varied and complex life forms. However, the many fossil types that appear suddenly suggest hardly any time at all. On the other hand, *those who favor the God hypothesis* see the Cambrian explosion as evidence of His creative ability. Some specifically interpret it as evidence of the first group of organisms buried during the catastrophic biblical Flood.[457]

Now for some striking details:

> Stephen C. Meyer, PhD: The Cambrian explosion represents an incredible quantum leap in biological complexity. . . . Without any ancestors in the fossil record, we have a stunning variety of complex creatures appearing in the blink of an

---

[450]William D. Stansfield, PhD, "Evidence of Evolution—Paleontology and Biogeography," *The Science of Evolution*, MacMillan Publishing Co., Inc. NY; Collier MacMillan Publishers, London, 1977, pp. 76-77.
[451]Richard Dawkins, PhD, "Puncturing Punctuationism," *The Blind Watchmaker*, W.W. Norton & Company; New York, London, 1987, p. 229.
[452]Richard B. Goldschmidt, MD, PhD, "Evolution, as viewed by one geneticist," *American Scientist*, Yale University, Vol. 40, January 1952, No. 1, pp. 97-98.
[453]Dr. Colin Patterson, *Evolution*, London: British Museum (Natural History), and NY: Cornell University Press, p. 133; Dr. Patterson lists this explanation but does not especially defend it.

[454]Ariel A. Roth, PhD, "What Fossils Say About Evolution," *Origins: Linking Science and Scripture*, Review and Herald Publishing Association, Hagerstown, MD, 1998, p. 188.
[455]Michael Pitman, PhD, "Programmes Typed in Dust," *Adam and Evolution*, Rider & Company, London, Melbourne, Sydney, Auckland, Johannesburg, 1984, p. 190.
[456]David C. Read, JD, "When Did the Dinosaurs Live? The 'Scientific' View," *Dinosaurs*, Clarion Call Books, Keene, TX, 2009, p. 47, (emphasis added).
[457]Ariel A. Roth, PhD, "So Little Time for Everything," *Science Discovers God: Seven Convincing Lines of Evidence for His Existence*, Autumn House Publishing, a division of Review and Herald Publishing, Hagerstown, MD, 2008, pp. 144-145.

eye, geologically speaking.

For example, the trilobite—with an articulated body [that is still in the same configuration as it was when it was alive], complicated nervous system, and compound eyes—suddenly shows up fully formed at the beginning of the explosion. It's amazing! And this is followed by stasis, which means the basic body plans remained distinct over the eons.

All of this totally contradicts Darwinism, which predicted the slow, gradual development in organisms over time. Darwin admitted the Cambrian explosion was "inexplicable" and "a valid argument against [his] theory." . . . He thought he would be vindicated; however, as more fossils were discovered, the picture has only gotten worse.

The big issue is where did the information come from to build all these new proteins, cells, and body plans? . . . Where did the genetic information come from to build those complicated molecules? This would require highly complex, specified genetic information of the sort that neither random chance, nor natural selection, nor self-organization can produce.[458]

Horatio H. Newman, PhD, pioneer in human genetics, wrote in 1932: Reluctant as he may be to admit it, honesty compels the evolutionist to admit that there is no absolute proof of organic evolution.[459]

Phillip E. Johnson, JD, almost 60 years later: After attending a geological conference on mass extinctions, [Dr. Stephen Jay] Gould wrote a remarkable essay reflecting on how the evidence was turning against Darwinism. He told his readers that he had long been puzzled by the lack of evidence of progressive development over time in the invertebrates with which he was most familiar. "We can tell tales of improvement for some groups, but in honest moments we must admit that the history of complex life is more a story of multifarious variation about a set of basic designs than a saga of accumulating excellence."[460]

Just about everyone who took a college biology course during the last sixty years or so [published in 1991] has been led to believe that the fossil record was a bulwark of support for the classic Darwinian thesis, not a liability that had to be explained away. And if we didn't take a biology class we saw [the film] *Inherit the Wind* [a story that fictionalized the 1925 Scopes "Monkey" trial] and laughed along with everybody else when [the attorney defending Scopes] made a monkey out of [the prosecuting attorney]. But I wonder if [the prosecutor, speaking then in favor of Tennessee's pro-creation law] would have looked like such a fool if he [had known what today's creation scientists have discovered, or, at least if he] could have found a distinguished paleontologist [like Gould] having one of those "honest moments," and produced him as a surprise witness to tell the jury and the theater audience that the fossil record shows a consistent pattern of sudden appearance followed by stasis, that life's history is more a story of variation around a set of basic designs than one of accumulating improvement, that extinction has been predominantly by catastrophe rather than gradual obsolescence, and that orthodox interpretations of the fossil record often owe more to Darwinist preconceptions than to the evidence itself.[461]

Do you understand why I engage in "quotation mining?" There is no way I, a non-scientist and historian for a university, could make such a compelling statement. Furthermore, the "scientific evidence" presented during the "movie trial" would be considered laughable by today's scientists.

Michael Pitman, PhD: "Two main lines of evidence [in the trial] for evolution were the Piltdown man [known today to be a scientific fraud] and Nebraska man (Hesperopithecus) [later found to be nothing more than the water-worn tooth of an extinct pig]. Nowhere in the trial did the scientific problems receive any sensible discussion. [The defense attorney] displayed ignorance both about the theory of evolution and the teachings of the Bible, and leveled a barrage of insult and vilification at [the creation-leaning prosecuting attorney who, after the barrage, was then blocked by the attorney general and the judge, and given no legal opportunity to defend his views].[462]

---

[458]Stephen C. Meyer, PhD, quoted by Lee Strobel, "The Evidence of Biological Information," *The Case for a Creator*, Zondervan, Grand Rapids, MI, 2004, p. 240, (bracketed copy added).
[459]Horatio H. Newman, PhD, *Evolution, Genetics, and Eugenics*, Third Edition Chicago University Press, Chicago, IL, 1932, p. 57.
[460]Phillip E. Johnson, JD, "The Fossil Problem," *Darwin on Trial*, Regnery Gateway, Washington, DC, 1991, pp. 57-58, (emphasis added).

[461]Phillip E. Johnson, JD, "The Fossil Problem," *Darwin On Trial*, Regnery Gateway, Washington, DC, 1991, pp. 58-59 [Bracketed added to condense pages of context.]
[462]Michael Pitman, PhD, "Monkey Business," *Adam and Evolution*, Rider & Company, London, Melbourne, Sydney, Auckland, Johannesburg, 1984, p. 100.

Chapter 19
Human Reproduction

Notice the complexity and impossibility of "evolving" simultaneously two interdependent human paths to all the interrelated and complementary anatomy and physiology involved in human reproduction.

In the beginning of a love relationship, both parties might feel a little shy. But as time goes by, the epithelial cells of his digits might stroke the epidermis on the interosseus muscles of her hand. The sensation of touch is one of the more important functions of the central nervous system. First of all, a caring, loving thought in his mind starts an amazing electrical circuit that energizes and sends a signal from the cerebral cortex by synapses of the toucher's nerves.

This signal travels through the cerebrum, cerebellum, pons, medulla oblongata, and vertebral canal to the brachial plexus in the shoulder, through the ulnar and median nerves in the arm, through the anterior interosseus nerve in the forearm and the carpal tunnel of the wrist, and finally by the Pacinian corpuscles and cutaneous receptors in the epidermis of the fingers, resulting in a little squeeze which says, "I like you."

The pressure sensation resulting from his little squeeze goes through the cutaneous receptors in the epidermis of the hand of the touchee. Then it travels by Pacinian corpuscles and synapses of the nerves, through the carpal tunnel in the wrist, etc., etc., etc., and finally to her cerebral cortex, culminating in a happy thought.

Amazingly, this multi-structure signal, often interpreted as the first vital activity in the three-step communication loop (encode, decode, feedback), causes a like response and similar happy thought in the cerebral cortex of the party of the second part, which says, "I like you, too."

This is only the beginning. A whole orchestration of other brain centers, neurotransmitters, neuropeptides, and hormones communicate this sensation throughout the whole body . . . which leads to human reproduction, one of the most spectacular illustrations of irreducible complexity. Without it, we wouldn't be having this discussion.

An incredible variety of anatomical and physiological components come together so that the male sperm can join the female egg to allow human reproduction, and the above illustrates the central nervous system designed for complementary, deep shared joyful love. "At its best" it is also a high spiritual union—not just physical.

First of all, the male, by his appreciation of the female, responds by creating a temporary "bridge" for his sperm cells to get from Point A to Point B. So, sexual response usually begins in the male's brain.

Even when this temporary bridge is in place, it won't "work" unless the male becomes 100 percent excited. Only then will he be able to move his sperm cells. And that is accomplished by an involuntary mechanism, without which fertilization of the female egg would not occur. Think about it. What would cause chance mutations to develop this intricate mechanism?

> Geoffrey Simmons, MD: A single male will produce about 100 million new sperm each day, essentially a thousand with each heartbeat, and over a trillion in a lifetime. These torpedo-like cells with flagellating tails have a head that is loaded with explosive enzymes. They must also have a homing device.[463]

Human males also need a means to manufacture sperm, store them, mobilize them on a moment's notice, mix them in a fluid that will . . . deliver them in the correct manner. In a similar fashion, the egg needs to know when it's time to ovulate, how to pop out, how to travel through the fallopian tubes, how to receive a single sperm and close out the other sperm, and how to eventually implant in the uterus. That's a lot to achieve by accident.[464]

---

[463]"Reproduction: Macroscopic," *What Darwin Didn't Know*, Copyright © 2004 by Geoffrey Simmons, MD. Published by Harvest House Publishers, Eugene, Oregon, 97402, www.harvesthousepublishers.com. Used by permission, p. 56.
[464]"Reproduction: Macroscopic," *What Darwin Didn't Know*,

Human reproduction is one of the most compelling examples of irreducible complexity anyone could possibly acknowledge. Like the mouse trap that wouldn't work if it didn't have a base, spring, hammer, trigger, and bait holder, we wouldn't be here if our mother didn't have a connection between her ovaries and her fallopian tubes. She also had to have a way to propel her egg the full length of the fallopian tube and a connection with her uterus. Then once the sperm cells, with their incredible amount of information, have been moved to the correct environment, they would not be able to fertilize the female egg if they also didn't somehow know how to "swim" to their goal and in which direction.

Here, when we consider the anatomy and physiology of the entire process of human reproduction, we can be reminded of Dr. Wernher Von Braun's question confirming his belief in the certainty of a Creator: "... *Must we really light a candle to see the sun?*" [465]

> Dr. Simmons: The egg is the size of a period [0.15 mm], and yet it must make a five-inch journey in 24 hours [a distance over 800 times its diameter].... Sperm are considerably smaller than the egg, yet they travel [over 11,000 times their length] ... [and] somehow the sperm know exactly where to go.[466]

Once contact has been made, a human life has been conceived.

> Dr. Simmons: Once inside, the head detaches, and a chemical reaction follows that makes the egg's outer wall impermeable to other sperm. The 23 chromosomes, or six billion chemical bases, in the sperm readily combine with the 23 chromosomes in the egg. This is a very quick and very exacting event. Millions of characteristics are decided at the moment of conception.[467]

But even then, this fertilized ovum would not mature, unless it was placed in a perfect environment. "Somehow" it knows where to go and how to attach itself for nourishment and maturation. During this process, this one cell goes through a process known as mitosis and divides into two cells, four cells, eight cells, sixteen cells, etc. The cells become either the placenta or embryo and start to "differentiate," some becoming digestive system, skeletal system, circulatory system, immune system (both cellular and humoral), a central nervous system, lymph system, and most wonderfully (and thankfully!), a new reproductive system.

> Dr. Simmons: Every cell has extremely specific instructions. Some bone marrow cells become white blood cells, and others become red blood cells. Some nerve cells form the brain and others the spinal column.... Our chromosomes can duplicate the six billion chemical bases in a matter of six to eight hours.... Thousands of enzymes help this duplication.... [468]

All of this new anatomy and physiology will start to work together in a complementary relationship that demands they all came into existence at the same time—as a *functional system*.

> Dr. Simmons: As a newly formed embryo divides, the [DNA and how it is to be used] Book of Life tells individual cells what shape to assume, what hormones to make, which minerals to absorb, how to communicate, which vitamins are needed, when to divide again, and much, much more. Parathyroid cells are told to monitor calcium concentrations—not sugar—and how to respond to high and low levels. Genes tell nerve cells how to sense pain, white cells how to fight infections, and gonad cells when to make testosterone or estrogen.... A person's blueprint is not only recorded in the DNA, but it is also linked to a timer. Barring something cataclysmic, growth and development will occur at predetermined times and predetermined rates.[469]

Copyright © 2004 by Geoffrey Simmons, MD. Published by Harvest House Publishers, Eugene, Oregon, 97402, www.harvesthousepublishers.com. Used by permission, p. 57.
[465]Wernher Von Braun, letter read by Dr. John Ford to California State Board of Education, September 14, 1972, quoted in Ann Lamont, *Twenty-One Great Scientists Who Believed the Bible*, Acacia Ridge, Queensland, Australia: Creation Science Foundation, 1995, p. 47, (emphasis added).
[466]"Reproduction: Macroscopic," *What Darwin Didn't Know*, Copyright © 2004 by Geoffrey Simmons, MD. Published by Harvest House Publishers, Eugene, Oregon, 97402, www.harvesthousepublishers.com. Used by permission, p. 61.
[467]"Reproduction: Macroscopic," *What Darwin Didn't Know*, Copyright © 2004 by Geoffrey Simmons, MD. Published by Harvest House Publishers, Eugene, Oregon, 97402, www.harvesthousepublishers.com. Used by permission, pp. 62-63.
[468]"Reproduction: Macroscopic," *What Darwin Didn't Know*, Copyright © 2004 by Geoffrey Simmons, MD. Published by Harvest House Publishers, Eugene, Oregon, 97402, www.harvesthousepublishers.com. Used by permission, pp. 70-71.
[469]"The Beginning," *What Darwin Didn't Know*, Copyright © 2004 by Geoffrey Simmons, MD. Published by Harvest House Publishers, Eugene, Oregon, 97402, www.harvesthousepublishers.com. Used by permission, pp. 72-74, (emphasis in the original).

Brain development begins early. By the embryo's fourth week, there is evidence of nervous tissue, and 250,000 cells per minute start migrating outward. . . . Each neuron knows exactly where to go, what to do, and how to link up with as many as 10,000 connections. . . . A special fluid with a precise concentration of nutrients and salts bathes the brain and spinal cord, buffering them from trauma and maintaining their health.[470]

The human brain is an extraordinary multitasking, multipurpose biological computer. . . . It can store between 100 trillion and 280 quintillion bits of information in three pounds of matter. . . . The heart, lungs, stomach, and kidneys are geared to keep it functioning. If calorie needs increase, the brain makes a person feel hungry; if oxygen levels fall, the lungs are told to breathe faster; if the blood pressure drops, the heart is told to pump harder; and if thyroid hormone is low, the thyroid gland is told to increase production.[471]

Published in 2004, the numbers in Dr. Simmons' overview of some of the wonders of the human body could no doubt be refined today given the advances in measuring such examples of complexity, and the instrumentation and discoveries unknown in 2004. However, his overview makes clear that we are "fearfully and wonderfully made," as David said in Psalm 139—and to believe that all these interdependent wonders have just happened requires more faith than David's.

Alexander Tsiaras, associate professor of Medicine and Chief of Scientific Visualization at Yale University, Department of Medicine, created a video featuring human embryology: "From Conception to Birth." In it, he shows that the fertilized ovum divides a few hours after fusion and divides anew every 12 to 15 hours. [Within 22 days, the heart is beating]. At four weeks it is developing one million cells per second. Within 32 days, arms and hands are developing. At day 36, a primitive vertebrae is developing. These weeks are the period of the most rapid development of the fetus. *If the fetus continues to grow at this speed for the entire 9 months, it would weigh 1.5 tons at birth.* At 52 days its retina, nose, and fingers are developing. At birth there are almost 60,000 miles of blood vessels, and only one mile is visible. . . . The complexity of building that within a single system is, again, beyond any comprehension of any existing mathematics today."[472]

> Alexander Tsiaras: The magic of the mechanisms inside each genetic structure saying exactly where that nerve cell should go—the complexity of these, the mathematical models of how these things are indeed done are beyond human comprehension. Even though I'm a mathematician, I look at this with a marvel of how did these instruction sets not make mistakes as they build what is us.[473]

After a baby is born, it dies unless the mother provides proper nutrition. And that is made possible by another example of the complementary relationships evident in human anatomy and physiology, but this time for the benefit of another. When a baby is born a message is sent, at just the right moment, to the mother's mammary glands which communicate a vital message that *now* is the time to make a special kind of milk that includes substances that stimulate the baby's immune system. All the anatomy and potential physiology to make this nutritional substance was present but dormant until the mammary glands received this critical message.

> Dr. Simmons: A combination of hormones secreted by the placenta, the maternal pituitary, and the mother's ovaries interact to change the mammary duct system so that it can make, store, and secrete milk. These cells already carry the recipes for how much water, salt, fat, sugar, minerals, vitamins, calcium, and antibodies will be needed.[474]

---

[470]"The Beginning," *What Darwin Didn't Know*, Copyright © 2004 by Geoffrey Simmons, MD. Published by Harvest House Publishers, Eugene, Oregon, 97402, www.harvesthousepublishers.com. Used by permission, pp. 94-95.

[471]"The Beginning," *What Darwin Didn't Know*, Copyright © 2004 by Geoffrey Simmons, MD. Published by Harvest House Publishers, Eugene, Oregon, 97402, www.harvesthousepublishers.com. Used by permission, p. 89.

[472]Alexander Tsiaras, "From Conception to Birth," TED, Ideas worth spreading, One of 1,000+ TED Talks, New ideas every weekday, TED.com, Recorded at INKConference, 2010, Youtube.com/watch_popup?v+fKyljukBE70_ DEC2010, LAVASA-INDIA, (emphasis added).

[473]Alexander Tsiaras, "From Conception to Birth," TED, Ideas worth spreading, One of 1,000+ TED Talks, New ideas every weekday, TED.com, Recorded at INKConference, 2010, Youtube.com/watch_popup?v+fKyljukBE70_ DEC2010, LAVASAINDIA.

[474]"The Beginning," *What Darwin Didn't Know*, Copyright © 2004 by Geoffrey Simmons, MD. Published by Harvest House Publishers, Eugene, Oregon, 97402, www.harvesthousepublishers.com. Used by permission, p. 79.

Not only do the mother's mammary glands make milk, but they also have very well-designed structures for the milk to exit. And in order to extract the milk, these exits are designed to respond to the infant's sucking mechanism. We would not exist today if we had all of our anatomy and physiology but the baby did not have the ability to learn how to suck, with all the coordinated muscular activity of the baby's tongue and lips.

A baby born with hypoplastic left-heart syndrome is a good example of the importance of precise/exact genetic instructions. It is a perfectly formed baby with one exception. Because it has a lethal underdevelopment of the left side of its heart it will die, 100 percent of the time, without intervention. Many of these babies have heart transplants in order to live a relatively normal life.

The baby with hypoplastic left-heart syndrome has—with that one exception—all of its anatomy and physiology, and all of its hormones, neuropeptides, and neurotransmitters to live a useful and productive life. It must have all of the above, including a normally functioning heart, all functional at the same time, to live. Could a perfectly formed baby develop by chance?

Dr. Simmons: The interior of the human body is a much busier place than New York City, London, Mexico City, Tokyo, and Bombay combined. Ten to 75 trillion cells participate in more than a quadrillion purposeful chemical interactions each day that help us walk, read, think, sleep, procreate, see, hear, smell, feel, digest food, eliminate waste, write, read, talk, make red cells, remove dead cells, fight infections, behave, misbehave, absorb nutrients, transport oxygen, eliminate carbon dioxide, maintain balance, carry on dialogue, understand instructions, argue, and make complex decisions, just to name a few common activities. In addition, each of these processes has dozens—and sometimes hundreds—of smaller, interacting steps, checks, counterbalances, balances, and regulatory mechanisms. . . . Darwin didn't have a clue how these mechanisms truly work. For the most part, he didn't even know they existed.[475]

Look at how we transfer sugar, minerals, proteins, fats, carbohydrates, and vitamins from our dinner plates to our mouths, down to the gastrointestinal tract, through the walls of the small bowel, into the bloodstream, through the liver, and ultimately to every cell in the body. Millions of macroscopic and microscopic processes are utilized. How does the body even know which sugar (and there are many types) to absorb, or which protein [and there are many thousands of proteins] . . . goes where, when, and in what quantity? How does it know which substances are safe to absorb, and which should be ignored, quickly eliminated, or destroyed? How does the small bowel know how to cooperate with the 500 different kinds of bacteria that live in it? These are incredibly complex functions that work together—and only together—to maintain the health of an individual.[476]

---

[475] "The Ever-Increasing Problems of Darwin's Theories," *What Darwin Didn't Know,* Copyright © 2004 by Geoffrey Simmons, MD. Published by Harvest House Publishers, Eugene, Oregon, 97402, www.harvesthousepublishers.com. Used by permission, pp. 33-34.
[476] "The Ever-Increasing Problems of Darwin's Theories," *What Darwin Didn't Know,* Copyright © 2004 by Geoffrey Simmons, MD. Published by Harvest House Publishers, Eugene, Oregon, 97402, www.harvesthousepublishers.com. Used by permission, pp. 33-34 [brackets added].

IV
Hypothesis and Theory vs. Fact

Chapter 20
Neo-Darwinism—The Theory of Imagination

There are a variety of fields of evidence that are better supported by creation science than by the theory of evolution. For example, take the irreducible complexity of the eye and ear.

Many leading scientists have commented on the staggering complexity of the human eye. What some have not studied are the many diverse types of non-human eyes, each of which adds to the problem of how each marvel could have evolved. One of the strangest is a multiple-lensed, compound eye found in fossilized worms.[477]

Also, the incredible eye of some *trilobites*, which evolutionists claim are very *early* forms of life. These trilobite eyes had compound lenses and sophisticated designs for eliminating image distortion (spherical aberration). Today, only the best cameras and telescopes contain compound lenses. According to physicist Riccardo Levi-Setti, PhD, trilobite eyes "represent an all-time feat of function optimization."[478]

How, then, can evolutionists claim that they are very early?

Lisa J. Shawver in *Science News*, described trilobites as having *"the most sophisticated eye and lenses ever produced by nature."* [479]

> Mark Whorton, PhD; Hill Roberts, MA: The abundance of trilobite fossils in the Cambrian strata . . . had a very modern arthropod body plan and the most complex *visual* system known in the entire animal kingdom, living or extinct.[480]

Evolutionist Stephen Jay Gould, PhD, admits that "the eyes of early trilobites, for example, have never been exceeded for complexity or acuity by later *arthropods*. . . . I regard the failure to find a clear vector of progress in life's history as *the most puzzling fact of the fossil record."* [481]

What do recent textbooks have to say about the human eye?

Merrill Biology: The complex structure of the human eye may be the product of millions of years of evolution.[482]

Glencoe Biology, Living Systems: When we marvel at a structure as complex as our eye, we are looking at a product of the ongoing process of evolution. You can better understand how an eye might have evolved if you picture a series of changes during the evolution of the eye.[483]

We must imagine how it "might have evolved?" Since we never observe such changes, we are told we must try to make it happen in our imagination?

> I. L. Cohen: Every single concept advanced by the theory of evolution (and amended thereafter) is imaginary as it is not supported by the scientifically established facts of microbiology, fossils, and mathematical probability concepts. Darwin was wrong. . . . The theory of evolution may be the worst mistake made in science.[484]

> Michael Denton: It was Darwin, the evolutionist, who admitted in a letter to Asa Gray that one's imagination must fill up the very wide blanks.[485]

Notice what the scientific community recently [2005] acknowledges about imagination.

---

[477]Donald G. Mikulic, "A Silurian Soft-Bodied Biota," *Science*, Vol. 228, May 10, 1985, pp. 715-717.
[478]Riccardo Levi-Setti, PhD, *Trilobites*, second edition, University of Chicago Press, 1993, pp. 29-74.
[479]Lisa J. Shawver, "Trilobite Eyes: An Impressive Feat of Early Evolution," *Science News*, Vol. 105, February 2, 1974, p. 72.
[480]Mark Whorton, PhD, Hill Roberts, MA, "Impossibilities in Chemistry: The Origin of Life," *Understanding Creation, a Biblical and Scientific Overview*, a Holman Reference Book, B & H Publishing Group, 127 Ninth Avenue, North Nashville, TN, p. 331, (emphasis added).
[481]Stephen Jay Gould, PhD, "The Ediacaran Experiment," *Natural History*, February 1984, pp. 22-23, still in 2016 unexplained, (emphasis added).
[482]*Merrill Biology*, 1983, p. 202.
[483]"Adaptation," *Glencoe Biology, Living Systems*, Glencoe/McGraw Hill, New York, NY; Columbus, OH; Mission Hills, California; Peoria, IL, 1998, p. 337.
[484]I. L. Cohen, "Summary," *Darwin Was Wrong: A Study in Probabilities*, New Research Publications, Inc., Greenvale, NY, 1984, pp. 209-210.
[485]Michael Denton, *Evolution: A Theory in Crisis*, Adler & Adler, Bethesda, MD, 1986, p. 117, citing Darwin., (1858) in a letter to Asa Gray, September 5, 1857, Zoologist, 16:1697-1699.

Adrienne Mayor, PhD: Seeking visions might seem light-years away from scientific inquiry. Yet the most creative paleontologists can be described as visionaries, and many respected scientists have described important theoretical breakthroughs that came to them as revelations while they slept or daydreamed.

This hyperaware dream state is called "lucid dreaming" or "power dreaming" by neuropsychologists. . . . Certainly paleontology, solving the mystery of stone creatures that will never be seen walking the earth, inspires dreams and theoretical narratives.[486]

Derek Isaacs: "[Adrienne Mayor's above] public admission about the true state of evolutionary-based paleontology is catastrophic for its integrity. Considering that paleontology is the cornerstone and lynchpin that holds together the evolutionary line of descent of creatures, this is a critical indictment about the nature of what they deem "science" to be.

. . . Considering the prominent paleontologists that [Dr. Mayor] consulted on this work, including the world-renowned Jack Warner, she is directly implying that this method of imagining and dreaming up lines of descent is a practice that goes to the very top of evolutionary paleontology. Her book [published by Princeton University Press] is from and approved by the circles of the evolutionary academic elite.

Therefore, by their own admission, they are living in a world of make-believe. This is an overriding theme in putting together timelines and ancestral descendents. Evidence cannot be found to truly document the descendent of creatures from one to another, so they have to make up where, when, and how things happened.

And do not be quick to forget that it is the paleontologists who tell other disciplines of academia what fossil is the evolutionary ancestor to what fossil. It is this discipline, no less, that is admitting that they are making things up—the discipline that builds the all-important evolutionary family tree.[487]

Warren L. Johns, JD: Lively imaginations grease the wheels of assumption's dreams. The imagination is a mediocre counterfeit for verifiable evidence. At best, assumptions provide cosmetic cover to unproven conjecture.[488]

Regarding the transitional fossils between invertebrates and vertebrates:

Homer W. Smith, ScD, in *From Fish to Philosopher*: As our present information stands, however, the gap remains unabridged, and *the best place to start the evolution of the vertebrates is in the imagination.*[489]

Evolutionist Stephen Jay Gould, PhD: The absence of fossil evidence for intermediary stages to major transitions in organic design, indeed our inability, *even in our imagination,* to construct functional intermediaries in many cases, has been a persistent and nagging problem for gradualistic accounts of evolution.[490]

David D. Riegle: When a scientist finds a single bone or tooth which supposedly dates back a few hundred thousand years, on what basis of measurement can he draw a picture of the whole creature? When the first fossil bones were discovered many years ago, there were no other bones with which to compare them, no other measurements by which to judge them, *so the first drawings of ancient men were the products of imagination.* The men who drew the first pictures imagined man rather ape-like in appearance, so they drew him with the facial features of a creature sort of halfway between a man and an ape. They gave him a slightly crouching stance, a long face with huge jaws, and a look of doubtful intelligence. This picture has *stayed with us* down through the years.[491]

Physical anthropologist Earnest A. Hooton, PhD: Put not your faith in reconstructions. . . . Some anatomists model reconstructions of fossil skulls by building up the soft parts of the head and face upon a skull cast and thus produce a bust

---

[486]Adrienne Mayor, *Fossil Legends of the First Americans*, Princeton Press, 2005, p. 329; quoted by Derek Isaacs, "Theory of Imagination," *Dragons or Dinosaurs? Creation or Evolution?* Bridge-Logos, Alachua, FL, 2010, pp. 111-112.
[487]Derek Isaacs, "Theory of Imagination," *Dragons or Dinosaurs? Creation or Evolution?* Bridge-Logos, Alachua, FL, 2010, pp. 112-113.
[488]Warren L. Johns, JD, "Superstitious Nonsense, Life from Spontaneous Generation?" *Genesis File*, www.GenesisFile.com, Lightning Source, LaVergne, TN, 2010, p. 18.
[489]Homer W. Smith, ScD, "The Protovertebrate," *From Fish to Philosopher*, Anchor Books, Doubleday & Company, Inc., Garden City, NY, 1961, p. 25, (emphasis added).
[490]Stephen Jay Gould, PhD, quoted by Dennis R. Petersen, "Some Famous Fossil 'Connections' To Evolution," *Unlocking the Mysteries of Creation: the Explorer's Guide to the Awesome Works of God*, Bridge-Logos Publishers, Alachua, FL, 2002, p. 105, (emphasis added).
[491]David D. Riegle, *Creation or Evolution?* Zondervan Publishing House, Grand Rapids, MI, 1971, pp. 47-48, (first emphasis added, second emphasis in the original).

purporting to represent the appearance of the fossil man in life. When, however, we recall the fragmentary condition of most of the skulls, the faces usually being missing, we can readily see that even the reconstruction of the facial skeleton leaves room for a good deal of doubt as to details. To attempt to restore the soft parts is an even more hazardous undertaking. The lips, the eyes, the ears, and the nasal tip leave no clues on the underlying bony parts. You can, with equal facility, model on a Neanderthaloid skull the features of a chimpanzee or the lineaments of a philosopher. These alleged restorations of ancient types of man have very little, if any scientific value and are likely only to mislead the public.[492]

Robert Martin in *New Scientist:* In recent years several authors have written popular books on human origins which were based more on fantasy and subjectivity than on fact and objectivity.[493]

According to *Webster's Dictionary*, **science** is systematized knowledge derived from observation. According to *The American Heritage Dictionary*, science is "the observation, identification, description, experimental investigation, and theoretical explanation of phenomena." Surprisingly, the British Naturalist himself acknowledged the following problem:

> Charles Darwin: To suppose that the eye, with all its inimitable contrivances for adjusting the focus to different distances, for admitting different amounts of light, and for the correction of spherical and chromatic aberration, could have been formed by natural selection, seems, I freely confess, absurd to the highest degree.[494]

But then he argues that natural selection could produce complex eyes.

We can substantiate the existence of a designer by the impossibility of such complexity having occurred by chance and natural selection. Dr. Simmons, in his book *What Darwin Didn't Know*, outlines the intricacies of the human eye:

> Geophrey Simmons, MD: Millions of cells lining the interior of each eye function as photochemical receivers that convert light waves into a myriad of electrical impulses, which are forwarded, at a speed of about 200 mph, to the brain—and then sorted, organized, and analyzed. This is accomplished in milliseconds. . . . Every anatomical, chemical, and physiological aspect of our eyes suggests design.
>
> Our eyes are kept moist and nearly sterile by tiny lacrimal glands along the outer edge of each upper eyelid. These glands secrete a viscous fluid that slowly moves downward and across the eye to an inner, lower tear duct and drains into the nasal passages. . . . Also, note how watery an eye becomes when irritated by a foreign object. The eye washes itself. The same fluid also brings oxygen to the cornea—which cannot have vision-blocking blood vessels on the surface—and brings proteins to coat the eye. Is that accidental or design?
>
> Blinking is another protective function that we each do about 14,000 to 20,000 times a day, or about 15 times a minute. . . . There are three kinds of eye blinks: protective (involuntary, in response to a loud sound or flash of light); voluntary (rocks seen coming at the eye); and cyclical (also involuntary), to clean the eye like a squeegee and move tears with debris toward the lacrimal ducts.[495]

Sylvia Baker, describing a university seminar: "We began . . . to discuss how that marvelous organ might have evolved. For an hour we argued round and round in circles. Its evolution was clearly impossible. All the specialized and complex cells that make up our eyes are supposed to have evolved because of advantageous mutations in some more simple cells that were there before. But what use is a hole in the front of the eye to allow light to pass through if there are no cells in the back of the eye to receive the light? What use is a lens forming an image if there is no nervous system to interpret that image? How could a visual nervous system have evolved before there was an eye to give it information?

We discussed the problem from every possible angle, but in the end had to admit that we had no idea how this might have happened. I then said that since we had found it impossible to describe how the eye had evolved, the honest and scientific thing was to admit the possibility that it had not

---

[492] Earnest A. Hooton, PhD, *Up From the Ape*, Revised Edition, Macmillan Company, NY, 1946, p. 329, (emphasis in the original).
[493] Robert Martin, "Man is Not an Onion," *New Scientist 75*, New Science Publications, London, August 4, 1977, p. 283.
[494] Charles Darwin, "Difficulties on Theory," *On The Origin of Species by Means of Natural Selection or the Preservation of Favoured Races in the Struggle for Life*, John Murray, Albemarle Street, London, 1859, p. 186.
[495] "The Beginning," *What Darwin Didn't Know*, Copyright © 2004 by Geoffrey Simmons, MD. Published by Harvest House Publishers, Eugene, Oregon, 97402, www.harvesthousepublishers.com. pp. 106-109. Used by permission.

evolved. My words were followed by a shocked silence. The lecturer leading the seminar then said that he refused to enter into any controversy, while the others in the group began to mock me for believing in God. I had not mentioned God! I had simply been trying to view the problem in an objective and scientific way.[496]

To illustrate the irreducible complexity of the eye, creationist Ray Comfort suggested what it would take for today's sophisticated medical science to create an eye:

> Ray Comfort: If you lost one eye, the very best medical science could do is replace it with a fake eye. It may look as good as your other eye, but it certainly won't look as *well*. You would be blind in that eye. This is because we don't know how to create its inter-related systems of about forty individual subsystems, including the retina, pupil, iris, cornea, lens, and optic nerve. Nor do we know how to make the retina's 137 million light-sensitive cells that send messages to the brain. The eye is a nightmare of complexity beyond words, for those who hope to imitate it.
>
> A special section of the brain called the visual cortex interprets the pulses as color, contrast, depth, etc., which then allows us to see "pictures" of our world. Incredibly, the eye, the optic nerve, the visual cortex are completely separate and distinct subsystems. Yet together they capture, deliver, and interpret up to 1.5 million pulse messages a millisecond![497]

> Michael J. Denton, MD, PhD: The human brain consists of about ten thousand million nerve cells. Each nerve cell puts out somewhere in the region of between ten thousand and one hundred thousand connecting fibers by which it makes contact with other nerve cells in the brain. Altogether the total number of connections in the human brain approaches . . . a thousand million million. . . . Even if only one hundredth of the connections in the brain were specifically organized, this would still represent a system containing a much greater number of specific connections than in the entire communications network on earth.[498]

According to James Perloff, "Each neuron contains about one trillion atoms. The brain can do the work of hundreds of supercomputers. Building a computer requires great intelligence. Who believes even a simple one could arise by chance? Indeed, the brain is more than a computer[499] — 'It is a video camera and library, a computer and communications center, all in one.'"[500]

> W. H. Yokel: Perhaps the most elusive questions surround the brain functions that make us human—the capacities of memory and learning. Transcending what might be called the hardware of the brain, there comes a software capacity that eludes hypothesis. The number that expresses this capacity in digital information bits exceeds the largest number to which any physical meaning can be attached.[501]

---

[496]Sylvia Baker, *Bone of Contention: Is Evolution True?* Evangelical Press, Welwyn, England, 1976, pp. 17-18.
[497]Ray Comfort, "The Evolution Illusion," *Nothing Created Everything*, A WND Books book, WorldNetDaily, Los Angeles, CA, 2009, p. 17.
[498]Michael Denton, PhD, "The Puzzle of Perfection," *Evolution: a Theory in Crisis*, Adler & Adler, Bethesda, MD, 1986, pp. 330-331.
[499]James Perloff, "Logic Storms Darwin's Gates," *Tornado in a Junkyard*, The Relentless Myth of Darwinism, Refuge Books, Arlington, MA, 1999, p. 36.
[500]Donald B. DeYoung, PhD, "Thinking about the Brain," *Impact #200*, Institute for Creation Research, El Cajon, CA, February 1990, p. 1.
[501]W. H. Yokel, promotional letter for Scientific American, 1979, quoted by John Whitcomb, *The Early Earth*, Baker Books, Grand Rapids, MI, 1986, p. 126.

Chapter 21
Vestigial Organs

Evolutionists have thought that organs or parts of the body that seem to serve no useful purpose must be remnants or "vestiges" of the long-distant past, when they may have then been useful. These "vestigial" organs have been considered by evolutionists to be evidence that the animal has evolved into a higher state where it no longer needs that part.

Even in the 1990s, "vestigial organs" were thought to include the tail bone in humans:

Modern Biology: The vestigial tail bone in humans is homologous [derived from the same ancestral form as] . . . the functional tail of other primates. Thus, vestigial structures can be viewed as evidence for evolution: organisms having vestigial structures probably share a common ancestry with organisms in which the structure is functional.[502]

Heath Biology: The coccyx is a small bone at the end of the human vertebral column. It has no present function and is thought to be the remainder of bones that once occupied the long tail of a tree-living ancestor.[503]

Even in 2005—

Biology Today and Tomorrow: At the end of your backbone is a coccyx, a few small bones that are fused together. Could the human coccyx be a vestigial structure? Or is it the start of a newly evolving structure?[504]

Actually, it is not a useless vestige of early times. The coccyx, commonly called the tail bone, is "the small bone at the end of the spine."[505]

It has several functions. It was designed to serve "as an attachment site for tendons, ligaments, and muscles. It also functions as an insertion point of some muscles of the pelvic floor. The coccyx also functions to support and stabilize a person while he or she is in a sitting position." It aids posture while we sit.[506]

In the past, evolutionist-oriented biology textbooks taught that the *appendix* is one of the supposed vestigial organs.

Ernst Mayr, PhD: Many organisms have structures that are not fully functional or not functional at all. The human caecal appendix is an example. . . . Such vestigial structures are the remnants of structures that had been fully functional in their ancestors. . . . They are informative by showing the previous course of evolution.[507]

Grolier Encyclopedia: Long regarded as a vestigial organ with no function in the human body, the appendix is now thought to be one of the sites where immune responses are initiated.[508]

National Institutes of Health: The caecal appendix functions fully in our immune system.[509]

Dr. Ron Allard: The appendix is required to activate killer B cells in your immune system like your thyroid activates "T" cells.[510]

Gill slits

In 1996, *Life* magazine described how human embryos grow "something very much like gills," which, among all the other embryos portrayed in the story is "some of the most compelling evidence of evolution since Darwin published *[On] The Origin of Species* in 1859."[511]

A 2009 textbook illustrates a human embryo with what it identifies as gill pouches. The accompanying text states confidently: "Patterns

---

[502]"Evolution: Evidence and Theory," *Modern Biology,* Holt, Rinehart and Winston, NY, 1989, p. 323.
[503]James E. McLaren, Lissa Rotundo, 'Laine Gurley-Dilger, PhD, "Human Evolution," *Heath Biology,* D. C. Heath and Company, Lexington MA, 1991, p. 264.
[504]Critical Thinking," *Biology Today and Tomorrow,* 2005, p. 292.
[505]https://www.merriam-webster.com/dictionary/coccyx.
[506]www.healthline.com/human-body-maps/coccyx.
[507]Ernst Mayr, PhD, "What is the Evidence for Evolution on Earth?" *What Evolution Is,* Basic Books, a member of the Perseus Books Group, New York, NY, 2001, pp. 30-31.
[508]Roy Hartenstein, *Grolier Encyclopedia,* 1998, see also: Jerry Bergman, PhD, and George Howe, PhD, *"Vestigial Organs" are Fully Functional,* Creation Research Society, St. Joseph, MO, 1990, pp. 43-46.
[509]Google NIH caecal appendix.
[510]Dr. Ron Allard, drronallard@yahoo.com.
[511]Kenneth Miller, Anne Hollister, "What Does It Mean to Be One of Us?" *Life,* November 1996.

of growth in the very earliest stages of life can also provide evidence of organisms' evolutionary past.... [It then makes reference to] observations that the embryos of fishes, amphibians, reptiles, birds, and mammals (including humans), all develop gill pouches."[512]

Another textbook states that the human embryo growing inside the mother has gills like a fish. According to the *American Heritage Dictionary*, gills are "the respiratory organ of most aquatic animals that obtain oxygen from water." The art illustrating the point showed a human embryo supposedly with gill slits.

Gill slits? Those are not fish-like gills, nor the gill slits of sharks nor the embryonic gill slits (pre-gills) of aquatic animals who *breathe in* oxygen from water. *Those little folds of skin on the human embryo grow into bones in the ear and glands in the throat.* They have nothing to do with breathing. The more we learn about what were thought to be vestigial organs, the more we recognize they have previously undiscovered functions important to life.

One of the greatest hoaxes in scientific history was perpetuated by Ernst Haeckel, a German embryology professor. Haeckel said the turning point in his thinking was when he read Charles Darwin's *On the Origin of Species* in 1860, a year after it was first published. He read the book and thought it was a great theory. If only science had some real evidence. Nine years later, there was still no evidence. So, Haeckel decided to make some. He noted that drawings of a human and dog embryo at four weeks looked the same—and they all had gill slits.

Actually, as an excellent artist he *drew* them to look exactly alike. He even depicted them as the same size. The similarities Haeckel exaggerated can be compared to striking differences he ignored in earlier stages of development.

In 1997, Michael K. Richardson, PhD, and an international team of experts compared Haeckel's art with photographs of actual embryos from all seven classes of vertebrates, demonstrating quite clearly that the drawings misrepresent the truth. Richardson concludes, "It looks like it's turning out to be one of the most famous fakes in biology."[513]

> James Perloff: At Jena, the university where he taught, Haeckel was charged with fraud by five professors and convicted by a university court. His deceit was thoroughly exposed in *Haeckel's Frauds and Forgeries* (1915), a book by J. Assmuth and Ernest J. Hull. They quoted nineteen leading authorities of the day. F. Keibel, professor of anatomy at Freiberg University, said that it clearly appears that Haeckel has in many cases freely invented embryos or reproduced illustrations given by others in a substantially changed form. L. Rutimeyer, professor of zoology and comparative anatomy at Basle University, called his distorted drawings a sin against scientific truthfulness deeply compromising to the public credit of a scholar.[514]

Haeckel had huge posters made of his fake drawings and gave them wide exposure throughout Germany. In so doing, he tricked people into believing in evolution. Fifteen years later, in 1875, Haeckel's own Jena University held the trial that convicted him of fraud. At the trial he confessed, "I should feel utterly condemned . . . were it not that hundreds of the best observers, and biologists lie under the same charge."[515]

The German scientist Wilhelm His accused Haeckel of shocking dishonesty in repeating the same picture several times to show the similarity among vertebrates at early embryonic stages in several plates of Haeckel's book.[516]

> James Perloff: Michael Richardson, an embryologist at St. George's Medical School, London, found there was no record that anyone ever actually checked Haeckel's claims by systematically comparing human and other fetuses during development. [Richardson] assembled a scientific team that did just that—photographing the growing embryos of 39 different species. In a 1997 interview in London's *The Times*, Dr. Richardson stated: "This is one of the worst cases of scien-

---

[512]Cain, Yoon, Singh-Cundy, "Organisms contain evidence of their evolutionary history," *Discover Biology*, W. W. Norton & Company, NY, Fourth Edition, 2009, p. 334.
[513]Michael K. Richardson, PhD, as quoted by Elizabeth Pennisi, "Haeckel's Embryos: Fraud Rediscovered," *Science*, Vol. 277, September 5, 1997, p. 1,435.
[514]James Perloff, "Old Myths Never Die—They only Fade Away," *Tornado in a Junkyard: The Relentless Myth of Darwinism*, Refuge Books, Arlington, MA, 1999, p. 112.
[515]Ernst Haeckel, *Records from the University of Jena trial*, 1875.
[516]Stephen Jay Gould, PhD, "Notes to pages 190-197," *Ontogeny and Phylogeny*, The Belknap Press of Harvard University Press, Cambridge, MA, 1977, p. 430.

tific fraud. It's shocking to find that somebody one thought was a great scientist was deliberately misleading. It makes me angry." . . .[517]

What [Haeckel] did was to take a human embryo and copy it, pretending that the salamander and the pig and all the others looked the same at the same state of development. They don't. . . . These are fakes. In the paper we call them "misleading and inaccurate," but that is just polite scientific language.[518]

Even though Haeckel's famous figures illustrate considerable inaccuracy, these drawings are sometimes still reproduced in textbooks and review articles, and continue to exert a significant influence on the development of ideas in this field.[519]

Ernst Haeckel's comparative drawings of embryos have often been described as among the best evidence for Darwinism. Not only did his drawings "illustrate" Darwinism, but so also did his written word:

> Ernst Haeckel: When we see that, at a certain stage, the embryos of man and the ape, the dog and the rabbit, the pig and the sheep, though recognizable as higher vertebrates, cannot be distinguished from each other, the fact can only be elucidated by assuming a common parentage.[520]

Lee Strobel outlined his first impression of the Haeckel drawings. "I was mesmerized by the 19th-century drawings when I first encountered them as a student. As I carefully compared the embryos at the earliest stage, looking back and forth from one to the other, I could see that they were virtually indistinguishable. I searched my mind, but I couldn't think of any logical explanation for this phenomenon other than a common ancestor. My verdict was swift: Darwin prevails."[521]

David C. Read, JD: Haeckel's drawings never fooled experts in embryology who immediately noticed that he was taking artistic license to support his pet theory.[522] Moreover, Haeckel's idea that "ontogeny [the biological development or course of development of an individual organism] recapitulates [restates or summarizes] phylogeny" [the evolution of a race or genetically related group of organisms][523] had by 1910 been "conclusively disproved and abandoned."[524]

According to Stephen Jay Gould, PhD, Haeckel's *"biogenetic law,"* the comparison between *ontogeny* and *phylogeny*, "provided an argument second to none in the arsenal of evolutionists during the second half of the nineteenth century."[525]

Even though the drawings were confirmed to be a fraud in 1875, the exact same art is still used in modern textbooks, more than 140 years later, to demonstrate evolution.[526]

When some biologists exposed the Haeckel drawings in an article a few years ago, Harvard University's evolutionist Stephen Jay Gould, PhD, complained that this was nothing new. He had known about it for 20 years. It was no secret to the experts. But he acknowledged that textbook writers should be ashamed of the way the drawings had been mindlessly recycled for over a century. At least he was honest enough to call it what it was: "the academic equivalent of murder."[527]

> Stephen Jay Gould, PhD: If so many historians knew all about the old controversy [over Haeckel's falsified drawings], then why did they not communicate this information to the numerous contemporary authors who use the Haeckel drawings in

---

[517]James Perloff, "Old Myths Never Die—They only Fade Away," *Tornado in a Junkyard: The Relentless Myth of Darwinism*, Refuge Books, Arlington, MA, 1999, p. 113.
[518]Michael K. Richardson, PhD, "An Embryonic Liar," *The Times*, London, August 11, 1997, p. 14.
[519]Michael K. Richardson, PhD, et al, "There Is No Highly Conserved Embryonic Stage in the Vertebrates," *Anatomy and Embryology*, Vol. 196, No. 2, August 1997, p. 104.
[520]Ernst Haeckel, *The Riddle of the Universe at the Close of the Nineteenth Century*, trans. Joseph McCabe, Harper and Brothers, NY, 1900, pp. 65-66.
[521]Lee Strobel, "Doubts About Darwinism," *The Case for A Creator*, Zondervan, Grand Rapids, MI, 2004, p. 47.
[522]Stephen Jay Gould, PhD, *I Have Landed*, Harmony Books, NY, 2002, pp. 305-320; as reported by David C. Read, JD, *Dinosaurs*, Clarion Call Books, Keene, TX, 2009, p. 372.
[523]Stephen Jay Gould, PhD, " . . . Although Haeckel was almost addicted to obfuscation by using fashionable words in meaningless contexts, it is important to recognize that when he said 'phylogeny is the mechanical cause of ontogeny' he really meant it." "Evolutionary Triumph, 1859—1900," *Ontogeny and Phylogeny*, The Belknap Press of Harvard University Press, Cambridge, MA, 1977, p. 78.
[524]Stephen Jay Gould, PhD, *I Have Landed*, Harmony Books, New York, 2002, p. 316; as reported by David C. Read, JD, *Dinosaurs*, p. 372.
[525]Stephen Jay Gould, PhD, "The Analogistic Tradition from Anaximander to Bonnet," *Ontogeny and Phylogeny*, The Belknap Press of Harvard University Press, Cambridge, MA, 1977, p. 13
[526]*Evolutionary Analysis*, 1998, p. 28. Used at the University of West Florida in 1999; Jonathan Wells, PhD, "Haeckel's Embryos," Icons of Evolution: Science or Myth? Regnery Publications, Washington, DC, 2000, p. 82.
[527]Stephen Jay Gould, PhD, "Abscheulich! (Atrocious!)," *Natural History*, March, 2000, pp. 42-50.

their books? I know of at least *fifty recent biology texts* which use the drawings uncritically.[528]

Even as late as 1996, *Life* magazine cited Haeckel:

> According to Haeckel's "biogenetic law," a pig, a human or any other vertebrate begins as a one-celled animal and on its way to birth becomes a fish, an amphibian and so on up the evolutionary ladder. Haeckel's successors took his concept further. Even in childhood, they theorized, we continue to relive the primitive stages of our species: Kids are prone to monkey business because they are biologically close to monkeys.[529]

---

[528] Stephen Jay Gould, PhD, "Abscheulich! (Atrocious!)," *Natural History*, March, 2000, pp. 42-50, (emphasis added).

[529] Kenneth Miller, Anne Hollister, "What Does It Mean to Be One of Us?" *Life*, November 1996.

Chapter 22
Deoxyribonucleic Acid —DNA

Intelligent man has never come close to developing a mechanism for the storage of information as efficient as that which we find in nature. In 1960, George Vandeman wrote a book, titled *Planet in Rebellion*, in which he told of a huge Mathematical Analyzer, Numerical Integrator And Computer (MANIAC) that had 40,000 characters of memory.[530] The first desktop computer I bought was a Radio Shack TRS-80 that had 48,000 characters of random access memory and no hard drive. I now carry in my pocket a portable USB "flash drive," smaller than a toothpick holder, that has 64 gigabytes of storage to be used on the laptop computer (which I am using to write this book) that has 232 gigabytes of hard-drive memory.

Microsoft chair Bill Gates once said, "DNA is like a software program, *but it's much more complex than anything we've been able to design."* [531]

In 2003, *Time* magazine stated: IBM models its newest . . . computers . . . after DNA. The quantity of information is so vast, we have to invent new numbers to measure it: not just a terabyte (a trillion bits of genetic data) but a petabyte (equal to half the contents of all the academic libraries in America), exabytes, yottabytes and zetabytes. All the words ever uttered by everyone who ever lived would amount to five exabytes.[532]

Monroe W. Strickland in *Genetics:* We each have 100 trillion cells in our bodies. We have 46 segments of DNA in most of our cells. We received 23 segments from our mothers and 23 from our fathers. DNA contains the unique information that determines what we look like, much of our personality, and how every cell in our bodies function throughout our lifetimes. If the 46 segments in our DNA from just one of our cells were uncoiled, connected, and stretched out, it would be about 7 feet long. It would be so thin its details could not be seen, even under an electron microscope. If all this very densely coded information from one cell were written in books, it would fill a library of about 4,000 books. If all the DNA in each of our bodies were placed end to end, it would stretch from here to the moon 552,000 times. In book form, that information would completely fill the Grand Canyon more than 75 times. ***If one set of DNA (one cell's worth) from every person who ever lived (estimated to be almost 50 billion) were placed in a pile, it would weigh less than an aspirin.*** Understanding DNA is just one small reason for believing we are "fearfully and wonderfully made."[533]

I. L. Cohen, mathematician, researcher, author, member of New York Academy of Sciences, and officer of the Archaeological Institute of America: In a certain sense, the debate transcends the confrontation between evolutionists and creationists. *We now have a debate within the scientific community itself; it is a confrontation between scientific objectivity and ingrained prejudice*—between logic and emotion—between fact and fiction. . . . In the final analysis, objective scientific logic has to prevail—no matter what the final result is—no matter how many time-honored idols have to be discarded in the process.[534]

After all, it is not the duty of science to defend the theory of evolution, and stick by it to the bitter end—no matter what illogical and unsupported conclusions it offers. . . . If in the process of impartial scientific logic, they find a creation by outside super-intelligence is the solution to our quandary, then let's cut the umbilical cord that tied us down to Darwin for such a long time. It is choking us and holding us back.[535]

Susanne Phillips, PhD, chair of a university earth sciences department, said recently in a

---
[530] George Vandeman, "The Hinge of Time," *Planet in Rebellion*, Southern Publishing Association, Nashville, TN, 1960, pp. 126-127.
[531] As quoted on Chuck Missler's tape, *In the Beginning There Was Information*.
[532] As quoted on Chuck Missler's tape, *In the Beginning There Was Information*.
[533] Monroe W. Strickland, *Genetics*, 2nd Edition, Macmillan Publishing Co., NY, 1976, p. 54, (emphasis added).
[534] I. L. Cohen, "Introduction," *Darwin Was Wrong: A Study in Probabilities*, New Research Publications, Inc., Greenvale, NY, 1984, pp. 6-8.
[535] I. L. Cohen, "In Retrospect," *Darwin Was Wrong: A Study in Probabilities*, New Research Publications, Inc, Greenvale, NY, 1984, pp. 214-215.

personal communication about the preceeding information in *Nature PLOS* and *Genetics Journal* websites that it is extremely damaging to the theor of relatedness between humans and chimps, and that it goes a long way to refute the theories of the evolution of man.

Mitochondrial Eve

Virtually all cells of every living thing (plants, animals, and humans) contain tiny strands of coded information called DNA. In human cells, the nucleus contains 95.5 percent of the DNA. Half of it came from the individual's mother and half from the father. . . . Each cell has, outside its nucleus, thousands of little energy-producing components called mitochondria, each containing a circular strand of DNA. Mitochondrial DNA (mtDNA) comes only from the mother.[536] Where did she get hers? From her mother—and so on. Normally, mtDNA does not change from generation to generation.

DNA is written with an alphabet of four letters: A, T, C, and G. One copy of a person's mtDNA is 16,559 letters long. Sometimes a mutation changes one of the letters in the mtDNA which a mother passes on to her child. These rare and somewhat random changes allow geneticists to identify families. For example, if your grandmother experienced an early mutation in her mtDNA, her children and any daughter's children would carry the same changed mtDNA. It would differ, in general, from that in the rest of the world's population.

In 1987, a team at the University of California at Berkeley published a study comparing the mtDNA of 147 people from five of the world's geographic locations. They concluded that all 147 had the same female ancestor. She is now called "the mitochondrial Eve." Researchers concluded that she lived up to 200,000 years ago.[537]

A greater surprise, even disbelief, occurred 11 years later in 1998, when it was announced that mutations in mtDNA occur 20 times more rapidly than previously thought. Mutation rates can now be determined directly, by comparing the mtDNA of many mother-child pairs. Using the new, more accurate rate, mitochondrial Eve "would be a mere 6,000 years old."[538]

The highly respected sources of this story *(Nature* and *Science)* are mainstream science journals—not creation-science publications. The *Science* headline acknowledged that this "Research News" section raised "troubling questions about the dating of evolutionary events."[539]

---

[536]Thomas J. Parsons, et al, "A High Observed Substitution Rate in the Human Mitochondrial DNA Control Region," *Nature Genetics*, Vol. 15, April 1997, p. 364.

[537]Rebecca L. Cann, et al, "Mitochondrial DNA and Human Evolution," *Nature*, Vol. 325, January 1, 1987, pp. 31-36.

[538]Ann Gibbons, "Calibrating the Mitochondrial Clock," *Science*, Vol. 279, January 2, 1998, p. 29; Walt Brown, PhD, In the Beginning, pp. 229-231.

[539]Ann Gibbons, "Calibrating the Mitochondrial Clock," *Science*, Vol. 279, January 2, 1998, p. 29.

V
Catastrophism

Chapter 23
The Biblical Flood

The entire earth was very different "in the beginning." Scientists who are open to the Biblical historical record believe that most of today's geologic column, continental drift, mountain building, lava distribution, and weather patterns resulted from a catastrophe of biblical proportions known today as the *Noachian Flood,* and then from its following catastrophes of flow, drying, and freezing of water on the surface of the earth. The original atmosphere may have included a greater percentage of oxygen, smaller temperature variation, and a higher concentration of atmospheric carbon dioxide. Pre-flood conditions would have supported longer life spans, minimal disease, rapid healing, and larger plants. If this were true, we should be able to find evidence in the fossil record, including catastrophes that occurred during and after the Flood.

First, we shall look at the biblical description of the Flood and then at the evidence:

> Genesis 6:5-7: And God saw that the wickedness of man was great in the earth, and that every imagination of the thoughts of his heart was only evil continually. And it repented ["grieved" NIV] the Lord that he had made man on the earth, and it grieved him at his heart ["It broke His heart" TLB.] And the Lord said, I will destroy man whom I have created from the face of the earth; both man, and beast, and the creeping thing, and the fowls of the air; for it repenteth me [I am very sad] that I have made them [because:]

> Genesis 6:11-14: The earth also was corrupt before God, and the earth was filled with violence. And God looked upon the earth, and, behold it was corrupt; for all flesh had corrupted his way upon the earth. And God said unto Noah, the end of all flesh is come before me; for the earth is filled with violence through them; and, behold, I will destroy them with the earth. Make thee an ark of gopher wood.

According to the book of Genesis, when Noah and his wife and their three sons and their wives entered the Ark, so did a representative number of earth's animals, male and female:

> Genesis 7:2, 14-15: Of every clean beast thou shalt take to thee by sevens, the male and his female [later to be used for sacrifices or food]: . . . They, and every beast after his kind, and all the cattle after their kind, and every creeping thing that creepeth upon the earth after his kind, and every fowl after his kind, every bird of every sort. And they went in unto Noah into the ark, two and two of all flesh, wherein is the breath of life.

> Genesis 7:22: All in whose nostrils was the breath of life, of all that was *in the dry land.*

From this text we can conclude there were no fish or other marine creatures on the Ark.

The whole world was covered with water, some from the firmament that was over the earth from the beginning, but mostly from the fountains of the deep, which possibly shot violently into the sky as huge geysers of water and added to the rainfall already coming from the water in the firmament [the atmospheric heavens].

> Genesis 7:11-12: In the six hundredth year of Noah's life, in the second month, the seventeenth day of the month, the same day were all the fountains of the great deep broken up, and the windows of heaven were opened. And the rain was upon the earth forty days and forty nights.

What are the scientific facts?

Leonard R. Brand, PhD: The earth contains a tremendous amount of sedimentary rock that was formed by the hardening of sediments deposited in basins by wind or water. In fact, there is enough sedimentary rock on the continents alone to cover them to an average depth of 1,500 meters [a little more than 4,921 feet].[540]

Water-deposited rock layers around the world contain billions upon billions of fossils. Fossilization usually requires *sudden death and immediate burial.* We do not see many fossils for the millions of

---
[540]Leonard R. Brand, PhD, "Faith and the Flood," *Ministry,* February, 1980, pp. 26-27.

bison that were killed in America. The red tides in the Gulf of Mexico can kill a million fish at a time. But they do not fossilize. Instead they simply decay:

> Immanuel Velikovsky, MD: When a fish dies its body floats on the surface or sinks to the bottom and is devoured rather quickly, actually in a matter of hours, by other fish. However, the fossil fish found in sedimentary rocks is very often preserved with all its bones intact. Entire shoals of fish over large areas, numbering billions of specimens, are found in a state of agony, but with no mark of the scavenger's attack.[541]

Ian T. Taylor: In England, one of the largest sedimentary rock deposits covering thousands of square miles is known as the Old Red Sandstone, and it contains many millions of fossilized fish in contorted positions indicating that they died in agony. In some of the Sandstone quarries the fossil fish are so densely packed it is estimated that there are more than a thousand per cubic yard. . . . The famous fossil bird, *Archaeopteryx*, found in the Solnhofen Limestone, east of Stuttgart, Germany, and which appears in most school biology textbooks, died with its neck contorted backwards.[542]

Scientists from the Natural History Museum in Milan, Italy, caught some fish, placed the bodies in wire cages to prevent other animals from destroying them, and lowered them onto the bottom of a warm Southern marsh. They returned six-and-a-half days later only to find that all the flesh had decayed and even the bones had become disconnected.[543]

> Giovanni Pinna: When an organism dies, the substances that compose its soft parts undergo more or less rapid decay, due to such factors as attack by bacteria. . . . If an organism is to be preserved, it must be protected from destructive agents as quickly as possible.[544]

> F. H. T. Rhodes, H. S. Zim, and P. R. Shaffer: To become fossilized a plant or animal must usually have hard parts, such as bone, shell or wood. It must be buried quickly to prevent decay and must be undisturbed throughout the long process.[545]

> Ian T. Taylor: The parlor aquarium was introduced to England during the 1850s and became a popular part of Victorian life. Had Lyell and his supporters been keepers of goldfish, they would have been well aware that expired individuals are not found on the bottom of the tank. When a living creature dies, internal bacterial action produces gas that, if the body is in water, keeps it from sinking, and in the case of a large animal, the body may remain suspended for weeks. During this time, it is picked clean by scavengers. . . . One can thus appreciate that fossil formation by the falling of sediment over the body on the ocean bottom must have been rare indeed.[546]

There is **no** microbiologist ever who would support the claim that an anoxic environment prevented decay of an organism. Lack of oxygen in an environment never, ever, prevents microorganisms from consuming dead animals!

How does Uniformitarianism explain fossils?

Uniformitarianism explains fossils by many small catastrophes over the eons in each particular locality. It cannot account for the rapid burial of the vast quantity of plants and animals that we see today in the geologic column. I repeat for emphasis: usually, when a plant or animal dies, it rots or is eaten by other animals. To become a fossil, the plant or animal has to be buried quite rapidly. The biblical Flood can account for this.

Fossils all over the world show evidence of rapid burial. Many fossils, such as fossilized jellyfish,[547] show by the details of their soft, fleshy portions[548] that they were buried rapidly, before they could decay. Many other animals buried in mass graves and in twisted and contorted positions, suggest violent and rapid burials over large

---

[541] Immanuel Velikovsky, MD, "Collapsing Schemes," *Earth in Upheaval*, Doubleday & Company, Inc., Garden City, NY, 1955, p. 222.
[542] Ian T. Taylor, "Science and Geology," *In the Minds of Men*, Minneapolis: TFE Publishing, 1996, p. 90.
[543] Harold G. Coffin, PhD, "Fossil Evidences in Support of Catastrophe," *Origin By Design*, Review and Herald Publishing Association, Washington, DC, Hagerstown, MD, 1983, pp. 33-34.
[544] Giovanni. Pinna (Deputy Director of the Museum of Natural History in Milan, Italy), *The Dawn of Life*, pp. 1-2.
[545] F. H. T. Rhodes, H. S. Zim, and P. R. Shaffer, "Introduction to Fossils," *Fossils: A Guide to Prehistoric Life*, Golden Press, NY, 1962, p. 10.
[546] Ian T. Taylor, "Science and Geology," *In the Minds of Men*, Minneapolis: TFE Publishing, 1996, pp. 88-89.
[547] Preston Cloud and Martin F. Glaessner, "The Ediacarian Period and System: Metazoa Inherit the Earth," *Science*, Vol. 217, August 27, 1982, pp. 783-792; see also the cover of that issue: Martin Glaessner, "Pre-Cambrian Animals," Vol. 204, March 1961, pp. 72-78.
[548] Donald G. Mikulic et al, "A Silurian Soft-Bodied Biota," *Science*, Vol. 228, May 10, 1985, pp. 715-717.

areas.[549] These observations, together with the occurrence of compressed fossils and fossils that cut across *two or more layers of sedimentary rock*, are *strong evidence that the sediments encasing these fossils were deposited rapidly*—not over hundreds of millions of years.

The fossilization process must have been quite rapid to have preserved a fish in the act of *swallowing another fish*.[550] Geologists now explain this away by saying the larger fish died because it tried to eat another fish that was too big. They call them "aspirations" though that still doesn't explain the lack of deterioration. The Stuttgart Museum of Natural History in Germany displays a fossil ichthyosaur (fish) preserved in the act of feeding her young.[551]

> Ian T. Taylor: In the Ludwigsburg Museum of Natural History in Germany, there is an even more spectacular specimen of an ichthyosaur fossilized in the process of giving birth with the young clearly visible in the birth canal.[552]

Creation *ex nihilo* displays a photograph of that ichthyosaur. This is clear evidence of its having been buried quickly by water-born sediments. The fossil record is consistent with creatures having been buried suddenly, otherwise most would either have rotted or have been devoured by scavengers.[553]

Fossilized while swallowing? Feeding? Birthing?

> Ian T. Taylor: In the Princeton Museum of Natural History there is a perch fossilized in the act of swallowing a herring. In each of these examples . . . their burial under fine sediment and subsequent fossilization had to have been sufficiently rapid to leave no trace of decomposition. . . . [The perch swallowing the herring was] found in the Eocene varves of Fossil Lake in Wyoming, where it is assumed that a foot of rock took two thousand years to form. . . . The fine details preserved in both [feeding and birthing] German specimens, each of which is almost six feet long, show no signs of decomposition, and the natural explanations proposed, without the appeal to a catastrophe, are strained to say the least.[554]

It is interesting to note that massive accumulations of well preserved fish, the extensive deposits of dinosaur and other reptilian remains, and large concentrations of plant materials in the form of coal beds represent striking examples of mass burial:[555]

Uniformitarianism cannot account for the rapid burial of the quantity of plants and animals that we see today in the geologic column.

Uniformitarianism—'The present is the key to the past'

Before *radiometric dating* was available, many people had estimated the age of the earth to be only a few thousand years old. But in the 1700s, Scottish scientist James Hutton estimated that the earth was much older. He was the principal architect of the theory of uniformitarianism. This principle states that processes occurring today are similar to those that occurred in the past. He observed that the processes that changed the rocks and land around him were very slow, and he inferred that they had been just as slow throughout earth's history. Hutton hypothesized that it took much longer than a few thousand years to form the layers of rock around him and to erode mountains that once towered kilometers high. John Playfair advanced Hutton's theories, but starting in 1830, an English geologist, Sir Charles Lyell, from his three-volume book *Principles of Geology*, is given the most credit for advancing uniformitarianism.

Evolutionists once believed in strict uniformitarianism; that the present is the key to the past (now redefined to state that the laws of nature remain invariant).

But recently strict uniformitarianism has fallen on hard times.

> Arthur V. Chadwick, PhD: With time and additional data, it became clear to geologists that Lyel-

---

[549] Presse Grayloise, "Very Like a Whale," *The Illustrated London News*, 1856, p. 116.
[550] Walt Brown, Jr., PhD, "Rapid Burials," *In the Beginning, Compelling Evidence for Creation and the Flood*, Center for Scientific Creation, Phoenix, AZ, 2001, p. 9.
[551] Ian T. Taylor, "Science and Geology," *In the Minds of Men*, TFE Publishing, Minneapolis, MN, 1996, p. 88.
[552] Ian T. Taylor, "Science and Geology," *In the Minds of Men*, TFE Publishing, Minneapolis, MN, 1996, p. 89.
[553] "The Greatest Catastrophe of all Time," *Creation ex nihilo*, Vol. 22, No. 1, December 1999, p. 12.
[554] Ian T. Taylor, "Science and Geology," *In the Minds of Men*, TFE Publishing, Minneapolis, MN, 1996, p. 89.
[555] Harold G. Coffin, PhD, "Fossil Evidences in Support of Catastrophe," *Origin By Design*, Review and Herald Publishing Association, Washington, DC; Hagerstown, MD, 1983, p. 33.

lian uniformitarianism did not stand up to careful scrutiny. For instance, a consideration of the magnitude of certain geological phenomena, such as giant debris flows, implied that past processes must at times have involved forces far above the range of forces experienced on the earth today. At the same time, a careful study of present day-to-day geological processes indicates that those forces are hardly involved at all. For example, on the occasion of the centennial celebration of Powell's historic traverse of the Grand Canyon, attempts were made to reoccupy the photographic sites in Grand Canyon used by Powell 100 years earlier. The results were unexpected. In about seventy percent of the cases, the sites, mostly at river level, appeared virtually unchanged. In photographic sites where changes from the original photographs were observed, the environment nearly always appeared catastrophically altered.

Gradually, geologists confronted with the absurdity of strict Lyellian uniformitarianism rejected it. But they were unwilling to be called "catastrophists," a phrase too closely linked with the Flood of Noah for the comfort of most. As a result, geologists reworked a related concept, *actualism*.

The modern concept of actualism rejects the strict Lyellian uniformitarianism and in its place acknowledges two specific premises: geologic processes have *varied in rates* and intensities over time, and there have been many processes operative in the past that are not occurring in the present.[556]

The geologic column

The geologic column was first described in the late 1700s and early 1800, (before the invention of radiometric dating techniques). Each layer of rock was associated with a group of index fossils. Again, we see two completely different interpretations from the same facts. Evolutionists still claim the sedimentary layers formed *slowly* over millions of years despite the now-adopted concept by geologists of actualism. Creationists claim the layers are mostly from Noah's Flood and that their fossils were buried *quickly*. They can't both be right.

What are index fossils?

Science Explorer, Earth Science: To be considered index fossils, fossils must meet certain requirements. First, they must be present in rocks scattered over a wide area of the earth's surface. Second, index fossils must have features that clearly distinguish them from all other fossils. Third, the organisms from which the index fossils formed must have lived during a relatively short span of geologic time. Fourth, they must occur in fairly large numbers within the rock layers.[557]

Because fossils determined to be older are usually lower in the geologic column, evolutionists determine the age of the fossils by the sedimentary layer in which they are found. So, the rock layers date the fossils. But then they say the fossils date the rock layers. This is, unfortunately, circular reasoning.

American Journal of Science: The intelligent layman has long suspected circular reasoning in the use of rocks to date fossils and fossils to date rocks. The geologist has never bothered to think of a good reply, feeling the explanations are not worth the trouble as long as the work brings results.[558]

Even the *Encyclopædia Britannica* weighed in: "It cannot be denied that from a strictly philosophical standpoint geologists are here arguing in a circle. A succession of organisms has been determined by a study of their remains embedded in the rocks, and the relative ages of the rocks are determined by the remains of organisms they contain."[559]

L. James Gibson, PhD: One example of *circular reasoning* is to date the fossils by the rocks and then use the rocks to date the fossils. The other example is to use the theory of evolution to explain the sequence of fossils and then use the sequence of fossils to prove the theory of evolution.[560]

Ronald R. West: Contrary to what most scientists write, the fossil record does not support the Darwinian theory of evolution because it is this theory (there are several) which we use to interpret the fossil record. By doing so we are guilty of circular reasoning if we then say the fossil record supports this theory.[561]

---

[556] Arthur V. Chadwick, PhD, "A Modern framework for Earth Sciences in a Christian Context," http://origins.swau.edu/papers/geologic/geology/index.html, 2004, p. 4, (emphasis added).

[557] "Finding the Relative Age of Rocks," *Science Explorer*, Earth Science, Prentice Hall, Needham, Massachusetts, 2002, p. 296.
[558] J. E. O'Rourke, "Pragmatism vs. Materialism in Stratigraphy," *American Journal of Science*, 1976, 276:51.
[559] R. H. Rastall, "Geology," *Encyclopædia Britannica*, Vol. 10, 1949, p. 168.
[560] L. James Gibson, PhD, Director, Geoscience Research Institute, note to author, July 23, 2013.
[561] Ronald R. West, "Paleoecology and Uniformitarianism,"

Niles Eldredge, PhD: There is no way simply to look at a fossil and say how old it is unless you know the age of the rocks it comes from. And this poses something of a problem: if we date the rocks by their fossils, how can we then turn around and talk about patterns of evolutionary change through time in the fossil record?[562]

But true believers in evolution continue to support this circular reasoning and also acknowledge its relationship to radioactive decay:

J. E. O'Rourke: Radiometric dating would not have been feasible if the geologic column had not been erected first. The rocks do date the fossils, but the fossils date the rocks more accurately.[563]
... The charge of circular reasoning in stratigraphy can be handled in several ways. It can be ignored, as not the proper concern of the public. It can be denied, by calling down the Law of Evolution. It can be admitted, as a common practice. . . . Or it can be avoided, by pragmatic reasoning.[564]

What?

So, what about radiometric dating? And is it as accurate as alleged? Although some creationists will acknowledge that there is some value in sequencing findings (their general order of age, not actual age), they agree with some evolutionists that this dating method has serious problems.

There is evidence against "millions of years" and radiometric dating. Dinosaurs lived in the Cretaceous Period supposedly 65 to 70 million years ago. However, a medical pathologist has documented unfossilized soft tissue and red blood cells inside dinosaur bones. (More recently, scientists have discovered measurable amounts of *carbon-14* in them—as is documented in the following pages. Both discoveries "should" be impossible.)

In 2005, the *New York Times* reported on the newly published scientific article that *Tyrannosaurus rex* bone fragments showed unfossilized dinosaur tissue, including vascular canals:

New York Times: A 70-million-year-old *Tyrannosaurus rex* recently discovered in Montana, scientists reported today, has apparently yielded the improbable: soft tissues, including blood vessels and possibly cells, that "retain some of their original flexibility, elasticity and resilience. . . . When fossilizing mineral deposits in the tissues were dissolved by a weak acid, the scientists were left with stretchy material threaded with what looked like tiny blood vessels. Further examination revealed reddish brown dots that the scientists said looked like the nuclei of cells lining the blood vessels."[565]

According to every biochemical assessment and the laws of physics applied to macromolecules like collagen fibers, soft elastic tissue could never be preserved for 65 to 70 million years. Is it possible that the tissue is not 70 million years old?

Mary H. Schweitzer and T. Staedter: A thin slice of *T. rex* bone glowed amber beneath the lens of my microscope . . . the lab filled with murmurs of amazement, for I had focused on something inside the vessels that none of us had ever noticed before: tiny round objects, translucent red with a dark center . . . red blood cells. The shape and location suggested them, but blood cells are mostly water and couldn't possibly have stayed preserved in the sixty-five-million-year-old *Tyrannosaur*.[566]

David C. Read, JD: The scientists began testing the *T. rex* tissue for hemoglobin, a blood protein that carries oxygen from the lungs to the body tissues. Hemoglobin consists of heme, which contains iron, and globin, which contains a long string of proteins. The evidence that hemoglobin had indeed survived in the *T. rex* bone is as follows:

• The tissue extracted from the *T. rex* was reddish brown—the color of hemoglobin.

• Chemical signatures unique to heme were found when certain wavelengths of laser light were applied to the *T. rex* tissue.

• Because it contains iron, heme reacts to magnetic fields differently from other proteins. Extracts from *T. rex* reacted the same way modern heme compounds react.

Compass 45, May, 1968, p. 216, quoted in *The Revised Quote Book,* Creation Science Foundation, Acacia Ridge, Queensland, Australia, 1990, p. 10.
[562]Niles Eldredge, PhD, *Time Framers: The Rethinking of Darwinian Evolution and the Theory of Punctuated Equilibria,* Simon and Schuster, NY, 1985, p. 52
[563]J. E. O'Rourke, "Pragmatism versus Materialism in Stratigraphy," *American Journal of Science,* Vol. 276, 1976, pp. 53-54.
[564]J. E. O'Rourke, "Pragmatism versus Materialism in Stratigraphy," *American Journal of Science,* Vol. 276, 1976, p. 54.

[565]John Noble Wilford, "Tissue Find Offers New Look Into Dinosaurs Lives," *New York Times,* March 24, 2005; see also: *Science,* March 25, 2005; see also: M. Schweitzer and T. Staedter, "The Real Jurassic Park," *Earth,* June 1997, pp. 55-57.
[566]M. Schweitzer and T. Staedter, "The Real Jurassic Park," *Earth,* June 1997, pp. 55-57.

- Extracts from the *T. rex* were injected into rats. If there was hemoglobin in the extracts, the rats' immune systems should build up antibodies against it. This is exactly what happened in carefully controlled experiments.[567]

- Schweitzer and her colleagues isolated a heme molecule, a ring-like structure that is found in hemoglobin and similar compounds.[568]

The team concluded that "the parsimonious explanation of this evidence is the presence of blood-derived hemoglobin compounds preserved in the dinosaurian tissues."

Derek Isaacs: For the documentary "Dragons or Dinosaurs?" we interviewed Dr. David Menton from "Answers in Genesis." From that interview, we learned that Jack Horner was offered a substantial amount of money to take a portion of that very *T. rex* bone and have a carbon-14 test run on it. The offer was refused.

Carbon-14 is used to date plant and animal fossils that are only thousands of years old. The significance of carbon-14 found in specimens estimated by evolutionists to be 65 to 70 million years old cannot be overemphasized. (See chapters 31 and 32 on radiometric dating.)

Certainly, a bone that was supposed to be 68 million years old, but still has soft tissue, warrants a carbon-14 test. However, such testing was denied.[569]

Schweitzer recently made an even more spectacular discovery, announced in March 2005, after Jack Horner found another *Tyrannosaurus rex* at Hell Creek, Montana (where Barnum Brown found the first *T. rex*). She soaked samples from the femur in a solution that dissolved the calcium.

What remained was "flexible vascular tissue that demonstrates great elasticity and resilience." After being stretched, it resumed its original shape. Schweitzer also found red blood cells (again) and what appear to be osteocytes, specialized cells that build and repair bone. Remarkably, nucleated red blood cells could be seen within the blood vessels. . . . The researchers were able to squeeze red blood cells out of the flexible blood vessels.[570]

The significance of the *T-rex* bone cannot be overstated. Now that soft tissue has been found, other scientists have started to actually look for it when they uncover dinosaur remains. Similar evidence is starting to pour in from around the world. It is as if a faucet had been turned on.

Another theropod from Argentina and a woolly mammoth have shown evidence of containing probable blood vessels, cells, and connective tissue[571] and researchers in Sydney found blood vessels and proteins in hadrosaur remains.[572] Samples from South America and Australia join the North American find in proclaiming these creatures are much younger than what has been taught.[573]

Soft, flexible blood vessels in dinosaurs (discovered in 2000, confirmed in 2015) are also documented in the *Geoscience Research Institute Newsletter*, Number 44, January 2016. The proteins that constitute the structure of elastic and collagen fibers, and the shape of cells could never remain intact for more than a few thousand years.

David C. Read, JD: Red blood cells in fossilized dinosaur bone, and the survival of soft tissue and hemoglobin, constitute powerful evidence against the notion that the dinosaurs lived sixty-five million years ago.[574]

---

[567] Mary H. Schweitzer and I. Staedter, "The Real Jurassic Park," *Earth*, June 1997, pp. 55-57; Mary H. Schweitzer, C. Johnson, T. G. Zocco, J. H. Horner, J. R. Starkey, "Preservation of biomolecules in cancellous bone of "Tyrannosaurus rex," *Journal of Vertebrate Paleontology*, 17:2, June 19, 1997; Mary H. Schweitzer, Mark Marshall, Keith Carron, D. Scott Bohle, Scott C. Busse, Ernst V. Arnold, Darlene Barnard, J. R. Horner, and Jean R. Starkey, "Heme compounds in dinosaur trabecular bone," *Proceedings of the National Academy of Sciences*, Vol. 94, pp. 6291-6296; Virginia Morell, "Dino DNA: The Hunt and the Hype," *Science*, 261 (5118):160-162, July 9, 1993, pp. 160-162; Jeff Hecht, "Dinosaur Bones Yield Blood Protein," *New Scientist*, Vol. 154, New Science Publications, London, June 21, 1997, p. 16. Schweitzer and Horner stood by their description of what they saw in the *T. rex* bone as "red blood cell-like intravascular structures." Mary H. Schweitzer, James G. Schmitt, John R. Horner, "Paleoenvironmental and taphonomic parameters of an exceptionally preserved Tyrannasaurus rex." GSA Annual Meeting, November 5-9, 2001, paper 160.
[568] *Science News*, Vol. 148, November 11, 1995, p. 314; quoted by David C. Read, JD, "When Did the Dinosaurs Live? The Biblical View, *Dinosaurs*, Clarion Call Books, Keene, TX, 76059, 2009, p. 150.
[569] Derek Isaacs, "Hell Creek," *Dragons or Dinosaurs? Creation or Evolution?* Bridge-Logos, Alachua, FL, 2010, p. 139.
[570] Mary H. Schweitzer, J. L. Wittmeyer, J. R. Horner, J.K. Toporski, "Soft-Tissue Vessels and Cellular Preservation in Tyrannosaurus rex," *Science*, Vol. 307, March 25, 2005, pp. 1952-1955.
[571] Helen Fields, "Dinosaur Shocker," *Smithsonian* magazine, May 2006, retrieved online June 8, 2009.
[572] Nick Evershed, "Blood tissue extracted from duck-billed dinosaur bone," *Cosmos*, May 1, 2009, retrieved on June 8, 2009.
[573] Derek Isaacs, "Hell Creek," *Dragons or Dinosaurs? Creation or Evolution?* Bridge-Logos, Alachua, FL, 2010, p. 139.
[574] David C. Read, JD, "When Did the Dinosaurs Live? The

In 2005, msnbc.com reported the soft tissue find. "It may be that this isn't a unique specimen," reported Jack Horner, SCD, Montana State Paleontologist. "The find could force scientists to reconsider how all fossils are formed."

And what did scientists who believe in creation recently find?

> Derek Isaacs: Another groundbreaking endeavor was well underway in Montana. In 2005, field paleontologist Otis Kline and his excavation team unearthed fossils from a Triceratops and a hadrosaur. These specimens came from the same Hell Creek formation as the Schweitzer *T-rex*. These fossils were in such fine condition that the scientific team was propelled to determine if these, like the *T-rex*, still contained some fresh remains that had not yet fossilized.
>
> To the excitement of the scientists, not one, but both the *Triceratops* and the hadrosaur still possessed collagen! The presence of such organic material flies in the face of evolutionary ideology that these creatures had been dead for 65 million years.

Again, carbon-14 is used to date plant and animal fossils that are only thousands of years old. The significance of carbon-14 found in specimens estimated by evolutionists to be 65 to 70 **million** years old cannot be overemphasized. The quotation continues:

> Due to the significance of this study, the industry-recognized Accelerator Mass Spectrometer (AMS) was used to test for carbon-14. Furthermore, not one but two internationally recognized labs, Geochron Laboratories and the University of Georgia Isotope Center, were used so that the results could be independently confirmed. The tests yielded paradigm-shifting results: *Both bones contained carbon-14* [a scientifically recognized impossibility for 65- to 70-million-year-old tissue]. . . .

This is the result that no evolutionist wants to see, and it is why the previous *T-rex* that we discussed, in my opinion, has not been tested.[575]

A study of nine dinosaurs from Texas to Alaska and one from China found measurable carbon-14 in dinosaur bones, where there should be none. Subjects, which reportedly are at least 65 million years old, included a *Psittacosaurus;* an *Allosaurus, Acrocanthosaurus,* and *Apatosaurus;* two *Triceratops;* and three hadrosaurs. An Accelerated Mass Spectrometer (AMS) was used for 20 of 22 samples, primarily at the University of Georgia. All samples were pretreated to remove contaminants. The two large samples were tested on conventional equipment as another cross check.[576]

This data was reported at the 2012 joint meeting of the American Geophysical Union (AGU) and the Asia Oceania Geophysical Society (AOGS) conference in Singapore. However, after the meeting, the abstract was removed from the conference website by two chairmen because they could not accept the findings. Unwilling to challenge the data openly, they erased the report from public view without a word to the authors or even to the AOGS officers. . . . It won't be restored.[577]

Biblical View," *Dinosaurs*, Clarion Call Books, Keene, TX, 2009, pp. 149-151

[575] Derek Isaacs, "Hell Creek," *Dragons or Dinosaurs? Creation or Evolution?* Bridge-Logos, Alachua, FL, 2010, pp. 141-142, (brackets added).
[576] Hugh Miller, Hugh Owen, Robert Bennett, Jean De Pontcharra, Maciej Giertych, Joe Taylor, Marie Clair Van Oosterwych, Otis Kline, Doug Wilder, Beatrice Dunkel, *Abstract*, "BG02-A012, A Comparison of δ13C & pMC Values for Ten Cretaceous to Jurassic Dinosaur Bones from Texas to Alaska USA, China, and Europe," see also: Paul Giem, MD, on You Tube, "The Missing Presentation," March 30, 2013. https://youtu.be/s_53hGlasuk
[577] Paul Giem, MD, on You Tube, "The Missing Presentation," March 30, 2013. https://youtu.be/s_53hGlasuk

Chapter 24
The Geologic Column and Fossil Record

The geologic column, with its literally millions of fossils, and their often clearly rapid burial, makes the most logical sense within the framework of a global catastrophe, like the Genesis Flood. First, if each layer took millions of years or even thousands or even hundreds of years to form, how were the fossils embedded in them without being eroded or scavenged or rotted away?

If the layers took eons to form, we should see major evidence of *erosion* from wind and moving water, as would be seen with occasional catastrophes in either the uniformitarianism or gradualism (actualism) geologic-column paradigm. Yet how can one look at the light-colored, Coconino Sandstone layer near the top of the Grand Canyon—which shows no serious evidence of erosion—and conclude that it took eons to form, when geologists can provide scientific documentation of solid rock being eroded from 3.15 to 748 inches, [i.e., 62 feet, or 80 to 19,000 millimeters] *per 1,000 years?*[578]

It is difficult to see the implications of lack of erosion in the Coconino Sandstone layer if one accepts evolution's prior assumptions that exclude the possibility of a Divine Creator or global Flood. Once one has accepted evolution's worldview and long-age presuppositions, it becomes more difficult to see a need to research whether Coconino's lack of erosion points more persuasively to layers laid down quickly in a Flood than to layers built up over 260 million erosionless years.

It is easy to overlook a phenomenon (like lack of erosion) if it doesn't fit within one's worldview.

Yet this lack of major erosion is viewed in the Grand Canyon by millions of visitors every year, not only in the tan Coconino sandstone, the most visible and obvious layer, but throughout most of the geologic sequence.

We already noted in chapter 3 that the existence of the geologic column (and its fossil record)—at least the wide-spread sedimentary strata containing fossils above the Precambrian layer—was shown from scientifically reported rates of erosion to suggest that the earth is not billions of years old.

> David C. Read, JD: The sedimentary rocks can be interpreted as the residue of the Genesis Flood. No one disputes that water deposited the great bulk of the geologic column, and almost all of the major divisions of sedimentary strata contain marine fossils. The disagreement between creationists and mainstream geologists relates primarily to when and how rapidly the strata were deposited.[579]

Fossils buried rapidly in worldwide, deep, watery-sediment layers

> Vine V. Deloria, Jr.: [After outlining the fact that the Upper Cretaceous Chalk strata, identified as "The White Cliffs of Dover" are also found in northern Ireland, England, northern France, through the Low Countries, northern Germany and southern Scandinavia to Poland, Bulgaria and eventually to Georgia in the south of the Soviet Union, Egypt, Israel, and in the United States in Texas, Arkansas, Mississippi, and Alabama, and finally in Australia,] "it is not possible to explain the *widespread* occurrence of this particular strata. In summary, we cannot begin to explain the origin of what we have called *sedimentary* strata because there are no processes of deposition that we can observe that would create anything resembling what we see in rock formations.[580]

One of the most intriguing examples of an interpretation bias involves a type of microscopic hard-shelled algae known as a *diatom*. Since this

---

[578] Ruxton BP, McDougall I. 1967, "Denudation rates in northeast Papua from potassium-argon dating of lavas," *American Journal of Science,* 265:545-561; Ollier CD, Bown MJF, 1971, "Erosion of a young volcano in New Guinea," *Zeitschrift für Geomorphologie,* 15:12-28.

[579] David C. Read, JD, "When Did the Dinosaurs Live? The 'Scientific' View." *Dinosaurs,* Clarion Call Books, Keene, TX, 76059, 2009, p. 76.
[580] Vine Deloria Jr., "Geomythology and the Indian Traditions," *Red Earth: White Lies,* Fulcrum Publishing, Golden, CO, 1997, pp. 163-165, (emphasis added).

tiny object has a shell that does not decompose when it dies, the shells of dead diatoms generally settle down to the ocean bottoms and, under appropriate conditions, form deposits called *diatomaceous earth*. Such deposits (such as the white cliffs of Dover) can be hundreds of feet thick.

All over the world, whales, dolphins, fish and even pterodactyls (thought to be extinct flying dinosaurs) have been found preserved in diatomaceous earth. The preservation of these fossils has created an interpretation dilemma for evolutionists, because these fossils were buried *quickly* rather than over many years. These discoveries are opening the door to a different interpretation of these fossil burials. Fossil fish are found by the millions packed in diatomaceous earth with their gills and fins extended as if they were without oxygen and terrified when they died. Near Lompoc, California, along the San Andreas Fault, diatomaceous earth hundreds of feet thick covers more than 100 square miles. This appears to be incontrovertible evidence of a catastrophic Flood.

Geologists often assume that the accumulation of thick layers of tiny microscopic *diatoms* such as the White Cliffs of Dover in England required lengthy periods of time. But such accumulation can occur rapidly. Along the coast of Oregon, a three-day storm of high winds and rain deposited a 10-15 centimeters [4-6 inches] layer of microscopic diatoms for a distance of 32 kilometers [20 miles].[581]

Peer-reviewed 'young' whales

At a diatomaceous-earth quarry in Lompoc, California, a remarkable discovery was made during mining operations in 1976.[582] Workers of the Dicalite Division of Grefco Corporation uncovered an 80-foot fossil skeleton of a baleen whale in the quarry that was being exposed gradually as the diatomite was mined. What does this mean? It means that this layer of diatomaceous earth could not have been built up gradually over millions of years. The billions of tiny diatom shells making up the formation had to be deposited in a very short period of time.

Here's why:

Again, in order for a creature to become a fossil after it dies it must be buried deeply and quickly in wet sediment to seal it off from the atmosphere. Otherwise it will simply rot. Thus, one can be confident that [even though the whale and sediment have both been lifted into a vertical position] the formation containing the whale had to be deposited quickly.[583]

The Miocene/Pliocene Pisco Formation of Peru also contains a large number of fossil whales that were buried in an accumulation of the minute silica skeletons of diatoms. According to Dr. Leonard Brand, "conventional geology assumes that fossil diatom deposits accumulated very slowly: only a few centimeters per thousand years." However, Dr. Brand's biblical worldview predicts that geological deposits like this formed in a much shorter time. He stated, "Our worldview, with its predictions of short chronology, opened our eyes to seeing things that others have not noticed or taken seriously." In fact, his team of geologists and paleontologists concluded, after several summers of excavations, that each of the whales was buried rapidly—"probably a matter of a few weeks or months, not thousand of years." The evidence also suggests some processes that help to explain how ancient diatomite may have accumulated much more rapidly than is usually assumed.[584]

Team members then presented several scientific papers at the national meetings of the Geological Society of America and at an international paleontological conference in Spain, and then published a scientific paper in *Geology*, a peer-reviewed scientific journal.[585]

---

[581] Campbell, A.S., Radiolaria, In: Moore R.C. editor, *Treatise on Invertebrate Paleontology*, Part D. (Protista 3), University of Kansas Press, Geological Society of America, Lawrence, KS, NY, p. D17.
[582] "Workers Find Whale in Diatomaceous Earth Quarry," *Chemical and Engineering News*, October 11, 1976, p. 40.
[583] Paul D. Ackerman, PhD, "Back Down to Earth," *It's a Young World After All: Exciting Evidences for Recent Creation*, Baker Book House, Grand Rapids, MI, 1986, p. 83.
[584] Leonard R. Brand, PhD, "Appendix A, Interventionist-inspired Research," *Beginnings: Are Science and Scripture Partners in the Search for Origins?* Pacific Press Publishing Association, Nampa, ID, 2006, pp. 153-154.
[585] R. Esperante-Caamaño, L. Brand, A. Chadwick, and O. Poma, "Taphonomy of Fossil Whales in the Diatomaceous Sediments of the Miocene/Pliocene Pisco Formation, Peru," in M. De Renzi, M. Alonso, M. Belinchon, E. Penalver, P. Montoya, and A. Marquez-Aliaga, eds. *Current Topics on Taphonomy and Fossilization*, International Conference Taphos, 2002, Third Meeting on Taphonomy and Fossilization, Valencia, Spain; L. R. Brand, R. Esperante, A. Chadwick, O. Poma, and M. Alomia, "Fossil

Water catastrophes

Some underwater landslides called *turbidity currents* can cover large areas in a short time. In 1929, twelve trans-Atlantic cables were cut by a turbidity current, which covered 40,000 square miles, flowing at 50 miles per hour. Considering the violent action of such underwater currents (think *Global Flood*), it's no wonder that fossil graveyards are found all over planet earth where the animal fossils, including dinosaurs, are found incomplete and ripped apart in tangled masses. In 1934, Barnum Brown, a famous dinosaur hunter, found a concentration of fossils in Wyoming at the foot of the Big-Horn Mountains that were piled like logs in a log jam.[586]

What was called "a mysterious bone bed" was discovered in eastern Africa in 1992. After studying the site for over 10 years, an international team of scientists was unable to solve what a textbook reported as one of the deepest mysteries: "How did the Kipsaramon bone bed form?"

> *Modern Earth Science:* Scientists have developed several ideas about this question and tested them against the fossil evidence. So far, every hypothesis has turned out to be incorrect. The fact that the fossils are disarticulated shows that the bones were disturbed after they were deposited. The fossils show no signs of wear from movement or of damage from predators, however. How did they get jumbled up without getting damaged? How did they get so closely packed together? Scientists are seeking answers to these and other questions as they continue their investigations of the incredible Kipsaramon bone bed.[587]

H. R. Siegler: There are so-called *fossil graveyards* in which is often found a rich conglomeration of organisms. One such graveyard found in Eocene lignite [coal] deposits of the Geiseltal in central Germany, contains more than six thousand remains of vertebrate animals together with an even greater number of mollusks, insects, and plants. So well-preserved are many of these animals that it is still possible to study the contents of their stomachs. It is easy to imagine how these could have been deposited in swirling and receding waters of a great Flood, but *not* how this could have happened under uniformitarian conditions.[588]

Ariel A. Roth, PhD: By the middle of the twentieth century some geologists had noticed that strict uniformitarianism conflicted with the data from the rocks themselves. . . . Under normal quiet conditions, changes over earth's surface proceed extremely slowly. However, many examples of catastrophic activity permit us to conceive of major changes in a short time.[589]

Professor Erich von Fange, PhD, of Concordia University, puts this into perspective:

> Great age is assumed from the successive strata in rocks that are believed to have been laid down very slowly and gradually. This is the heart of uniformitarianism in establishing the dating of the geologic column from Precambrian to modern times. This principle, however, is not faring well in [new] geological research. As just one example, the trunk of a tree at Craigleth Quarry in England was found to intersect from ten to twelve successive strata of limestone. It became fossilized without even losing its bark. Presumably the tree and its bark survived patiently for many millions of years as the strata covered it millimeter by millimeter.[590]

Non-creationists sometimes emphasize that the geologic column is far too thick to have been deposited in the single year of the Deluge.[591] But, while most creationists would exclude the lowest (Precambrian) and highest portions of the geologic column from the Flood, some *present* [geologic] rates of deposition are so rapid that it would be possible to deposit the whole column in a few weeks.[592]

Evolutionists claim that the animals found in

Whale Preservation Implies High Diatom-Accumulation Rate in the Miocene-Pliocene Pisco Formation of Peru," *Geology*, Vol. 32, 2004, pp. 165-168.
[586] Edwin H. Colbert, PhD, "Jurassic Giants of the Western World," *Men and Dinosaurs: The Search in Field and Laboratory*, E. P. Dutton & Co., Inc., NY, 1968, pp. 171-173.
[587] "The Mysterious Bone Bed," *Modern Earth Science*, Holt, Rinehart, and Winston, Austin, New York, Orlando, Atlanta, San Francisco, Boston, Dallas, Toronto, London, 2002, pp. 384-385.
[588] H. R. Siegler, *Evolution or Degeneration—Which*, 1972, pp. 78-79, (emphasis added).
[589] Ariel A. Roth, PhD, "Catastrophes: The Big Ones," *Origins: Linking Science and Scripture*, Review and Herald Publishing Association, Hagerstown, MD, 1998, p. 200.
[590] Erich A. von Fange, PhD, "Ancient Plant Oddities and Mysteries," *In Search of the Genesis World*, Concordia Publishing House, St. Louis, MO, 2006, p. 223.
[591] E. G. Ecker, "Geology," *Dictionary of Science & Creationism*, Prometheus Books, Buffalo, New York, 1990, p. 102.
[592] Ariel A. Roth, PhD, "Catastrophes: The Big Ones," *Origins: Linking Science and Scripture*, Review and Herald Publishing Association, Hagerstown, MD, 1998, p. 201, (italics and bracketed copy added).

the geologic column were buried in the order in which they evolved. They claim that clams in the bottom layers of sediment and birds in the top layers are proof that clams evolved first and birds evolved last. But there is another equally convincing theory. They could have died and were buried in sequence according to the progress of the biblical Flood. Dr. Morris provides a sound hypothesis for why various animals are found in the water-laid sediments as they are:

> Henry M. Morris, PhD: The general order from simple to complex in the fossil record in the geologic column, considered by evolutionists to be the main proof of evolution, is thus likewise predicted by the rival model, only with more precision and detail.
>
> 1. As a rule, there would be many more marine invertebrate animals trapped and buried in the sediments than other types, since there are many more of them and, being relatively immobile, they would usually be unable to escape.
>
> 2. Animals caught and buried would normally be buried with others living in the same region. In other words, fossil assemblages would tend to represent ecological communities of the pre-cataclysmic world.
>
> 3. In general, animals living at the lowest elevations would tend to be buried at the lowest elevations . . . with elevations in the strata thus representing relative elevations of habitat or ecological zones.
>
> 4. Marine invertebrates would normally be found in the bottom rocks of any local geologic column, since they live on the sea bottom.
>
> 5. Marine vertebrates (fishes) would be found in higher rocks than the bottom-dwelling invertebrates. They live at higher elevations and also could escape burial longer.
>
> 6. Amphibians and reptiles would tend to be found at still higher elevations, in the commingled sediments at the interface between land and water.
>
> 7. There would be few if any terrestrial sediments or land plants or animals in the lower strata of the column.
>
> 8. The first evidence of land plants in the column would be essentially the same as that for amphibians and reptiles, when the rafts of lowland vegetation were brought down to the seashore by the swollen rivers.
>
> 9. In the marine strata, where invertebrates were fossilized, these would tend locally to be sorted hydrodynamically into assemblages of similar size and shape. Furthermore, as the turbulently upwelling waters and sediments settled back down, the simpler animals, more nearly spherical or streamlined in shape, would tend to settle out first because of lower hydraulic drag. Thus, each kind of marine invertebrate would tend to appear in its simplest form at the lowest elevation. . . .
>
> 10. Mammals and birds would be found in general at higher elevations than reptiles and amphibians, both because of their habitat and because of their greater mobility. However, few birds would be found at all, only occasional exhausted birds being trapped and buried in sediments.
>
> 11. Because of the instinctive tendency of the higher animals to congregate in herds, particularly in times of danger, fossils of these animals would often be found in large numbers if found at all.
>
> 12. Similarly, these higher animals (land vertebrates) would tend to be found segregated vertically in the column in order of size and complexity, because of the greater ability of the larger, more diversified animals to escape burial for longer periods of time.
>
> 13. Very few human fossils or artifacts would be found at all. Men would escape burial for the most part and, after the waters receded, their bodies would lie on the ground until decomposed. The same would apply to their lighter structures and implements, whereas heavier metallic objects would sink to the bottom and be buried so deeply in the sediments they would probably never be discovered.
>
> 14. All the above predictions would be expected statistically but, because of the cataclysmic nature of the phenomena, would also admit of many exceptions in every case. In other words, the cataclysmic model predicts the general order and character of the deposits but also allows for occasional exceptions.

... And Morris concludes: Now there is no question that all of the above predictions from the cataclysmic model are explicitly confirmed in the geologic column.[593]

I reiterate the above descriptions. Clams would be the first ones buried because that's where they live; at the bottom. Birds are going to be the last ones buried because they can fly around until they get tired. The same comparison can be made with fish, amphibians, and mammals. If one were to take a snapshot of all living forms of life on earth today, or at any time in the past, one would find that it would reflect the fossil record from the lowest levels of the ocean to the highest. The lower levels in the geologic column would contain the animals that are naturally found on the bottom of the sea. Then, although marine fossils are found in every part of the geologic column, as they were buried by underwater landslides during the Flood, in general they would be buried in sequence: fish, amphibians, reptiles, birds and mammals, including man. Dr. Morris's hypothetical model fits the data as well as or better than evolutionary hypothetical models.

Due to the enormous size of the events predicted, this model can never be scientifically tested, only "modeled" based on the evidence in the rocks. (Other equally qualified creationists have alternative models that fit the evidence equally well, but also cannot be tested.) But Morris's model creates "reasonable doubt" about the evolutionary model or at least a feasible alternative to long-age evolution and geology.

---

[593] Henry M. Morris, PhD, "Uniformitarianism or Catastrophism?" *Scientific Creationism*, CLP Publishers, San Diego, CA, 1974, pp. 118-120.

Chapter 25
Where Did the Floodwaters Come From, and Where Did They Go?

The Bible indicates that originally there were waters under the earth. "The earth is the Lord's, and the fullness thereof. . . . For he hath founded it upon the seas, and established it upon the floods" (Psalm 24:1). "To him that stretched out the earth above the waters" (Psalm 136:6). The Bible tells us that the water creating the Flood in Noah's time came from the *firmament above and from below* the surface of the earth. "In the six hundredth year of Noah's life, . . . the same day were all the *fountains of the great deep* broken up, and the windows of heaven were opened. And the rain was upon the earth forty days and forty nights" (Genesis 7:11-12).

If all the water in today's atmosphere were to fall as rain it would measure about one inch.[594] Obviously it would not cover the earth much less the tops of the highest mountains. Therefore, to cover the mountains in existence at that time would require water from the "deep" (the oceans and underground fountains and erupting geysers [Genesis 7:11 and 8:2; Revelation 14;7] before and during the flood) in addition to the water in the firmament (atmospheric heavens where the rain clouds are, and beyond, i.e., the sky).[595] The mountain building now observed by geologists is likely the result of geologic forces set in motion by the yearlong biblical Flood. Our mountains today are much taller than the pre-Flood hills. Mt. Ararat, itself, where the Ark reportedly rested, is a volcano that did not exist before the Flood.

Where did the Deluge water go? Some Bible-believing geologists hypothesize that during the last few months of the Flood the unstable *tectonic plates* of the earth's crust shifted. Some plates went down in places, creating the ocean basins. Some went up, creating the high mountain ranges. This movement in the earth's crust probably continued for hundreds, possibly even thousands of years after the Flood. It continues to this day by earthquakes, and volcanoes, albeit with less intensity.

It seems likely that the Flood buried most of the tropical plants and animals scientists have documented finding around the world, including at the North and South Pole regions where there are no trees today. Within the Arctic Circle, the New Siberian Islands and Spitzbergen Islands have been discovered to contain immense frozen gravel mounds entombed with entire fruit trees with fruit still on them.[596]

On one of his major expeditions, Admiral Richard E. Byrd discovered near the South Pole a prehistory that was inconsistent with today's cold:

> Here, at the most southern-known mountain in the world . . . , scarcely 200 miles from the South Pole, was found conclusive evidence that the climate in Antarctica was once temperate or even subtropical.[597]

Dolph Earl Hooker in *National Geographic:* The rock fragments from this mountainside invariably included plant fossils, leaf and stem impressions, coal and fossilized wood.[598]

---

[594]Howard E. Brown, Victor E. Monnett, J. Willis Stoval, "The Atmosphere," *Introduction to Geology*, Gin and Company, Boston, New York, Chicago, 1958, p. 128.
[595]The "deep" in Genesis 2:2 is from the Hebrew word *teham*, literally, per *Young's Analytical Concordance to the Bible*, "the deep (sea);" and Isaiah 51:9-10 says, "Lord . . . art thou not [He] which hath dried the sea, the waters of the great deep; that hath made the depths of the sea . . . ?"
[596]D. G. Whitley, "The Ivory Islands in the Arctic Ocean," *Journal of Philosophical Society of Great Britain*, XII, p. 49; Immanuel Velikovsky, MD, "In the North," *Earth in Upheaval*, Doubleday & Company, Inc., Garden City, NY, (Dell ed., 1955), p. 8.
[597]Admiral Richard E. Byrd, "Exploring the Ice Age in Antarctica, *National Geographic*, National Geographic Society, Washington, DC, October 1935, p. 457.
[598]Dolph Earl Hooker, *Those Astounding Ice Ages*, Exposition Press, NY, 1958, p. 44, as taken from Admiral Richard E. Byrd, "When a South Pole Suburb was Semitropical: Exploring the Ice Age in Antarctica," *National Geographic*, National Geographic Society, Washington, DC, October 1935, pp. 456-457.

Chronicle of Higher Education: In January, Mr. Webb—with David Harwood, an assistant professor of geology at the University of Nebraska at Lincoln, and Barrie McKelvey, a professor of geology at the University of New England in Australia—found the deposit of leaves on the side of a cliff in a desolate stretch of the Transantarctic mountains, about 250 miles from the South Pole.[599]

Secular and Bible-believing geologists agree that the theory of *"plate tectonics"* explains how the original land mass of the earth became separated into the various continents we have today.

For example, the mountain ranges and fossils on the East Coast of South America match the mountain ranges and fossils on the West Coast of Africa. On a globe the two continents could fit into each other like pieces of a giant jigsaw puzzle.

Some commentators believe plate tectonics is acknowledged in the Bible: "And unto Eber were born two sons: the name of one [was] Peleg; for in his days was the earth divided" (Genesis 10:25). Others believe that these words refer to the people of the earth being divided into different language groups during the confusion of the languages at the Tower of Babel. "From thence did the Lord scatter them abroad upon the face of all the earth" (Genesis 11:9).]

---

[599] Chris Raymond, "Scientists Report Finding Fossils of Dinosaurs in Antarctica's Interior," *Chronicle of Higher Education*, March 20, 1991, p. A11.

Chapter 26
Water, Water Everywhere

While ignoring supporting evidence to the contrary, Darwin wrote, "We may feel certain . . . that no cataclysm has desolated the whole world."[600]

> Warren L. Johns, JD: This rash, arbitrary pronouncement came from a [person] lacking credentials as hydrologist, paleontologist, or geologist. Nor did he claim to have explored personally all dry land continents.[601]

Genesis tells us not only that God created life "in the beginning," but also that He judged His creation and destroyed most of it by a global Flood. It tells us that only Noah and his family chose to enter the "life boat"—Noah's huge Ark that God had instructed Noah to build to save whosoever would believe in His rescue plan [see also John 3:16]. And a representative number of animals were saved from the Flood. Rejection of this story as a myth is a major component of the creation/evolution controversy. What are the facts?

We have focused on scientific evidence of the Flood. However, there is also much rich data from the study of ancient cultures; we can analyze these cultures for additional knowledge that will contribute to informed decisions on this issue.

> Steven D. Peet in *American Antiquarian:* Today there are 270 flood legends that have been identified around the world in different countries and cultures. . . . There are many descriptions of the remarkable event [the Genesis Flood]. Some of these have come from Greek historians, some from the Babylonian records, others from the cuneiform tablets, and still others from the mythology and traditions of different nations, so that we may say that no event has occurred either in ancient or modern times about which there is better evidence or more numerous records, than this very one which is so beautifully but briefly described in the sacred Scriptures. It is one of the events which seems to be familiar to . . . nations—in Australia, in India, in China, in Scandinavia, . . . in the various parts of America"[602] and around the world.

> Donald Wesley Patten: Von Humboldt, one of the outstanding geographers of the early 19th century . . . [concluded] (1) that the traditions of the Flood were global in geographical distribution, and (2) that the more primitive the culture, the more vivid or emphatic the traditions of an ancient Flood were apt to be.[603]

> Colin W. Mitchell, PhD: There is no tradition so widespread among all the peoples of the world. Strickling[604] reported that anthropologists have collected fifty-nine Flood legends from the aborigines of North America, forty-six from Central and South America extending from Alaska to Tierra del Fuego, thirty-one from Europe ranging from Greece to Ireland and Iceland, seventeen from the Middle East, twenty-three from Asia, and thirty-seven from the South Sea Islands, Australia and New Zealand. . . . As the tribes migrated farther and farther from Ararat, the stories become more and more distorted. This substantiates the view that they have a *common origin* and are *not* exaggerated tales of *local catastrophes.*[605]

These 200-plus Flood legends generally agree on three main features:

1. That water destroyed all the human race and almost all other living things. [Some water-creatures may have survived the uniquely stormy seas.]

---

[600] Charles Darwin, "Recapitulation and Conclusion," *On the Origin of Species by Means of Natural Selection or the Preservation of Favoured Races in the Struggle for Life,* 150th Anniversary Edition, Bridge Logos Foundation, Alachua, FL, 2009, p. 275.
[601] Warren L. Johns, JD, "Mother Earth's Facelift: Global Gully-Washer," *Genesis File,* www.GenesisFile.com, Lightning Source, LaVergne, TN, 2010, p. 181.
[602] Steven D. Peet, "The Story of the Deluge," *American Antiquarian,* Vol. 27, No. 4, July-August 1905, p. 203.
[603] Donald Wesley Patten, "Astral Catastrophism in Ancient Literature," *The Biblical Flood and the Ice Epoch,* Pacific Meridian Publishing Co., Seattle, WA, 1966, p. 166.
[604] J. A. Strickling, "A statistical analysis of Flood legends," *Creation Research Society Quarterly,* 9 (3) 1972, pp. 152-155.
[605] G. J. Kean, *Creation Rediscovered,* Credis Pty Limited, PO Box 451, Doncaster, Australia, 1991, p. 58; as quoted by Colin Mitchell, PhD, "Evidences for human antiquity," *The Case for Creationism,* Autumn House Limited, Alma Park, Grantham, Lincolnshire, England, 1995, p. 160, (emphasis added).

2. That an ark or boat provided a means of escape.

3. That a family was preserved to perpetuate the human race.

A wide variety of names are given to the man, but Noh in the Sudan, Nu-u in Hawaii, and Nu-Wah in China, are similar to the biblical Noah. . . . The written character for "ship" in the Chinese language is made up of a boat with eight mouths, showing that the first ship carried eight people.[606] [Noah and his wife and his three sons and their three wives.]

Jesus' disciple the Apostle Peter, a New Testament prophet, predicted in his second letter that there would be those living "in the last days," just before Jesus' return to earth (II Peter 3:3-13), who discount the evidences that there was a worldwide Flood:

Knowing this first, that there shall come in the last days scoffers . . . saying, Where is the promise of his [Jesus' promised Second] coming? For since the fathers fell asleep, all things continue as *they were* from the beginning of the creation. For this they willingly are ignorant of, that by the word of God the heavens were of old, and the earth standing out of the water and in the water: Whereby the world that then was, being overflowed with water, perished (II Peter 3:3-6 KJV; "was deluged and destroyed" verse 6, NIV).

Rather than a global Flood, the words "all things continue as they were from the beginning" (II Peter 3:4), foretell the theory of *uniformitarianism*—the belief that "the present is the key to the past." To be more specific, uniformitarianism states that slow geological processes observed today, experienced over millions of years, account for what we see in the Grand Canyon. On the other hand, *catastrophism* sees these same geological evidences as having taken place rapidly, during a catastrophe, such as the biblical Flood.

Why is the Flood controversial? The countless Bible verses that speak about Creation also tells us that the Creator is the true God. (Jeremiah 10:10-12). Because pre-flood humans had chosen to become so wicked, so malicious and violent, only evil continually (Genesis 6:5-17), He decided the kindest thing to do was to do a start over for those who did not deserve to live in such an evil society. Anyone was welcome on the Ark who, like Noah, wanted to know and be like their gracious Creator.

First, God mercifully warned all that a flood was coming and said that all who were disobeying Him, in rebellion against Him, if they believed His Spirit's warning message through Noah (2 Peter 2:5 and 1 Peter 3:18-20), were welcome to join Noah and his family on the Ark. None did. All perished in the Flood. And it deposited millions of fossils in water-laid strata around the world. But the Flood let God start over with those He hoped would reveal His plan to recreate planet earth and whoever entrusts their lives to Him (John 3:16).

Donald W. Patten: These abundant fossil finds demanded an explanation. Many, such as Cuvier, felt that some sort of gigantic, watery cataclysm or cataclysms had indeed engulfed the past. This possibility immediately suggested the biblical Flood. Yet others cast about for an alternative explanation. Modern humanists, increasingly anti-Genesis in outlook, were growing in numbers and in positions of importance, especially in academic circles.

. . . The doctrine of uniformitarianism was born and nurtured from the mother principle of *humanism*, as was the daughter principle of evolution. . . . Thus, our century has received an almost pure heritage of uniformitarianism, and as a consequence, is leaving a legacy of anti-spiritual humanism in various forms.

Modern uniformitarianism was conceived 200 years ago [writes Patten in 1966], and about 100 years ago it became the dominant theory of earth history: . . . *No great, sudden cataclysms ever occurred*. But is this theory defensible in light of new evidence? Was it ever really defensible in light of former evidence? . . .

There are abundant evidences of a watery, global cataclysm—evidences which are not easily refuted. They are so universal, so astounding, and so interrelated that they require re-examination.[607]

---

[606] R. Noorbergen, *Secrets of the Lost Races*, New English Library, Times-Mirror, London, 1978, p. 6; as quoted by Colin Mitchell, PhD, "Evidences for human antiquity," *The Case for Creationism*, Autumn House Limited, Alma Park, Lincolnshire, England, 1995, pp. 160-161.

[607] Donald Wesley Patten, "Global flood or local flood," *The Biblical Flood and the Ice Epoch*, Pacific Meridian Publishing Company,

Some geologists concede that there appear to be many cataclysmic flood events in the fossil record, but say that these were likely local events.

> David C. Read, JD: There are a number of serious problems with the local flood theory. . . . [It] cannot be reconciled with a straightforward reading of Scripture. Genesis tells of a Flood that covered "all the high mountains under the entire heavens." Moses was at pains to describe, in the most categorical language he could muster, a universal Flood with universal consequences. . . .

Some theologians, wishing to be more in harmony with secular science, also try to allow the biblical flood to be local only. However:

> [D. C. Read, continuing from a Biblical perspective]: After examining the Hebrew syntax of [Genesis 7:18-23], theologian Gerhard F. Hasel concluded: "There is hardly any stronger way in Hebrew to emphasize total destruction of 'all existence' of human and animal life on earth than the way it has been expressed. The writer of the Genesis Flood story employed terminology, formulae, and syntactical structures of the type that could not be more emphatic and explicit in expressing his concept of a universal, worldwide Flood."[608]

If the Flood were merely local, it would have been absurd for Noah to spend 120 years constructing a giant, 150-yards long barge to save him, his family, and the animals, and it would have been absurd for God to order him to do so. Noah and his family could simply have moved out of the way. In fact, [with 120 years warning] anyone could have moved out of the way, defeating the purpose of the Flood.[609]

And from a scientific perspective, the abundance of evidence—including the extinction of species, changes in the climates world wide, and the fossil record in the geologic column—demonstrates the results of major catastrophes. Remains of coral reefs have been found in the Canadian Arctic. Fig trees and magnolias have been discovered in Greenland. Plant remains have been found within a few hundred miles of the South Pole.[610]

How was coal formed?

Geologists usually ascribe coal formation to the slow burial of peat deposits on land. However, the necessary amounts of starting material for conversion to coal seem unrealistic for land deposition. A more likely origin is rafting of organic material (vegetation), and burial under flood conditions.[611]

> Erich A. von Fange, PhD: One of the sacred tenets of geology is that vegetable matter covered in swampy areas converts into peat. So far, so good. As time passes, and under the right conditions, peat changes into lignite, and later into bituminous coal. Under great pressure the bituminous coal is metamorphosed into anthracite coal.[612]

> Henry M. Morris, PhD: There are many existing peat bogs, of course, but none of these grade vertically downward into a series of coal seams. The uniformitarian *peat-bog theory* of coal seam origin seems quite unrelated to the real world.[613]

> Harold G. Coffin, PhD, Robert H. Brown, PhD, L. James Gibson, PhD: The thickness of peat supposedly needed to produce one foot of coal depends on a number of factors, such as the type of peat, the amount of water in the vegetable matter, and the type of coal that results. The scientific literature on coal gives figures ranging from a few feet to as many as 20 feet of peat for the formation of one foot of coal. To produce thick coal seams, this compression factor requires the accumulation of a great thickness of vegetable debris, far more than we can find in modern situations.[614]

> Colin Mitchell, PhD: The thickness of coal seams is too great for them to be derived from terrestrial peat. A seam 10 metres [32.8 feet] thick would,

---

Seattle, WA, 1966, pp. 2-3, (emphasis added).
[608]David C. Read, JD, *Dinosaurs*, Clarion Call Books, Keene, TX, 2009, p. 98, see also: Gerhard F. Hasel, "The Biblical View of the Extent of the Flood," *Origins* 2(2): Geoscience Research Institute, Loma Linda, CA, 1975, p. 78.
[609]David C. Read, JD, "Can a Christian Follow Lyell?" *Dinosaurs*, Clarion Call Books, Keene, TX, 2009, pp. 97-99 (emphasis added).
[610]Erich A. von Fange, PhD, "Ancient Plant Oddities and Myster-ies," *In Search of the Genesis World*, Concordia Publishing House, St. Louis, MO, 2006, p. 236,
[611]Colin Mitchell, PhD, "Geological arguments for a young earth," *The Case for Creationism*, Autumn House Limited, Alma Park, Grantham, Lincs, England, 1994, p. 86.
[612]Erich A. von Fange, PhD, "Ancient Plant Oddities and Mysteries," *In Search of the Genesis World*, Concordia Publishing House, St. Louis, MO, 2006, p. 228; Whitcomb and Morris, *Genesis Flood*, pp. 163, 278; Otto Stutzer, Geology of Coal, trans. A. Noe, University of Chicago Press, Chicago, IL, 1940, passim; Philip LeRiche, "Scientific Proofs of a Universal Deluge," *JVI* Vol. 61, 1925, p. 95.
[613]Henry M. Morris, PhD, "Uniformitarianism or Catastrophism?" *Scientific Creationism*, CLP Publishers, San Diego CA, 1974, p. 107.
[614]Harold G. Coffin, PhD, Robert H. Brown, PhD, L. James Gibson, PhD, *Origin by Design*, Review and Herald Publishing Association, Hagerstown, MD, 2005, p. 200.

for instance, require 100 metres [328 feet] of peat. This is nowhere to be found on earth. In the Permo-Carboniferous of India, the Barakar Series of Damuda Series, overlying the Talchir Boulder Bed, includes numerous coal seams, some up to 100 feet thick. Some coal beds in Australia are even thicker—up to 240 metres [787 feet], impossibly thick for an accumulation of peat. Also, coal seams often cover a wider geographical area than could reasonably have been occupied by vegetated swamps. Drift accumulation [of plant debris] under flood conditions seems the only reasonable explanation.[615]

Harold G. Coffin, PhD, Robert H. Brown, PhD, L. James Gibson, PhD: The drift, or transport, idea of coal originated with Flood geologists. Modern creationists would generally consider that all or most coal originated this way. The major evidences for this are as follows: Marine fossils often accompany coal. They may consist of fish or sea creatures mixed directly with carbonaceous material or as marine fossils in thin shale partings (thin layers) in the midst of [and separating thicker] coal seams. Marine organisms often appear in the shales or sandstones directly above or below the coal. Comparison with modern bogs or marshes will not explain the great depths of carbon accumulation, especially in the Cretaceous-Eocene beds.[616]

How long does it take to form oil?

Does it really take millions of years? A 1988 textbook tells us that it takes millions of years of heat and pressure to make oil and natural gas from organisms that once lived in the sea.[617] However, in 1971 scientists learned a process that can make oil in the laboratory in 20 minutes.[618] And Argonne National Laboratory has shown that coal can be formed by heating wood, clay and water at 150°C for 36 weeks.[619]

Duane Gish: It has now been demonstrated that cellulosic (plant derived) material such as garbage or manure, can be converted into a good grade of petroleum in twenty minutes. . . . The experiments of Bureau of Mines scientists in which cow manure was converted to petroleum are described in *Chemical and Engineering News,* May 29, 1972, page 14. . . . The manure was heated at 716° F, at 2000 to 5000 pounds per square inch for twenty minutes in the presence of carbon monoxide and steam. The product was a heavy oil of excellent heating quality. The yield was about three barrels of oil per ton of manure.[620]

A factory in Texas claimed that it can turn 600 million tons of turkey guts and other waste into 4 billion barrels of light Texas crude [oil] each year.[621]

William D. Stansfield, PhD: Some geologists find it difficult to understand how the great pressures found in some oil wells could be retained over millions of years. [Though some geologists find it puzzling, to Creationists the "problem" is not a problem: it is] evidence that oil was formed [not millions of years ago, but] less than 10,000 years ago.[622]

Stephen Grocott, PhD: What do you see in the geology of the world? Massive sedimentary deposits. How did they form? Primarily through moving water. Belief that these formed through gradual erosion over millions of years does not fit with common sense or good science.[623]

Could widespread deposition result from merely localized floods?

Worldwide sedimentary strata laid by water

Clyde L. Webster, Jr., PhD: "The search for oil has produced extensive underground mapping that reveals the massive size of some of these sedimentary [oil] deposits. As more and more oil exploration takes place around the world, a picture of the earth's rock layers becomes clearer.

---

[615] Colin Mitchell, PhD, "Geological arguments for a young earth," *The Case for Creationism*, Autumn House Limited, Alma Park, Grantham, Lincs, NG31 9SL, England, 1994, pp. 86-87, see also: S. E. Hollingsworth, "The Climate Factor in the Geological Record," *Quarterly Journal*, Geological Society of London, Vol. 118 (March 1962), p.13.
[616] Harold G. Coffin, PhD; Robert H. Brown, PhD; L. James Gibson, PhD, "The Autochthonous Origin of Coal," *Origin by Design*, Review and Herald Publishing Association, Hagerstown, MD, 2005, pp. 198-200.
[617] *Holt General Science*, 1988, p. 294; Leigh Price, *Journal of Petroleum Geology*, 1983, p. 32.
[618] Hadden R. Appell, *Converting Organic Wastes to Oil*, US Department of Interior, Bureau of Mines, 1971.
[619] See *Organic Chemistry*, Vol. 6, 1984, pp. 463-471.

[620] Duane Gish, quoted in John C. Whitcomb, ThD, *The World That Perished*, Baker Books, Grand Rapids, MI, 1991, p. 124.
[621] Brad Lemley, "Anything Into Oil," *Discover*, Vol. 24, No. 5, May 2003, p. 50.
[622] William D. Stansfield, PhD, "Evidence of Evolution—Paleontology and Biogeography," *Science of Evolution*, MacMillan Publishing Co., NY; Collier MacMillan Publishers, London, 1977, p. 82.
[623] Edited by John Ashton, PhD, Stephen Grocott, PhD, "Stephen Grocott," *In Six Days, Why Fifty Scientists Choose to Believe in Creation*, Master Books, Inc., Green Forest, AR, 2000, p. 150.

[Of their various unique features] the most unique feature of these stratified sediments is not their thickness but their similarities and their wide range of distribution. Many strata can be traced over thousands of square miles and from continent to continent. Some extend over much of the world!

Strata identical to the English "White Cliffs of Dover" can be found in France, Germany, Scandinavia, Poland, Bulgaria, and Russia. Similar chalk deposits of the same geologic age and characteristics can also be found in Texas, Arkansas, Mississippi and Alabama in North America. And, last but not least, similar chalk deposits can be found also in Western Australia! All of these deposits are resting on the same type of glauconitic sandstone![624]

These chalk deposits are not the only deposits that are widespread. Other widespread deposits include various types of sandstone, limestone, and shale. *Surely very similar conditions had to exist at nearly the same time in all of these places for these similar deposits to form!*[625]

John R. Baumgardner, PhD: Just as there has been glaring scientific fraud in things biological for the past century, there has been a similar fraud in things geological. The error, in a word, is uniformitarianism. This outlook assumes and asserts that the earth's past can be correctly understood purely in terms of present day processes acting at more or less present-day rates. Just as materialist biologists have erroneously assumed that material processes can give rise to life in all its diversity, materialist geologists have assumed that the present can fully account for the earth's past. In so doing, they have been forced to ignore and suppress abundant evidence to the contrary that the planet has suffered major catastrophe on a global scale.

Only in the past two decades has the silence concerning global catastrophism in the geological record begun to be broken. Only in the past 10-15 years [published in 2000] has the reality of global mass extinction events in the record become widely known outside the paleontology community. . . . But the huge horizontal extent of Paleozoic and Mesozoic sedimentary formations and their internal evidence of high energy transport represent stunning testimony for global catastrophic processes far beyond anything yet considered in the geological literature. Field evidence indicates catastrophic processes were responsible for most, if not all, of this portion of the geological record. The proposition that present-day geological processes are representative of those which produced the Paleozoic and Mesozoic formations is utter folly. . . .

That the catastrophe was global in extent is clear from the extreme horizontal extent and continuity of the continental sedimentary [water-laid] deposits. That there was a single large catastrophe and not many smaller ones with long gaps in between is implied by the lack of erosional channels, soil horizons, and dissolution structures at the interfaces between successive strata. The excellent exposures of the Paleozoic record in the Grand Canyon provides superb examples of this vertical continuity with little or no physical evidence of time gaps between strata. . . .[626]

---

[624] D. V. Ager, *The Nature of the Stratigraphic Record*, 2nd ed., MacMillan Press, London, 1983.

[625] Clyde L. Webster, Jr., PhD, "Fossils and the Flood," *A Scientist's Perspective on Creation and the Flood*, Geoscience Research Institute, Loma Linda, CA, 1995, pp. 15-16, (emphasis in the original).

[626] Stephen Grocott, PhD, "John R. Baumgardner," *In Six Days, Why 50 Scientists Choose to Believe in Creation* [Baumgardner is one of the 50 scientists].

Chapter 27
The Worldwide Distribution of Coal and Other Water-laid Strata

We will now examine the evidence which further suggests that during the Flood, plant life buried under great pressure formed the massive layers of coal we see around the planet. Most have never heard this evidence for a biblical Flood.

> Harold G. Coffin, PhD: Perhaps the most convincing evidence of a universal flood is to be found in the extensive coal reserves of the world. No processes going on today approach in magnitude the action necessary to account for this phenomenon.... The condition required to satisfy the picture obtained from the examination of most coal deposits is that of extensive forests and vegetable debris buried suddenly.[627]

The thickness of certain coal beds is well known. Thirty feet is not uncommon.... In the western United States, seams sixty to ninety feet in thickness are known. Like the vertical extent, the horizontal extent is also tremendous. The Pittsburgh bed covers parts of Pennsylvania, Ohio, and West Virginia, an area of 2,100 square miles, and averages about seven feet thick.[628] The Appalachian coal basin extends over some 70,000 square miles.[629] The extent of mineable coal runs into the thousands of millions of tons. Custer County, Montana, is said to have 1.5 billion tons of mineable lignite [brown coal]. The Latrobe Valley in Australia is estimated to be able to yield 70 billion tons.[630]

The concept of a *global deluge* that eroded out the forests and plant cover of the pre-flood world, collected [them] in great mats of drifting debris, and eventually dropped [them] on the emerging land or on the sea bottom is the most reasonable answer to this problem of the great extent and uniform thickness of coal beds.[631]

Scientists believe that the Powder River Basin in Montana and Wyoming contains 56 billion tons of low-sulfur coal. They estimate that to be enough to power the United States at once-current consumption rates for about 50 years. Some of it is packed in immense layers more than 200 feet thick, hundreds of miles long, and 50 miles wide.[632]

Many who read for the first time this evidence find it compelling. But it leads to some good questions. How do we explain huge canyons like the Grand Canyon? Where does the ice age fit into the Bible?

And why did God destroy the whole world with a Flood (except the few who accepted the escape plan He offered—Noah's Ark)? Why? A hard question. He gives no answer except His anguish over man's selfish, violent inhumanity to his fellow humans—so vile, so demonic that He had to destroy this evil. He has left lasting evidence around the world, which we can plainly see today in the fossil-bearing, water-laid strata of the geologic column. The results are here for us to study. *Because of mountain uplift*, scientists have found evidence that objects on top of mountains were once below sea level:

> *Science Digest:* A fossil fish has been unearthed 17,000 feet up the slopes of the Andes, and marine fossil limestone has been spotted in the Himalayas at an altitude of 20,000 feet![633]

---

[627]Harold G. Coffin, PhD, "Fire in the Earth," *Creation—Accident or Design?* Review and Herald Publishing Association, Washington, DC, 1969, pp. 75-76.
[628]W. A. Tarr, *Introduction to Economic Geology*, McGraw-Hill Book Co., NY, 1930, p. 664.
[629]Otto Stutzer, Adolph C. Noe, *Geology of Coal*, The University of Chicago Press, Chicago, 1940, p. 461.
[630]H. Herman, *Brown Coal*, The State Electricity Commission of Victoria, 1952, p. 88.
[631]Harold G. Coffin, PhD: "Evidences of the Flood—II," *Creation—Accident or Design*, Review and Herald Publishing Association, Washington, DC, 1969, pp. 75-77.
[632]Doug McInnis, "Powder River Coal," *Earth* magazine, Vol. 2, Issue 3, May, 1993, p. 54.
[633]Lyall Watson, "The Water People," *Science Digest*, 90[5]44, May, as cited by Eric Lyons and Kyle Butt, *The Dinosaur Delusion*, Montgomery, AL: Apologetics Press, Inc., 2008, p. 140.

Did it rain enough to cover the 29,000-foot Mount Everest?

Mount Everest was probably not there at the time of Noah's Flood. *If the earth were much flatter then, it would have required much less water to cover the highest mountains.* Both Flood and non-Flood geologists agree that Mount Everest and many other mountains represent uplift after the deposition of their sedimentary layers. We should not use present topography to estimate the volume of water required for a worldwide Flood.[634]

The Himalayas, including Mt. Everest, are

[634] Ariel A. Roth, PhD, "Catastrophes: The Big Ones," *Origins: Linking Science and Scripture*, Review and Herald Publishing Association, Hagerstown, MD, 1998, p. 208.

The Coconino sandstone exposed in the Grand Canyon (the most noticeable, light-colored layer near the top) continues across Northern Arizona and has an estimated volume of 10,000 cubic miles. Such widespread, erosionless layers imply not long ages for evolution, they imply more recent creation.

said to be recent, and these mountains reportedly are still rising.[635]

Creationists believe that mountain building began in the earth upheavals that occurred during the Flood, and have continued after the Flood, though much more slowly (like dying tremors of a major earthquake).

> *Modern Earth Science:* Climbers reaching the top of Mount Everest plant their victory flags over the remains of animals that once lived in the sea. Rocks at the top of the highest mountain in the world are made of limestone from ancient sea-floor sediments. Material that was once far below the sea level had been thrust up to an altitude of 9,524 m (29,028 feet).[636]

If the biblical records of the Flood are accurate, and if it were as violent as depicted in Genesis, then physical evidence to support such a claim must exist worldwide. Such evidence does exist. It includes *widespread sedimentary* deposits, and massive burial and preservation of plants and animals *[coal and oil]*[637] that we don't see happening today:

> John R. Baumgardner, PhD: "This sort of dramatic global-scale catastrophism documented in the Paleozoic, Mesozoic, and much of the Cenozoic sediments implies a distinctively different interpretation of the associated fossil record. Instead of representing an evolutionary sequence, the record reveals a successive destruction of ecological habitat in the global tectonic and hydrologic catastrophe. This understanding readily explains why Darwinian intermediate types are systematically absent from the geological record—*the fossil record documents a brief and intense global destruction of life, and not a long evolutionary history!* The types of plants and animals preserved as fossils were the forms of life that existed on the earth prior to the catastrophe. The long span of time in the intermediate forms of life that the evolutionist imagines in his mind are simply illusions.[638]

The profound witness of paraconformities

What is the geologic column? Is it the evolutionary explanation of sedimentary layers of rock? It is said to be the historical geological record of living things, layer by layer, dating back to the origin of life on earth. To account for the presumed absence of a Divine Creator, the ages of each layer must be millions of years old.

If the worldwide geologic column said the bottom layer is A, and above that is B, and above that is C, then one would expect to find layer B between layers A and C in a certain region. If the life forms fossilized in layer A are, let's say, thought to be 75 million years old, and those in B 100 million, and in C 50 million, if B is entirely missing, then one would expect to see evidence for about 100 million years of activities that eroded away layer B. But if there is no erosion, if like two flat pancakes C sits on A, then how did 100 million years eat B and not leave a crumb to account for the theft?

> Or as Ariel A. Roth, PhD says: What is a paraconformity? It is a substantial [time] gap [between two of] the sedimentary layers [of the geologic column], where the surfaces above and below are flat. You can tell you have a [time] gap when a part of the geologic column [a sedimentary layer] is missing. What is the significance of paraconformities? Paraconformities challenge the geologic time scale. The usual lack of evidence at the underlayer for the long ages postulated for the gaps, especially the lack of erosion, suggests that the long geologic ages never occurred.[639]

According to Dr. Roth, "Geologists usually identify paraconformities by noting a gap in the fossil sequence. There are places where paraconformities are found, and there are other places that *have* the missing layers and fossil sequences of the geologic column that are not present at the gaps. For instance, the Ordovician and Silurian periods are missing in the Grand Canyon but are found in sequence between the Cambrian and Devonian in Wales."[640]

---

[635]Erich A. von Fange, PhD, "The Great Flood," *Noah to Abram: The Turbulent Years*, Living Word Services, Syracuse, IN, 1994, p. 62.
[636]William L. Ramsey, Clifford R. Phillips, Frank M. Watenpaugh, "Movement of the earth's crust," *Modern Earth Science*, Holt, Rinehart and Winston, Publishers, New York, Toronto, Mexico City, London, Sydney, Tokyo, 1983, p. 207.
[637]Harold G. Coffin, PhD, Robert H. Brown, PhD, L. James Gibson, PhD, "The Autochthonous Origin of Coal," *Origin by Design*, Review and Herald Publishing Association, Hagerstown, MD, 2005, pp. 198-200.
[638]Edited by John Ashton, PhD, John R. Baumgardner, PhD,

"John R. Baumgardner, PhD," *In Six Days, Why Fifty Scientists Choose to Believe in Creation*, Master Books, Inc., Green Forest, AR, 2000, pp. 230-233, (emphasis added).
[639]Ariel A. Roth, PhD, "Flat Gaps in the Rock Layers," www.scienceandscriptures.com
[640]Ariel A. Roth, PhD, "Implications of paraconformities," *Geoscience Reports*, No. 36, fall 2003, Number 36. p. 2.

Dr. Roth has discussed paraconformities at the University of California, Riverside, the University of Utrecht, and at the Goddard Space flight Center. The following question is typical: "Could these be flat areas where there is just no erosion or deposition? Answer: Not unless we can suspend the world weather pattern for millions of years over hundreds of thousands of square miles. On our restless planet, over the millions of years postulated, there is either erosion or deposition."[641]

According to Dr. Roth, these gaps at the Grand Canyon (which are common over the whole world) represent from 6 million years of missing sediment (according to the traditional geological timescale), to 600 million years. Question: Why are they flat? Are we to believe that there was no weathering or erosion for 6 million to 600 million years? According to Dr. Roth, based on conservative scientific measurements of the rates of erosion (one foot per thousand years), instead of being flat, these paraconformities in the Grand Canyon should exhibit from 1,000 to 6,000 feet of erosion. This missing erosion profoundly challenges long geological ages.[642]

> Dr. Roth: An outstanding feature of erosion is the highly irregular surface (topography) it creates as streams and rivers keep cutting deeper gullies, canyons, and valleys into the landscape. . . . These flat gaps are so common that they pretty much challenge the validity of the whole geologic time scale. Over the many millions of years postulated for these gaps, you would expect pronounced irregular erosion, and the gaps should not at all be flat.[643]

Because paraconformities are so abundant over the earth, they represent an important component for the interpretation of earth history. Paraconformities pose a serious challenge to the standard geologic time scale, radiometric dating, and interpretation of extended time for the development of life on earth. All of this raises a more fundamental question, namely: Do the assumed time gaps between sedimentary layers show the effects of time, or do they suggest rapid deposition as would be expected if they were laid down during the catastrophic flood described in the book of Genesis?[644]

Paraconformities, or flat gaps, are what would be expected from the rapid deposition of sediments during the Genesis Flood.[645]

And what did a traditional scientist have to say about this phenomenon?

> N. D. Newel, PhD: The origin of paraconformities is uncertain, and I certainly do not have a simple solution to this problem.[646]

The presuppositions of materialist/naturalist evolution today are that there is no spiritual/Supernatural Designer Creator God who gave humans the account of Creation; and of a global flood during Noah's day. Paraconformities—layers of sediments laid down quickly as flood waters ebbed and flowed and receded—are therefore given various alternate explanations. Because these layers "couldn't" be from a global flood.

Or could they? There is much in the scientific literature—on paraconformities (and on many other topics)—that is brought together in these pages. Take a fresh look; I believe the muddy waters will clear.

And what was Dr. Roth's conclusion?

> Ariel A. Roth, PhD: One does not have to give up his scientific integrity in order to believe the biblical account of beginnings. While the scientific literature overwhelmingly rejects the biblical account, there is good data that supports it. Instead of focusing too much on the question: how should we live with the dissonance between science and Bible; we should be focusing more on: how can we live with the dissonance within science?[647]

So, under our feet there is worldwide evidence for a global flood: not only the *mega-layers of coal*

---

[641] Ariel A. Roth, PhD, "Implications of paraconformities," *Geoscience Reports*, No. 36, fall 2003, Number 36. p. 5.
[642] Ariel A. Roth, PhD, "Flat Gaps in the Rock Layers," www.scienceandscriptures.com
[643] Ariel A. Roth, PhD, Jonathan Sarfati interviews biologist and geologist Ariel Roth, "Millions of years are missing," *Busting Myths, 30 PhD Scientists Who Believe the Bible and its Account of Origins*, Creation Book Publications, Powder Springs, Georgia, 2015, pp. 143-147.
[644] Ariel A. Roth, PhD, "Implications of paraconformities," *Geoscience Reports*, No. 36, fall 2003, Number 36. p. 9.
[645] Ariel A. Roth, PhD, "Flat Gaps in the Rock Layers," www.scienceandscriptures.com
[646] N. D. Newel, PhD, "Paraconformities, In: Teichert C. Yochelson F. L. editors: *Essays in paleontology and stratigraphy*, Department of Geology, University of Kansas, Special publication 2, University of Kansas Press, 1967, p. 164.
[647] Ariel A. Roth, PhD, "Flat Gaps in the Rock Layers," www.scienceandscriptures.com

from floating-then-settling rafts of forests and other plant debris—the weight being enough pressure to create coal; and the "missing" *sedimentary layers* (such as the *erosionless* paraconformities in, for example, the Grand Canyon); but there are also *mega-layers of sediments* (shales, sandstone, sedimentary deposits filled with land fossils—some formation so wide they are even joined across continents).

Harold W. Clark, MS: The Colorado Plateau covers much of Utah and bordering states where we find extensive cross-bedded shales and sandstones. Some of these beds are from ten to a hundred feet thick, but spread out quite regularly over a vast area of 100,000 square miles or more. How such deposits could have taken place without violent and widespread waves of water is impossible to imagine.[648]

Ariel A. Roth, PhD: In the United States, the Dakota Formation of the western United States, with an average thickness of 30 meters [about 100 feet], covers some 815,000 square kilometers [about 315,000 square miles]. The widespread nature of special sedimentary deposits with land-derived fossils offers evidence of a kind of catastrophic activity on the continents for which we have no contemporary analogs.[649]

H. E. Gregory: An outstanding example is the Triassic fossil-wood-bearing Shinarump conglomerate, a member of the Chinle Formation found in the southwestern United States. This conglomerate, which occasionally grades into a coarse sandstone, usually has a thickness of less than 30 meters [about 100 feet], but spreads as an almost continuous unit over nearly 250,000 square kilometers [about 96,500 square miles].[650]

Harold G. Coffin, PhD: Many indications of a major flooding of the world by water remain. Any serious study of geological literature, or personal observations in the field, turns up facts that are difficult to interpret as having been produced by any other means.

Sedimentary rocks constitute about 75 percent of the rocks exposed on the surface of the earth. Only in recent years has the depth of sedimentation in certain areas been shown by drillings or soundings. In North America some of the most extensive and deep sediments are located in the Midwest plains, the Colorado Plateau, the California coastal plains, the Gulf of Mexico coast, and the northern Rocky Mountains. Other areas of the world have similarly deep and numerous sedimentary deposits, with India presenting what may be the deepest known sedimentary basin—60,000 feet or more.

Many of these deposits are so deep that one of the most difficult problems facing the geologist is that of determining their source. Such processes as gradual submergence and the slow accumulations of sediments by erosion seem inadequate to account for the great quantities of water-and-wind deposited materials. . . . The layered nature of earth's crust is one of the most evident observations. Beds of rock and sediment lying one above another often consist of entirely different materials—limestone above sandstone, shale overlaying conglomerate, for example.[651]

Examples of thick sediments are seen the world over. The beautiful faultblock and overthrusted mountains of the northern Rockies show exposures of more than 4,000 feet of sedimentary strata, some of which give clear evidence of having been deposited rapidly. The well-preserved ripple marks, the excellent preservation of even delicate fossils, the burial of vast numbers of trilobites and other invertebrates that frequently show no sign of decay or disintegration, are some features that indicate that these sediments did not form by a process of gradual accumulation over millions of years.[652]

Scott M. Huse, PhD: Great sheets of fresh-water-laid sand that change only slightly in thickness and texture, mile after mile, cannot be observed forming anywhere in the world today. Flowing rivers, meandering streams, and trickling rivulets could not have produced these great deposits of sand.

Polystratic trees are fossil trees that extend through several layers of strata, often twenty feet or more in length. There is no doubt that this type of fossil

---

[648] Harold W. Clark, "The Delusion of Uniformity," *Fossils, Flood, and Fire*, Outdoor Pictures, 1968, p. 32
[649] Ariel A. Roth, PhD, "Geologic Evidence for a Worldwide Flood," *Origins: Linking Science and Scripture*, Review and Herald Publishing Association, Hagerstown, MD, 1998, p. 218.
[650] H. E. Gregory, "Geology and Geography of the Zion Park Region, Utah and Arizona," *U.S. Geological Survey Professional Paper*, 1950, Vol. 220, p. 65.
[651] Harold G. Coffin, PhD: "Evidences of the Flood—I," *Creation—Accident or Design*, Review and Herald Publishing Association, Washington, DC, 1969, pp. 64-65.
[652] Harold G. Coffin, PhD: "Evidences of the Flood—I," *Creation—Accident or Design*, Review and Herald Publishing Association, Washington, DC, 1969, pp. 67-68.

was formed relatively quickly; otherwise it would have decomposed while waiting for the strata to slowly accumulate around it.... Obviously the more reasonable interpretation calls for the simultaneous transportation and deposition of the trees and their surrounding sediments.... Once again, the evidence supports the biblical catastrophism of the Genesis Flood.[653]

According to Paul D. Ackerman, PhD, "The earth's sedimentary-rock layers are not a testimony of life's long struggle upward, but a witness to sudden terror."[654]

It is reasonable to conclude that by far most of them were buried during Noah's Flood.

> John C. Whitcomb, ThD, and Henry M. Morris, PhD: "Some kind of catastrophic action is nearly always necessary for the burial and preservation of fossils.... Nothing comparable to the tremendous fossiliferous beds of fish, mammals, reptiles, etc. that are found in many places around the world is being formed today.[655]

Decades—not millions of years: A time miscalculation

For years it was claimed that the fastest growing stalactites in Carlsbad Caverns took 250 million years to grow. This is based on unreliable assumptions. In 1975, the *Bulletin of the National Speleological Society* declared that the fastest stalactites can grow is 2.5 inches every 1,000 years.[656]

Yet, some 50-inch stalactites grew under the Lincoln Memorial in 40 years. *National Geographic* pictured a bat embedded in flow-stone before it could rot.[657]

The next statement adds a personal touch:

> Keith H. Wanser, PhD: An example of a supposed fact demonstrating an ancient age of the earth is the rate of growth of stalactites and stalagmites in limestone caverns. As a young boy I toured Carlsbad Caverns in New Mexico and remember the tour guide informing us matter-of-factly that the limestone caverns and formations were formed over many millions of years, which did not seem to agree with what I had been taught in Sunday school. A sign above the entrance until 1988 said the caverns were at least 260 million years old. In recent years, the age on the sign was reduced to 7-10 million years, then 2 million years, and now the sign is gone—perhaps as a result of observations that stalactite growth rates of several inches a month are common.[658] ... In May 1998 I observed stalactites longer than six inches growing from the edge of the concrete boarding platform at the Arlington, Virginia, Metro rail station, which was only completed in June 1991 [about ten years earlier].[659]

And the evidence available to the scientific community is not limited to the above recent observations. Notice what I found in *Origin of the World According to Revelation and Science*, by J. W. Dawson, LLD, FRS, FGS, *published in 1887:*

> Professor Hughes [in a lecture before the Royal Institution of London] refers, as a case of rapid deposition of matter akin to stalagmite, to the deposit of travertine in the old Roman aqueduct of the Pont du Gard, near Avignon, where a thickness of fourteen inches seems to have accumulated in about 800 years. Mr. J. Carey has given in *Nature*, December 18, 1873, another instance where a deposit 0.75-inch-thick was formed in fifteen years in a lead mine in Durham. Mr. W. B. Clarke in the same journal gives a case where in a cave in Brixton, known as Poole's Hole, a deposit one eighth of an inch in thickness was formed in six months. Such examples show how unsafe it is to reason as to the rate of deposit in past times, and when [climatic] and local conditions may have been very different from those at present subsisting.[660]

---

[653] Scott M. Huse, PhD, "Geology," *The Collapse of Evolution*, Grand Rapids, Michigan, Baker, 1997, p. 96. See photo, Ian T. Taylor, "Science and Geology," *In the Minds of Men*, TFE Publishing, Minneapolis, MN, 1996, p. 114.
[654] Paul D. Ackerman, PhD, "Back Down to Earth," *It's a Young World After All: Exciting Evidences for Recent Creation*, Baker Book House, Grand Rapids, MI, 1986, p. 84.
[655] John C. Whitcomb, ThD, and Henry M. Morris, PhD, "Modern Geology and the Deluge," *The Genesis Flood*, Presbyterian and Reformed Publishing Company, Phillipsburg, NJ, 1961, p. 203.
[656] *Bulletin of the National Speleological Society*, 1975, Vol. 37, p. 21.
[657] E. Tex Helm, photographer, "Carlsbad Caverns in Color," *National Geographic*, October 1953, p. 442.
[658] G. W. Wolfram, "Carlsbad 'Signs Off,'" *Creation Research Society Quarterly*, Vol. 31, June 1994, p. 34; E. L. Williams, "Rapid Development of Calcium Carbonate Formations," *Creation Research Society Quarterly*, Vol. 24, June 1987, pp. 18-19.
[659] Edited by John Ashton, PhD, Keith H. Wanser, PhD, "Keith H. wanser, PhD," *In Six Days: Why Fifty Scientists Choose to Believe in Creation*, Master Books, Inc., Green Forest, AR, 2000, pp. 103-104.
[660] J. W. Dawson, LLD, FRS, FGS, "Appendix," *Origin of the World according to Revelation and Science*, Dawson Brothers Publishers, 159-161 St. James Street, Montreal, 1977, pp. 388-389.

Chapter 28
Fossil Evidence for a Worldwide Flood?

A broad, worldwide selection from the fossil record offers impressive evidence for a worldwide Flood:

Joe White and Nicholas Comninellis: Remnants of this worldwide Flood can be found all over the earth. Marine crustaceans have been discovered on 12,000-foot mountaintops. . . . Hundreds of dinosaurs have been found buried together with other creatures that *did not share the same habitat*. The Norfolk forest beds in England contain fossils of northern cold-climate animals, tropical warm climate animals, and temperate zone plants all mixed together. . . . How were such diverse creatures transported and buried thousands of miles from their normal environments, at unexplainable elevation, except by a devastating universal Flood?[661]

Another challenge for evolution: What did the animals eat in order to survive through the "millions of years" of evolutionary development? Incomplete ecosystems are found in the geologic column. The fossil record yields evidence that animals existed without any corresponding indication of sufficient plants to provide them nourishment.

The Morrison Formation in the western United States, for example, has been one of the world's richest sources of dinosaur fossils, yet plants are rare, especially in the vicinity of dinosaur remains.[662] Paleontologist Theodore White comments that "although the Morrison plain was an area of reasonably rapid accumulation of sediment, identifiable plant fossils are practically nonexistent."[663]

Question: If dinosaurs lived there for millions of years, what did they eat if plants were so rare? Other investigators have also commented on this lack of plant fossils. One states that the Morrison Formation in Montana "is practically barren of plant fossils throughout most of its sequence."[664]

And others comment that the "absence of evidence for abundant plant life in the form of coal beds and organic-rich clays in much of the Morrison Formation is puzzling."[665]

It appears that the Morrison Formation was not a place where dinosaurs lived, but a Flood-created dinosaur burial ground, and that plants were sorted and transported elsewhere during the Global Deluge of Genesis. Flood geologists believe that these animals were transported here from their customary habitats and that the plants were washed elsewhere.[666]

Another interesting discovery is that well-pre-

---

[661] Joe White and Nicholas Comninellis, "Not As Old As You've Been Told," *Darwin's Demise*, Master Books, Green Forest, AR, 2001, p. 57, (emphasis added).
[662] P. Dodson, A. K. Behrensmeyer, R. T. Bakker, J. S. McIntosh, "Taphonomy and Paleoecology of the Dinosaur Beds of the Jurassic Morrison Formation," *Paleobiology*, 6(2):208-232.
[663] T. White, "The Dinosaur Quarry," *Guidebook to the Geology and Mineral Resources of the Uinta Basin*, Intermountain Association of Geologists, Salt Lake City, UT, pp. 21-28.
[664] R. W. Brown, "Fossil Plants and Jurassic-Cretaceous Boundary in Montana and Alberta," *American Association of Petroleum Geologists Bulletin*, 1946, 30:238-248.
[665] P. Dodson, A. K. Behrensmeyer, R. T. Baker, and J. S. McIntosh, "Taphonomy and Paleoecology of the Dinosaur Beds of the Jurassic Morrison Formation," *Paleobiology*, 1980, 6(2):208-232.
[666] Ariel A. Roth, PhD, "Geologic Evidence for a Worldwide Flood," *Origins: Linking Science and Scripture*, Review and Herald Publishing Association, Hagerstown, MD, 1998, p. 220.

served and abundant animal trackways left in the Coconino sandstone, the light-colored rock layer near the top of the Grand Canyon, indicate [from fossilized tracks made by the feet, tail, or other body parts] that the animals were going uphill. By the way, this layer has an average thickness of 150 meters [492 feet], and spreads across many thousands of square miles. If simple footprints are well preserved, we would also expect to find the imprints or casts of roots, stems, and leaves of plants. But we find no evidence for the presence of plant foods.[667]

Walt Brown Jr., PhD: Evolutionists believe amphibians evolved into reptiles, with either the *Diadectes* or *Seymouria* as a claimed **transition.** Actually, by the evolutionist's own timescale, this "transition" occurs 35 million years **after** the earliest reptile, *Hylonomus* (a cotylosaur). A parent cannot appear 35 million years after its child![668]

---

[667] Gilmore, C.W., "Fossil Footprints from the Grand Canyon," Second contribution, *Smithsonian Miscellaneous Collections*, 1927, 80(3):1-78.

[668] Walt Brown Jr., PhD, "The Scientific Case for Creation," *In the Beginning: Compelling Evidence for Creation and the Flood*, Center for Scientific Creation, Phoenix, AZ, 2001, p. 53, (emphases in the original). See also: Steven M. Stanley, PhD, *Earth and Life Through Time*, pp. 411-415. See also: Robert H. Dott, Jr., and Roger L. Baten, *Evolution of the Earth*, 2nd edition 1976, p. 311.

Chapter 29
Evidence That Demands an Explanation

We need to know what the originally created world was like and what the Flood did to the world, so that we can understand how life's challenges today—thorns, thistles, sweat, pain, tooth and claw, "and unto dust shalt thou return," (Genesis 3:18-19)—have resulted from the first man's defying his Creator. Do we have scientific evidence to illustrate a change in earth's atmosphere? And could such a change impact the size and lifespan of living things?

Gary Landis of the U.S. Geological Survey and Robert Berner of Yale University, reporting to the *Geological Society of America*, documented increased air pressure and 50 percent more oxygen from air bubbles found in amber (the aged and solidified resin of coniferous trees or petrified tree sap).[669]

Other research shows it had 35 percent oxygen (compared to 21 percent today).[670]

It is interesting to note that today hospitals transfer certain patients into hyperbaric oxygen chambers with 100 percent pure oxygen under double or even triple pressure. The procedure is useful in treating acute carbon monoxide poisoning, burns, compromised skin grafts and flaps, crush injury/reattachment of severed limbs, cyanide poisoning, decompression sickness, gas gangrene, lower-extremity diabetic wounds, and a few others.

Although there is no way to verify its existence today, according to scripture a water (or water-vapor or thick cloud) canopy existed in the atmosphere above the earth, between the creation of the earth "in the beginning" and the Noachian Flood. "And God made the firmament, and divided the waters which were under the firmament from the waters which were above the firmament: and it was so" (Genesis 1:7).

Could enough water in this atmosphere have caused a "green-house" effect on earth? Genesis 1:29 says that there were herbs [vegetation, plants] "upon the face of all the earth."

> Lambert Dolphin, physicist, theologian: While classical evolutionary theory presupposes earth's early atmosphere was a reducing atmosphere (devoid of oxygen) newer scientific evidence confirms what Bible scholars had previously suspected: the earth's ancient atmosphere probably contained a larger fraction of oxygen than it does at present. It is even possible that photosynthesis in plant life was more efficient than it is now. A warmer average climate in ancient times would also mean a higher rate of oxygen generation by the more numerous plant life.[671]

Historical evidence correlates that a greater level of oxygen before the Flood supported man, animals, plants, and even insects to not only to live longer, but also to maintain a larger size. Even a scientific publication, although not endorsing a global Flood, agrees that an oxygen-rich environment once existed on earth:

> *New Scientist:* An oxygen-rich atmosphere allowed *gigantic insects* to grow 300 million years ago. Dragonflies with wingspans of up to 70 centimeters [27 and a half inches] flourished in the Carboniferous period. Researchers had long suspected that the giant insects and lush forests needed more oxygen than the level of 21 percent found in today's atmosphere.[672]

Evolutionary scientists see the evidence, but still interpret life within the context of millions of years:

> *National Geographic* states that the *chambered nautilus* is essentially the same today as its ancestors were 180 million years ago, and called it, "a living link with the past." Not only did it flourish, but a long-extinct, *nine-foot nautiloid fossil* turned up in Arkansas. Now, with fewer than half a doz-

---

[669] Gary Landis, Robert Berner, "Putting on Ancient Airs," *Time*, November 9, 1987, p. 82.
[670] "Bug Breath," "In Brief," *New Scientist*, New Science Publications, London, March 11, 2000, p. 23.
[671] Lambert Dolphin, "The Antediluvian World," World Population Since Creation, April 8, 1998, ldolphin.org/popul.html, p. 1.
[672] "Bug Breath," "In Brief," *New Scientist*, New Science Publications, London, March 11, 2000, p. 23, (emphasis added).

en species in existence, "time has whittled these descendents to about 8 inches."[673]

The Bible says that before the Flood "There were *giants* on the earth in those days" (Genesis 6:4). Could their descendents confirm the biblical account?

1 Samuel 17:4 reports that *Goliath* of Gath was six cubits and a span, or about nine feet, five or six inches tall.

When Hernando Cortez conquered Mexico in the early 16th century, the people told him that in ancient times a race of men and women lived there who were of immense stature with heavy bones, and who were a very bad and evil-disposed people. They reported exterminating these giants by continual war, and that the few who remained had gradually died.[674]

In Brush Creek Township, *Ohio*, a mound was opened by the Historical Society of the township, under the immediate supervision of Dr. J. F. Everhart, of Zanesville. The story was reported in *Scientific American*, August 14, 1880.

> It measured sixty-four by thirty-five feet at the summit, gradually sloping in every direction, and was eight feet in height. There was found in it a sort of clay coffin including the *skeleton of a woman measuring eight feet in length.* In another grave was found the skeleton of a man and woman, the former measuring *nine* and the latter measuring *eight* feet in length. In a third grave occurred two other skeletons, male and female, measuring respectively *nine feet four inches* and *eight* feet. Seven other skeletons were found in the mound, the smallest of which measured *eight* feet, while others reached the enormous length of *ten* feet.[675]

If water in the firmament before the Flood contained a greater concentration of oxygen, might there also be more oxygen in the ocean and might that not support larger fish? *Eighty-foot shark fossils* have been documented. The Yale Peabody Museum in New Haven, Connecticut, displays a *turtle skeleton* that is *more than 13 feet long and 16 feet wide*, flipper to flipper.[676]

Mount Everest clams?

Petrified clams have been found in limestone sediments *on top of Mount Everest*, the tallest mountain in the world, which is 29,028 feet tall and about 450 miles from the ocean.[677]

The *clams* are not only petrified, but they are also closed, which is evidence that they were trapped and literally held closed while they died.

If one walks along a beach he will notice that every dead clam is open because clams always open when they die. The muscle inside relaxes, and the ligaments on the outside pull the shell open, almost instantly. What can explain the fact that billions of closed and fossilized clams are found on mountaintops around the world? It is reasonable to propose that they were buried alive during the biblical deluge and covered with water-laid sediment, which not only killed them, but also prevented them from opening. Then, these oceanic sediments were raised into mountain ranges toward the end of the biblical Flood.

Erosion on a grand scale

One of the best places to view exposed, stratified sedimentary deposits is at the rim of Arizona's Grand Canyon. There, water has cut a huge channel down through numerous sedimentary layers of rock of differing color and texture. The Global Flood is another interesting example of how two different groups of people can look at the same evidence and come to completely different conclusions. Most evolutionists say the Grand Canyon formed slowly by a little water over millions of years. Creationists look at the same Grand Canyon and say it was caused by lots of water over a little time.[678]

Given the findings in the previous chapters, the simplest, most elegant explanation is that it was caused by a world of water over a brief time as a result of Noah's Flood. If a giant dam were to be built across the Grand Canyon, huge lakes, covering several states, would fill in behind it.

---

[673]"The Chambered Nautilus," *National Geographic*, January 1976, pp. 38-41.
[674]Bernal Díaz del Castillo, *Discovery and Conquest of Mexico*, Read Books, 2008, p. 158.
[675]Bernal Díaz del Castillo, *Discovery and Conquest of Mexico*, Read Books, 2008, p. 158.
[676]"Archelon," *BBC – Science & Nature*, retrieved 2009-05-25, see also: "Archelon," *Wikipedia, the free encyclopedia*.
[677]William L. Ramsey, Clifford R. Phillips, Frank M. Watenpaugh, "Movement of the earth's crust," *Modern Earth Science*, Holt, Rinehart and Winston, Publishers, New York, Toronto, Mexico City, London, Sydney, Tokyo, 1983, p. 207.
[678]One of those creationists is Dr. Ariel Roth. His website—www.sciencesandscriptures.com—challenges the belief that it is impossible to harmonize science and the Bible.

Consider these facts:
1. The Grand Canyon is a breach in the Kaibab uplift, which is from 6,900 feet to 8,500 feet above sea level.
2. The Colorado River enters the area at an elevation of 2,800 feet.
3. The river drops 1,800 feet as it goes 277 miles through the canyon.
4. The top of the canyon is more than 4,000 feet higher than where the river enters.
5. Most of the canyon has no erosion between layers supposedly millions of years old.

One explanation of these five observations is that the Grand Canyon is a washed-out spillway from two giant lakes that covered parts of what is now Arizona, Utah (including Monument Valley), Colorado, and New Mexico, whose beach lines are still evident in places even though the lakes are long gone. Water flowed off this nature-made dam toward the east causing the wrong-way canyons we see there. Then, following its breach, the canyons on the west end eroded very quickly in the opposite direction to the west.

> Duane T. Gish, PhD: If these ancient lakes did exist, they would have covered a large portion of the "Colorado Plateau." The lakes could easily have covered an area of more than 30,000 square miles, and contained 3,000 cubic miles of water.[679]

Canyons on Mars are much larger than the Grand Canyon. Yet scientists who would scoff at the idea of the Grand Canyon being created in one year believe these Mars canyons *were formed in a few weeks:* [680]

> Smithsonian Geologist Ross Irwin: Water pouring out of an overfilled lake carved *an instant Grand Canyon*—a valley more than a mile deep—on the surface of Mars 3.5 billion years ago, according to a new analysis of pictures taken by spacecraft. Researchers at the National Air and Space Museum said the flood of water originated from a huge lake—large enough to flood both Texas and California—that overflowed into a nearby impact crater.

When that crater filled up, the water eroded away a ridge-like barrier and was sent rampaging across the plain. Within a short time, a deep and wide gully called Ma'adim Vallis was carved from the Martian surface. A ridge on the edge of the crater gave way, suddenly releasing the flood that created Ma'adim Vallis.[681]

Four processes help explain the destructive power of floodwater: cavitation, hydraulic plucking, abrasion, and liquefaction. *Cavitation* was seen at the Glen Canyon Dam in Arizona: water got going too fast and sucked rock right off the canyon walls. Within about 20 seconds it made an area about the size of a basketball court four feet deep. During *hydraulic plucking*, moving water during a flood picks up debris, such as gravel and mud and tree stumps which together can erode or *abrade* through solid rock. And *liquefaction* of soil takes place as the floodwater compresses, stresses (as in earthquake shaking), and releases the sediments.

In 1976, the newly built Teton Dam in Idaho sprung a leak that could not be stopped, and the rushing water cut through sediment to a depth of 100 meters [328 feet] in less than one hour.[682]

From Genesis to Revelation many, including Jesus, referred to Noah's flood as a global, literal, historical event. Today, some regard Noah's Flood as simply a local or regional event caused by a long, heavy rain. However, Genesis chapters 6-9 record that it was far more; it was worldwide. It had never before rained. Much of the water was still stored high in clouds and vapor in the earth's atmosphere, and also in the preflood oceans and "fountains of the deep." And then "The floodgates of the sky broke open" (Genesis 7:11—The Hebrew-English Translation).[683]

> "And the waters prevailed exceedingly upon the earth; and all the high hills, that were under the whole heaven, were covered. Fifteen cubits [over 20 feet above the highest mountains] upward did the waters prevail; and the mountains were covered.... And every living substance

---

[679] Duane Gish, PhD, "How Long Ago Did Dinosaurs Live?," *Dinosaurs By Design*, Master Books, Green Forest, AR, 1992, p. 14.
[680] Matthew G. Golombek, "The Mars Pathfinder Mission," *Scientific American*, July 1998, p. 45; MSNBC News, June 20, 2002.
[681] Smithsonian Geologist Ross Irwin, MSNBC, June 20, 2002, (emphasis added).
[682] For details from an eyewitness, see, "Teton: Eyewitness to Disaster," *Time*, June 21, 1976, p. 56.
[683] *Tenakh: A New Translation of the Holy Scriptures According to the Traditional Hebrew Text*, Philadelphia: the Jewish Publication Society, 1985.

was destroyed which was upon the face of the ground, both man, and cattle, and the creeping things, and the fowl of the heaven; and they were destroyed from the earth. . . . For the waters were on the face of the whole earth" (Genesis 7:11 and 23; and 8:9).

Does that sound like a local or regional flood? Scripture is so specific it tells us on what day it all started: "the six hundredth year of Noah's life, in the second month, the seventeenth day of the month." However, it was far more than just 40 days of rain. Genesis 7:11 also reads: "subterranean waters"[684] "burst forth;"[685] "all the fountains of the great deep burst apart" (the traditional Hebrew Bible translated into English, by the Jewish Publication Society). Imagine all over the earth, geysers and perhaps volcanoes from fissures in the earth's crust during earthquakes. Mountains rising, valleys forming. Continents moving. Considering how inhumane humanity had become, God nearly had to start over. Only eight people accepted His offer of a lifeboat. The rest had been ridiculing Noah for 120 years. Genesis chapters 6 through 9 tells the whole story.

The global Flood storms were so violent they may have caused the land to break up into continents, accompanied by earthquakes, tsunamis, thousands of volcanoes, and continued mountain building around the world:

> Andrew Snelling, PhD: As a geologist I am also interested in the biblical account of earth history, particularly the record of the year-long global catastrophic Flood in the days of Noah that must have totally reshaped the entire surface of the globe. In fact, based upon the biblical description of the Flood event, it is logical to predict that it would leave behind billions of dead animals and plants buried in sediments eroded and deposited by the moving Flood waters, that would all end up being fossils in rock layers laid down by water all over the globe. And that's exactly what we find—layers of water-deposited sedimentary rocks containing fossils all over the earth.

There is impressive evidence that fossil deposits and rock strata were formed catastrophically. There are also many indications that there were not millions of years, or even thousands, between various rock units. The rock sequence in the Grand Canyon is a case in point. Not only can it be shown that each of the rock units exposed in the walls of the canyon must have formed very rapidly under catastrophic watery conditions, but there are not sufficient time gaps [no erosion nor other evidence of time gaps] between the various rock layers. Thus, the total time involved to put in place [about] 4,000 feet (1,200 meters) thickness of rock strata is well within the time constraints the Bible stipulates for the Flood event.[686]

Some believe there is evidence that the atmosphere was different than today, perhaps because of the "waters which were above" spoken of in Genesis 1:7. The change in the atmosphere of the long ago may be implied in oxygen levels and air pressure measured today in amber (petrified tree sap).[687]

Evidence for the global Flood can be concluded from a combination of scientifically documented facts: the washed-out spillway and stratified sedimentary deposits of the Grand Canyon, and petrified and closed clams on top of the world's tallest mountains. The world's tallest mountains may have been upthrust during the Flood's geologic perturbations. The rapidly and catastrophically formed layers of water-deposited sedimentary rocks containing billions of plant and animal fossils all over the earth are nothing less than compelling.

---

[684] TLB: *The Living Bible*.
[685] ESV: English Standard Version.
[686] Edited by John Ashton, PhD, Andrew Snelling, PhD, "Andrew Snelling, PhD," *In Six Days: Why Fifty Scientists Choose to Believe in Creation*, Master Books, Inc., Green Forest, AR, 2000, pp. 295-296.
[687] Gary Landis, Robert Berner, "Putting on Ancient Airs," *Time*, November 9, 1987, p. 82.

Chapter 30
Noah's Ark

After reading the manuscript for this book, one of my valued consultants suggested that I include information on Noah's Ark. As I started to research the subject, I became quite discouraged because of the different "facts" I uncovered (including fraud), and the lack of hard-core scientific evidence. I had to delete some information I originally thought was credible.

First of all, there is a question whether today's Mount Ararat is the same as the Mount Ararat of Genesis. Not only are there two Mount Ararats (Greater and Lesser Mount Ararat in Turkey), but other mountains have been identified in ancient texts as the landing site of Noah's Ark. Mount Cudi (pronounced Mount Judi), for example, is located in the ancient kingdom of southern Urartu. However, Greater Mount Ararat (at an elevation of nearly 17,000 feet) and Lesser Mount Ararat (at approximately 13,500 feet), both inactive volcanoes permanently covered with glaciers, are considered to meet the plural definition of "the mountains of Ararat."

> Joe White and Nicholas Comninellis: Pillow lava is formed only under water, and yet geologists have found a field of pillow lava as high as 15,000 feet on Mount Ararat.[688]

Therefore, Mount Ararat must have been formed while the Flood waters still covered the continent, and then volcanic activity pushed the mountain up to its present elevation.

Greater Mount Ararat has a permanent ice cap which is approximately 17-20 square miles in size. Being perpetually frozen in ice would prevent Noah's Ark from decaying for thousands of years.[689]

Of great help was *The Explorers of Ararat, and the Search for Noah's Ark*, a collaborative effort of alleged eyewitnesses—accounts from experienced explorers on nearly 100 expeditions (some of which have been debunked). The book, including 175 photographs, edited by B. J. Corbin, avoids "misleading sensationalism," and includes "critical thought and objectivity" based on the writings of Bill Crouse and his "detailed research of historic sources." Publisher Rex Geissler provides valuable insight:

> Rex Geissler: It should be noted that there is absolutely no convincing photographic or hard evidence to indicate that Noah's Ark has survived to the current date on modern-day Mount Ararat or anywhere else. In fact, if it were not for the purported sightings and its name, Mount Ararat would have very little to link it with the biblical account. . . . The only major reason to consider Mount Ararat is because of the few documented eyewitnesses.[690]

As I continued my research, I found information, later confirmed in this multi-authored, 1999 book, which I concluded to be quite intriguing to say the least. It included involvement spanning from United States astronaut Colonel James Irwin (who walked on the moon with Apollo 15); and Ralph Havens, a two-star United States Air Force general and commander of U.S. forces at NATO headquarters in Ankara, Turkey; to Elfred Lee, a personal friend of mine. After participating in four expeditions to the Greater Mount Ararat and interviewing credible eyewitnesses and hearing their firsthand accounts, Elfred has become a world expert on the search for Noah's Ark. And, as a professional artist, he was able to convey through his careful art renditions that which could not have been done with mere words.

Therefore, after discarding some of my earlier discoveries, I eventually found information I felt was profound—information I felt my readers should evaluate for themselves. Geissler agrees:

---

[688] Joe White and Nicholas Comninellis, "Not As Old As You've Been Told," *Darwin's Demise*, Master Books, Inc., Green Forest, AR, 2001, p. 57.
[689] Bill Crouse, "Bill Crouse," *The Explorers of Ararat, and the Search for Noah's Ark*, Great Commission Illustrated Books, Long Beach, CA, 1999, p. 48.
[690] Publisher Rex Geissler: "Preface," *The Explorers of Ararat, and the Search for Noah's Ark*, Great Commission Illustrated Books, Long Beach, CA, 1999, pp. 7, 8.

"... only one valid sighting of an object is enough to make the study worthwhile." Then he summarizes conclusions: "While some believers in the Bible debate the age of the earth, most of the Ark searchers . . . believe the Ark could have survived in petrified and frozen conditions for thousands of years."[691]

There are some today who believe Noah's Ark is usually under deep ice but has been seen at various times (during rare thaws) on Greater Mount Ararat, where some believe the Bible and tradition say it came to rest (in Eastern Turkey, near the borders of Iran and Armenia).

Many fascinating stories about the Ark reach far back in history and continue into recent decades. Legend says that those who lived nearby knew where it rested and that some climbed this Mount Ararat to worship there. Others, including the United States and Russian militaries claim to have photographed it. Then, why can't we see their pictures?

I leave it to the reader to evaluate whether the Ark has been found.

Nobody knows exactly how some of the Biblical details of Noah's Ark are defined. We don't know today for example what "gopher wood" was or what were the exact measurements of the cubit used by Noah. We also don't know if some of the animals hibernated in the ark during the Flood (as some animals do during times of stress).

Could Noah's Ark carry millions of earth's species? This question is often asked by critics. The answer is "No." As you will see, it wouldn't have to. To determine whether Biblical statements might be correct, all we need is to do some logical calculations. Let's start with dimensions provided in the Bible. According to Genesis 6:15 the Ark was to be 300 cubits long, 50 cubits wide, and 30 cubits high. Inside were to be three floors. It was to be a huge, seaworthy vessel.

So, how long was the Ark in dimensions that we can understand? Most of the time, the cubit is considered to be a measurement from the elbow to the tip of the middle finger. In ancient days, there was never any definite length of the cubit. There also was the cubit measure of the whole arm. Naturally both of these measurements would differ depending on the size of the person being measured.

The cubit also varied by country. Ancient Egypt had different lengths at different times. One was the length of a newborn child. Another was the length of the king at a certain age. The standard measure of the Egyptian, Assyrian, Chaldean, and Babylonian cubit was about 20.7 inches.[692]

"The Babylonians [also] had a 'royal' cubit of about 19.8 inches; the Egyptians had a longer and a shorter cubit of about 20.65 inches and 17.6 inches respectively, while the Hebrews apparently had a long cubit of 20.4 inches and a common cubit of about 17.5 inches."[693]

From these figures we can calculate that, depending on which definition is used, the Ark could have been from 438 to 516 feet long, 73 to 86 feet wide, and 44 to 52 feet high. If we use the smallest dimensions, "it had a total deck area of approximately 95,700 square feet, . . . and its total volume was 1,396,000 cubic feet. The gross tonnage of the Ark . . . was about 13,960 tons. . . . "[694]

However, we can't know the Ark's actual dimensions unless and until the day enough of the Ark is exposed for investigation, and exact measurements taken and made public.

A longer cubit would also provide much more space for the Ark's cargo, as would the shape of the Ark. As described by people claiming to be eye witnesses, it had vertical sides and ends. It was the shape of a giant shoe box, not of a ship. It is estimated that this shape would have provided almost 35 percent more space than would the steeply slanted walls of a typical ship.

> Bernard L. Ramm, PhD: The shape of the ark was boxy or angular, and not streamlined nor curved. With this shape it increased its carrying capacity by one third. It was a vessel *designed for floating, not sailing.* A model was made by Peter Jansen of Holland, and Danish barges called *Fleuten* were modeled after the ark. These models proved

---

[691]Publisher Rex Geissler, "Preface," *The Explorers of Ararat, and the Search for Noah's Ark,* Great Commission Illustrated Books, Long Beach, CA, 1999, pp. 8, 9.

[692]Rene Noorbergen, "Chapter 3," *The Ark File,* Pacific Press Publishing Association, Mountain View, CA, 1974, p. 72.
[693]R. B. Y. Scott, "Weights and Measures of the Bible," *The Biblical Archaeologist,* Vol. XXII, No. 2, May, 1959, pp. 22-27.
[694]John C. Whitcomb, Jr., ThD, and Henry M. Morris, PhD, "The Size of the Ark," *The Genesis Flood,* Presbyterian and Reformed Publishing Company, Philadelphia, Pennsylvania, 1967, p. 10.

that the ark had a greater capacity than curved or shaped vessels. They were very seaworthy and almost impossible to capsize. . . . The stability of such a barge is great, and it increases as it sinks deeper into the water. The lower the center of gravity the more difficult it is to capsize. . . . Wherever the center of gravity may have been in the ark, it certainly was a most stable vessel.[695]

In addition to Noah and his wife, their three sons, and their three wives, Noah was instructed to take into the Ark at least two of each animal, a male and female, "wherein is the breath of life." There were no fish or other marine creatures on the Ark; therefore, it did not carry all of the species on earth at that time. Here is God's directive to Noah:

> Genesis 7:2-3: Take with you seven pairs of all clean animals, the male and his mate, and a pair of the animals that are not clean, the male and his mate, and seven pairs of birds of the heavens also, male and female, to keep their offspring alive on the face of all the earth.

These words, "on the face of all the earth," also confirm that the flood was to be worldwide. The enormous size of the Ark supports the belief that the Flood was not local. Otherwise, God could have simply asked Noah and his family to move out of the area. And to suggest that Noah and his three sons would not have been able to construct such a vessel assumes that *none* of the population of the time was hired to help.

Some of the problems of the story proposed by critics are simply based on the pre-suppositions of uniformitarianism that there could not have been a worldwide Biblical Flood.

The purpose of the Flood was to destroy "both the man, and the beast, and the creeping thing, and fowls of the air; for it repenteth me that I have made them" (Genesis 6:7), and "to destroy all flesh, wherein is the breath of life, from under heaven; and every thing that is in the earth shall die" (Genesis 6:17). And this was accomplished when "all flesh died that moved upon the earth, both of fowl, and of cattle, and of beast, and of every creeping thing that creepeth upon the earth, and every man: All in whose nostrils was the breath of life, of all that was in the dry ground, died. And every living substance was destroyed that was upon the face of the ground, both man, and cattle, and creeping things, and the fowl of the heaven; and they were destroyed from the earth: and Noah only remained alive, and they that were with him in the ark" (Genesis 7:21-23).

> John C. Whitcomb, Jr., ThD, and Henry M. Morris, PhD: For all practical purposes, one could say that, at the outside, there was need for no more than 35,000 individual vertebrate animals on the Ark . . . assuming the average size of these animals to be about that of a sheep (there are only a very few really large animals, of course, and even these could have been represented on the Ark by young ones).[696]

It is unwarranted to insist that all the [then] present species, not to mention all the varieties and sub-varieties of animals in the world today, were represented in the Ark. Nevertheless, as a gigantic barge, with a volume of 1,396,000 cubic feet (assuming one cubit = 17.5 inches), the Ark had a carrying capacity equal to that of 522 standard stock cars as used by modern railroads [1950s], or of eight freight trains with sixty-five such cars in each.[697]

And what about hibernation? "Hibernation is generally defined as a specific physiological state in an animal in which normal functions are suspended or greatly retarded, enabling the animal to endure long periods of complete inactivity."[698]

> John C. Whitcomb, Jr., ThD, and Henry M. Morris, PhD: We suggest the reasonable possibility, however, that the mysterious and remarkable factor of animal physiology known as *hibernation* may have been involved. There are various types of dormancy in animals, with many different types of physiologic and metabolic responses, but it is still an important and widespread mechanism

---

[695] Bernard L. Ramm, PhD, *The Christian View of Science and Scripture*, Wm. B. Eerdmans Publishing Company, Grand Rapids, MI, 1954, pp. 230, 231, (emphasis added).

[696] John C. Whitcomb, Jr., ThD, and Henry M. Morris, PhD, "The Capacity of the Ark," *The Genesis Flood*, The Presbyterian and Reformed Publishing Company, Philadelphia, Pennsylvania, 1967, p. 69.

[697] John C. Whitcomb, Jr., ThD, and Henry M. Morris, PhD, "The Size of the Ark," *The Genesis Flood*, The Presbyterian and Reformed Publishing Company, Philadelphia, Pennsylvania, 1967, pp. 67, 68; quoting Lionel S. Marks, ed., *Mechanical Engineers' Handbook*, Hill Book Co., Inc., 1958, pp. 11-35, states that the standard stock car contains 2,670 cubic feet, effective capacity; *see also* the *Car Builders' Cyclopedia of American Practice*, Simmons-Boardman Publishing Company, 1949-51, p. 121.

[698] Marston Bates, "Hibernation," *Collier's Encyclopedia*, 1956, Vol. 7, p. 11.

in the animal kingdom for surviving periods of climatic adversity.[699]

It is a little-known fact that over the years approximately 80,000 books pertaining to the Deluge have been written in 76 different languages—and these are only the ones that can be traced through the . . . indexes of the great libraries of the world. Many more have no doubt been started, edited, and written but never catalogued. Most of these are concerned with the archaeological and geological aspects and not with the legends and folklore underlying the basic history of the ancient civilizations. Yet these tales are of the utmost importance, as their very existence among widely separated tribes is what generally might be expected if the Flood was indeed a universal one.[700]

> Rene Noorbergen: Not until 1972 did the first meaningful piece of evidence [surface] supporting the Biblical flood. The evidence was inscribed in cuneiform among 20,000 clay tablets discovered in that year by George Smith, the British orientalist, among the ruins of the palace of the Assyrian King Ashurbanipal (669-626 B.C.). Startling was the fact that not only did the cuneiform inscriptions reveal names and events *already* known from the Bible, but they also brought to light a Babylonian version of the deluge now known as the Gilgamesh Epic.[701] [This deluge is *first* described in Genesis chapters 6-9 as the worldwide flood that occurred in the days of Noah. It is referred to by Jesus, Paul, Peter, etc. The references are available in any standard Bible Concordance.]

After careful comparison of these two major accounts—the Bible story and the Gilgamesh Epic—Merril F. Unger points out that:

1. Both accounts state that the Deluge was divinely planned.

2. Both accounts agree that the impending catastrophe was divinely revealed to the hero of the Deluge.

3. Both connect the Deluge with the defection of the human race.

4. Both tell of the deliverance of the hero and his family.

5. Both assert that the hero of the Deluge was divinely instructed to build a huge boat to preserve life.

6. Both indicate the physical causes of the Flood.

7. Both specify the duration of the Flood, although differing in the elapsed time.

8. Both name the landing place of the boat.

9. Both tell of the sending forth of birds at certain intervals to ascertain the decrease of waters.

10. Both describe acts of worship by the hero after his deliverance.

11. Both allude to the bestowment of special blessings upon the hero after the disaster.[702]

In 1996, John Woodmorappe, published *Noah's Ark: A Feasibility Study*. The book is a definitive, 306-page work regarding the size of Noah's Ark, its capacity, and a variety of realistic possibilities. He deliberately chose to assume a very conservative view that God did not perform any miracles during the great Flood. To help support this position, he simply asks, "Why was the Ark needed at all if God could have judged and destroyed the world miraculously?"

Woodmorappe shows the possibility that eight people could have cared for all the animals, (feeding, watering, and waste-disposal) without any miraculous, Divine intervention. He documents his position that the Bible presents an accurate picture of what happened (and that the Flood was world-wide, not a local or regional event as proposed by some).[703]

He states plainly that the Ark did not have to accommodate every type of animal life in existence today. Sea creatures and microorganisms,

---
[699] John C. Whitcomb, Jr., ThD, and Henry M. Morris, PhD, "The Capacity of the Ark," *The Genesis Flood*, The Presbyterian and Reformed Publishing Company, Philadelphia, Pennsylvania, 1967, p. 71.
[700] Rene Noorbergen, "Chapter 2," *The Ark File*, Pacific Press Publishing Association, Mountain View, CA, 1974, pp. 38-39.
[701] Rene Noorbergen, Introduction, *The Ark File*, Pacific Press Publishing Association, Mountain View, CA, 1974, pp. 7-8.
[702] Merril F. Unger, *Archaeology and the Old Testament*, third ed., Zondervan Publishing House, Grand Rapids, MI, 1960, pp. 55-65.
[703] John Woodmorappe, "Introduction," *Noah's Ark: A Feasibility Study*, Institute for Creation Research, Dallas, TX, 1996, Fifth Printing, 2014, pp. xi, xii.

for example, could have survived outside of the Ark. He proposes that the majority of land animals seen in the world today resulted from *micro*evolution—variation within each basic kind that had been saved in the Ark.[704]

Woodmorappe addresses criticisms of the size of the Ark by documenting reports that wooden ships approaching its size were built by pre-modern Chinese and Greeks (one capable of 4,000 tons of cargo).[705]

Has anybody seen the Ark? And if so, what does it look like?

All one has to do is Google "Search for Noah's Ark," to learn that many expeditions and claims have been made about the success or failure of recent efforts to find Noah's Ark. Some of these claims have been debunked as fraud. Even conservative Christian organizations have concluded that, although if found it would be an unparalleled archaeological find, there are conflicting opinions and no current scientific proof of its existence that would satisfy unbelievers.

There was a monastery called Saint Jacob's near the base of Mount Ararat,[706] which claimed to house relics from the Ark. It reportedly was common for pilgrims to hike to the Ark when the weather permitted, and they could stop at the monastery on their way. However, a major earthquake destroyed the monastery in 1840.

So, what evidence is there, in addition to rumors and folklore, that can provide creationists with hope for its eventual discovery?

Greater Mount Ararat, at an elevation of almost 17,000 feet, has been described as a great mountain of indescribable majesty. As you will see, over the years, many people claim to have seen Noah's Ark, including the United States and Russian militaries. Especially during the years of World War II, American pilots flew hundreds of missions over Mount Ararat, flying from allied bases in Tunisia to Russia in a major attempt to keep the allied Red Army supplied. On several occasions during that time, reports surfaced about American, Australian, and Russian airmen seeing the Ark locked in a glacier below. One of these sightings was published in a summer 1943 edition of the *Stars and Strips* Army newspaper.[707] Various reasons have been given as to why most of the documentation has been hidden or lost over time: politics, prejudice, earthquake, fire, theft, local fighting between tribes or countries.

The first modern explorer to conquer the summit of Mount Ararat in recent historical time was Dr. J. J. Friedrich W. Parrot, a Russian-born German physician. [The Parrot Glacier was later named after him.] The year was 1829. He was inspired by an artifact he saw during his visit to the Armenian monastery at Etchmiadzin. It was said to have been made from a piece of sacred wood from Noah's Ark.

At the 7,000-foot level of his journey, he visited the Saint Jacob's monastery at Ahora. He was told it, too, contained priceless remains of Noah's Ark. [This monastery was said to be 800 years old, and was destroyed in the June 20, 1840, earthquake.][708]

But later, in 1876, Sir James Bryce had discovered a piece of hand-tooled timber from the rocky upper slopes of Mount Ararat. When the noted British statesman, author, world traveler, and mountaineer brought his treasure back to England, he was ridiculed after claiming what he sincerely believed: that he had found a piece of what he called the "true Ark."

Complicating the possibility of successfully documenting the finding of Noah's Ark are a number of challenging situations: Mount Ararat is in a very sensitive military zone (on the borders of Turkey, Russia, and Iran). Some Ark seekers have suffered severe injuries or death. Earth and sky are not friendly and make discovery difficult. There are moving glaciers, earthquakes, landslides, avalanches, lightning, major thunderstorms (including smothering snow and

---

[704] John Woodmorappe, "Introduction," *Noah's Ark: A Feasibility Study*, Institute for Creation Research, Dallas, TX, 1996, Fifth Printing, 2014, pp. 3, 5, 13.
[705] John Woodmorappe, "Introduction," *Noah's Ark: A Feasibility Study*, Institute for Creation Research, Dallas, TX, 1996, Fifth Printing, 2014, p. 50.
[706] According to *The Imperial Cyclopedia*, page 133, "Ararat has retained the same name throughout all ages since the deluge."
[707] Dave Balsiger, Charles E. Sellier, Jr., "20th Century Expeditions and Sightings," *In Search of Noah's Ark, The Greatest Discovery of Our Time*, Sun Classic Pictures, Inc., Los Angeles, CA, 1976, p. 155.
[708] Dave Balsiger, Charles E. Sellier, Jr., "20th Century Expeditions and Sightings," *In Search of Noah's Ark, The Greatest Discovery of Our Time*, Sun Classic Pictures, Inc., Los Angeles, CA, 1976, pp. 84-87.

pelting hail), ice beds sometimes 600 feet thick, dangerous crevasses, bitter cold, and the potential for hypothermia, disorientation, and altitude sickness . . . not to mention snakes, ferocious wolf-like dogs, and even bears and cougars. And the difficulty today of obtaining access from local governments and military police due to martial law and terrorists.

> John McIntosh: About midnight that night we heard noise, saw lights shining and next we saw the barrels of AK-47s being shoved through our tent openings. The next three hours were a nightmare as we were held at gunpoint by masked outlaws as they searched for money, photo equipment, and passports. After they got what they wanted, they threw gasoline on everything and torched thousands of dollars worth of equipment.[709]

Atheists find Noah's Ark

The following account in the book *Noah's Ark, Fact or Fable?* by Violet M. Cummings wasn't reported until 1952, but occurred in 1856. It is a story of how several atheist scientists arranged with a Mount Ararat local to take them to Noah's Ark, in an effort to prove the Bible wrong. It is a story written by Harold N. Williams about the memories of his personal friend, Haji Yearam, an Armenian who had moved to the United States.

Haji Yearam asked Mr. Williams to get a "composition book" and carefully write down the personal story he was until that time afraid to tell.

Yearam's family had lived at the foot of Greater Mount Ararat in Armenia. According to family traditions, they were direct descendants of Noah's family and had never moved from the area.

> Harold N. Williams: For several hundreds of years after the flood his forebears had made yearly pilgrimages up to the ark to make sacrifices and to worship there. They had a good trail and steps in the steep places. Finally, the enemies of God undertook to go to Ararat and destroy the ark, but as they neared the location there came a terrible storm that washed away the trail, and lightning blasted the rocks. From that time on, even the pilgrimages ceased, because they feared to betray the way to the ungodly. . . .[710]

Tribesmen handed down these legends throughout many generations, and periodically lonely shepherds or hunters came back with stories that they had reached a little valley in very hot summers, and had seen one end of the Ark protruding from the ice and snow.

When Yearam was a teenager, several atheist scientists hired his father to be their guide. Their intent was to disprove what they believed to be a legend of Noah's Ark. A very hot summer had melted the snow and glaciers more than usual. Yearam's father agreed to be their guide because he wanted to prove to them that the Biblical account of the Flood and Noah's Ark was true.

As they reached the little valley near the top of the mountain, "they found the prow of a mighty ship protruding out of the ice." They entered the Ark and found that it was divided into many compartments with animal cages. The entire Ark was covered with a very thick varnish or lacquer. The door was missing from a great doorway. At finding what they hoped to prove nonexistent, these scientists were dumbfounded, and became so angry that they tried to destroy the ship. But, they had to give up, because they had no tools or means to wreck so mighty a ship. They found the wood was more like stone. It was so hard it was almost impossible to burn.

There were cages of all sizes inside, many of them having great strong bars like great animal cages. They could not see far inside because they did not have lanterns or torches. But they saw enough to know it was none other than the mighty ship called Noah's Ark.[711]

> Harold N. Williams: "They held a council, and then took a solemn and fearful death oath. Any man present who would ever breathe a word about what they had found would be tortured and murdered.
>
> They told their guide and his son that they would keep tabs on them and that if they ever told anyone and they found it out they would surely be tortured and murdered.[712]

---

[709] John McIntosh, "John McIntosh," *The Explorers of Ararat, and the Search for Noah's Ark*, Great Commission Illustrated Books, Long Beach, CA, 1999, p. 39.
[710] Harold N. Williams, quoted by Violet M. Cummings, "The Atheists and the Ark," *Noah's Ark: Fact or Fable?* Creation-Science Research Center, San Diego, CA 92116, 1972, pp. 188-193.
[711] Harold N. Williams, quoted by Rene Noorbergen, "Chapter 5," *The Ark File*, Pacific Press Publishing Association, Mountain View, CA, 1974, p. 104.
[712] Harold N. Williams, quoted by Violet M. Cummings, "The

The atheists reported to all whom they met that there were no evidences of any ship on the mountain, or any remains of any such thing, and that the tradition was only vain imagination.[713]

When Yearam was an old man of about 75 years of age and living in America, he wanted Williams to record his personal story before he died. He felt quite certain that these men who had threatened his life would no longer be alive.

One evening in about 1918, Williams saw a short story in the daily newspaper of Brockton, Massachusetts. It was the story of a dying man. He was an elderly scientist on his deathbed in London. He said he was afraid to die before making a terrible confession.

> Harold N. Williams: He said his two companions [the two scientists who had hiked with him to Noah's Ark] were dead, so he was responsible to no one but God. He confessed that he was convinced that there is a God and that the Bible is His Word.
>
> It gave briefly the very date and facts that Haji Yearam had related in his story. I got out the composition book containing the story he had me write. It was identical in every detail. . . . We had never for one moment doubted Haji's story but when this scientist on his death bed on the other side of the world confessed the same story in every detail, we knew positively that the story was true in every detail.[714]

Discovery of Noah's Ark—First public report in recent history

In August 1883, the account of a Turkish discovery of Noah's Ark was announced to the world. The setting was a group of Turkish officials who were investigating avalanches on the mountain. Fortunately, an otherwise unidentified Englishman with the group, known as "Gascoyne," was familiar with the Biblical story of the Flood, or else he just might not have recognized what was found.

The story was released in Constantinople, published in the London press, and appeared in leading newspapers across America.

> *CHICAGO TRIBUNE:* A paper at Constantinople announces the discovery of Noah's Ark. It appears that some Turkish commissioners appointed to investigate the question of avalanches on Mount Ararat suddenly came upon a gigantic structure of very dark wood protruding from a glacier. . . .
>
> They recognized it at once. There was an Englishman among them who had presumably read his Bible, and he saw that it was made of the ancient gopher wood of Scripture, which as everyone knows, grows only on the plains of the Euphrates. Effecting an entrance into the structure . . . they found . . . that the interior was divided into partitions fifteen feet high. Into three of these only could they get, the others being full of ice, and how far the ark extended into the glacier they could not tell. If, however, on being uncovered, it turns out to be 300 cubits long, it will go hard with disbelievers.[715]

> *The New York WORLD:* "Considerable competition has recently been shown by the discoverers of ancient manuscripts and the finders of ancient relics. The latter has suddenly come to the front by the discovery of Noah's Ark in that part of the Armenian plateau still known as Mount Ararat. . . .
>
> An extraordinary spell of hot weather had melted away a great portion of the Araxes glacier, and they were surprised to see sticking out of the ice what at first appeared to be the rude façade of an ancient dwelling. On closer examination it was found to be composed of longitudinal layers of gopher wood, supported by immense frames, still in a remarkable state of preservation.
>
> Assistance having been summoned from Nakhchevan, the work of uncovering the find commenced under the most extraordinary difficulties, and in a week's time the indefatigable explorers had uncovered what they claimed to be Noah's Ark, as it bore indisputable evidence of having been used as a boat.[716]

Atheists and the Ark," *Noah's Ark: Fact or Fable?* Creation-Science Research Center, San Diego, CA 92116, 1972, p. 192.
[713] Harold N. Williams, quoted by Rene Noorbergen, "Chapter 5," *The Ark File*, Pacific Press Publishing Association, Mountain View, CA, 1974, p. 105.
[714] Harold N. Williams, quoted by Violet M. Cummings, "The Atheists and the Ark," *Noah's Ark: Fact or Fable?* Creation-Science Research Center, San Diego, CA 92116, 1972, pp. 192, 193; Harold N. Williams, quoted by Rene Noorbergen, "Chapter 5," *The Ark File*, Pacific Press Publishing Association, Mountain View, CA, 1974, p. 105.

[715] *Chicago Tribune*, August 10, 1883, quoted by Violet M. Cummings, "The Journalists and the Ark," *Noah's Ark: Fact or Fable?* Creation-Science Research Center, San Diego, CA 92116, 1972, pp. 152, 153.
[716] *The New York World*, August 10, 1883, quoted by Violet M. Cummings, "The Journalists and the Ark," *Noah's Ark: Fact or*

Fraud?

In 1935, *New Eden* magazine published a controversial and widely-circulated article, "Noah's Ark Found," by Vladimir Roskovitsky. It told of a Russian aviator who claimed to have seen the remains of a great ship on the side of Mount Ararat as he was making a routine flight. When the pilot returned to base, his captain asked to be shown, and identified the object as Noah's Ark. In response, the Russian Czar sent two companies of soldiers to search for the Ark. One had 100 men, the other had 50. When they found the Ark, they took complete measurements, drew plans, and took pictures. A few days later, the Russian government was overthrown during the Bolshevik Revolution (March 8, 1917), and the records were never made public. They probably were destroyed to discredit all religion and belief in the Bible.

The article, identified as perhaps the most widely-publicized and distributed of any story ever printed about the discovery of the Ark, was based on a true story, but greatly embellished—so much so that the publisher eventually acknowledged being the author, using Vladimir Roskovitsky as a pen name.[717]

But later discoveries confirmed the most important parts of the story. In correspondence, Mr. James Frazier of Malotte, Washington, verified the story: "Your letter, dated March 30, is at hand, in regard to the Ark of Noah. Yes, my father-in-law John Schilleroff, told me at different times about the Ark of Noah. . . . Mr. John Georgeson, a Dane, formerly my neighbor here, now also deceased, told me the same story, he also having served in the Russian Army in the Ararat region. They had never met, though their accounts fully agree. They belonged to different expeditions and went at different times. They were both sober and reliable men, and therefore I believe their story. The following is the story as they both told it to me.

Mr. James Frazier: While in the Russian army, they were ordered to pack for a long tramp up into the Mountains of Ararat. A Russian aviator had sighted what looked to him like a huge wooden structure in a small lake. About two thirds of the way up, probably a little farther, they stopped on a high cliff, and in a small valley below them was a dense swamp in which the object could be seen. It appeared as a huge ship or barge with one end under water. And only one corner could be clearly seen from where these men stood. Some went closer, and especially the Captain. . . .[718]

Then on October 6, 1945, *Rosseya*, a White Russian paper published another story about Noah's Ark, written by a Colonel Alexander A. Koor. In addition to confirming the main points in the Russian aviator's story, *it included additional details.* The two divisions of 150 infantrymen, army engineers, and specialists reached the Ark after a month of most difficult effort. They had braved severe snowstorms and falling ice.

Violet M. Cummings: But the rigors of the difficult campaign were all forgotten, said the story, when they finally reached the object of their search. As the huge ship at last loomed before them, an awed silence descended, and "without a word of command everyone took off his hat, looking reverently toward the Ark; everybody knew, feeling in his heart and soul," that they were in the actual presence of the Ark. Many "crossed themselves and whispered a prayer," said the eye-witness account. It was like being in a church, and the hands of the archaeologist trembled as he snapped the shutter of the camera and took a picture of the old boat as if she were "on parade."

The investigating party found that the ship was, indeed, of a "huge size." Measurements disclosed that it was about 500 feet in length, with a width of about 83 feet in the widest place, and about 50 feet high. These measurements, of course, when compared with a 20-inch cubit, fitted "quite proportionately" with the size of the Noah's ark as described in Genesis 6:15.[719]

One end of the ship was in ice. But the investigating party was able to enter through a broken hatchway near the front. They found a very

---

Fable? Creation-Science Research Center, San Diego, CA 92116, 1972, p. 156.
[717] "Great Scientific Find," *The New York World*, August 13, 1883, quoted by Violet M. Cummings, "The Journalists and the Ark," *Noah's Ark: Fact or Fable?* Creation-Science Research Center, San Diego, CA 92116, 1972, p. 156.
[718] James Frazier, quoted by Violet M. Cummings, "The Long, Long Trail Starts Winding," *Noah's Ark: Fact or Fable?* Creation-Science Research Center, San Diego, CA 92116, 1972, pp. 106-115.
[719] Violet M. Cummings, "The Bolsheviks and the Ark," *Noah's Ark: Fact or Fable?* Creation-Science Research Center, San Diego, CA 92116, 1972, p. 136.

narrow upper room with a high ceiling. Next to it stretched rooms of various sizes, both small and large. On the walls they found cages, from the floor to the ceiling. Because much of the boat was filled with ice, they estimated there to be several hundred rooms.

The Ark was covered inside and out, according to the story, "with some kind of a dark brown color" resembling "wax and varnish." Most of the wood was excellently preserved.

> Violet M. Cummings: As in the Roskovitsky story . . . this very similar but far more detailed account stated that the description and measurements of the Ark, both inside and out, together with photos, plans, [and] samples of wood, were sent at once by special courier to the office of the chief commandant of the Army "as the Emperor had ordered."[720]

But as stated earlier, the Bolshevik Revolution had just taken place, and, according to the article, the courier was intercepted and shot. Some members of the royal household reportedly escaped, but the Czar himself, "did not live to see the results of the investigation he had ordered."[721]

Carveth Wells examines petrified wood

In 1933, Carveth Wells, a popular radio commentator from KFI in Los Angeles, made an unsuccessful attempt to find Noah's Ark. His interest was based on his reading James Bryce's account of finding hand-tooled wood on Mount Ararat in 1876. Although Wells and his party were stopped from proceeding to the majestic mountain, they secured permission to visit Etchmiadzin, reported to be the oldest Armenian monastery in the world.

According to Mrs. Cummings, the group was met by the aged Armenian Patriarch, Archbishop Mesrop, who welcomed the Americans with "true Oriental graciousness and charm." Wells later reported that the man was obviously a great scholar and author. A magnificent oil painting hung on his library wall, depicting Noah and his family descending Ararat, followed by a long line of animals.

> Violet M. Cummings: The highlight of the occasion, of course, came when Wells finally ventured to ask the Patriarch about the relic of Noah's Ark reputed to be in his possession. The genial Archbishop smiled: "We have [some of] the remains of the Ark here in the church. . . . It is the most prized possession of the monastery. We do not class it merely as a relic. This piece of Noah's Ark is in quite a different category and no other church in the world possesses or even claims to possess such a thing."[722]

A special meeting of the monks was called in order to obtain their permission to even glimpse the relic.

> Violet M. Cummings: At last, however, to the accompaniment of [what sounded to Wells like Armenian] chanting and the swishing of many brooms, the church was thoroughly cleansed and certain rituals performed in preparation for the special event. A stately procession appeared, led by a bearded, mitered priest bearing the sacred chest in his hands. Carveth Wells was graciously granted permission to open the casket and examine the wood. "You may examine it as much as you like," smiled the archbishop. "This is the portion of Noah's Ark which was brought down from Ararat by one of our monks named Jacob" [after whom Saint Jacob's Monastery was named].[723]

The wood, as reported by other eyewitnesses, was petrified. It was reddish-colored, and measured about twelve by nine inches and about one inch thick. Even though it was petrified, he later wrote, the grain was plainly visible. This unusual experience with the monks convinced Wells that Noah's Ark was still, "without a doubt," somewhere on the mountain.[724]

In 1952, an American pipeline and mining engineer, working in the Middle East, saw the Ark on Mount Ararat's northeast face. George Jefferson Greene was on a helicopter reconnaissance mission for his company when he saw the

---

[720] Violet M. Cummings, "The Bolsheviks and the Ark," *Noah's Ark: Fact or Fable?* Creation-Science Research Center, San Diego, CA 92116, 1972, p. 137.
[721] Violet M. Cummings, "The Bolsheviks and the Ark," *Noah's Ark: Fact or Fable?* Creation-Science Research Center, San Diego, CA 92116, 1972, p. 138.
[722] Violet M. Cummings, "The Expedition That Went 'Kapoot,'" *Noah's Ark: Fact or Fable?* Creation-Science Research Center, San Diego, CA 92116, 1972, p. 150.
[723] Violet M. Cummings, "The Expedition That Went 'Kapoot,'" *Noah's Ark: Fact or Fable?* Creation-Science Research Center, San Diego, CA 92116, 1972, p. 150.
[724] Violet M. Cummings, "The Expedition That Went 'Kapoot,'" *Noah's Ark: Fact or Fable?* Creation-Science Research Center, San Diego, CA 92116, 1972, p. 150.

massive vessel protruding from the ice. After directing the pilot to maneuver the craft as close as possible to the huge structure, sometimes less than 100 feet away, Greene took a variety of pictures. Six were extraordinary photographs. Enlargements clearly showed joints and parallel, horizontal timbers.

One of Greene's colleagues, an oilman named Fred Drake, met Greene in 1954, two years after Greene had photographed the Ark, and reports seeing Greene's pictures, including "the prow of a great ship, parallel wooden side planking and all.... I will admit that I had never been much of a Bible believer, but these pictures sure made a believer out of me!"

However, Greene failed to interest his American friends in financing an expedition to Mount Ararat, and instead transferred to a mining operation in British Guyana, where he was murdered for his gold. His personal belongings, including the valuable photos, disappeared.[725]

U.S. Navy photographer William Todd photographs Ararat anomaly

Between 1951 and 1955, The Navy Hydrographic Office in Washington, D.C. and the Army Map Service assisted the Turkish Air Force in making the first accurate topographic maps of Turkey. William Todd, a Photographer's Mate Chief flew in a U.S. Navy squadron AJ2P photographic-configured Savage aircraft, the steadiest platform the Armed Forces had at the time.

One afternoon, while descending to Diyarbakir, he and his crew spotted an anomaly sticking out of the ice at an altitude of 14,500 to 15,000 feet on Greater Mount Ararat. Because they had plenty of fuel and it looked interesting, they descended to take a long "look-see."

> William Todd: The very thought of the ark would have been ridiculous, as there was no reason to believe that it existed to modern times. But there appeared this structure lying on a little shelf protruding from ice and snow. At the base were snow patches and large to small rocks. The upper part had additions that looked like a railing or a roof that was in extreme disrepair....

> We witnessed the structure from about 2,000 feet above the object. However, once spotted, we flew around the mountain a number of times along the side of the object. It was a rectangular, slate-colored boat that we all claimed was Noah's Ark.... The object appeared to be of huge size and was surrounded by ice and snow. This anomaly looked like a craft of some sort.... As a Naval Photographer and Chief Petty Officer, it did not look like a typical boat and was completely out of place. It appeared more like a barge.

> Rails or openings ran all the way from the bow to the ice. About 35 feet of the structure was sticking out of the ice. As we were professional photographers, we measured the photos to determine the width and height of the object.... It turned out to be exactly 75 feet wide by 45 feet high using an 18-inch cubit. When we found that it precisely matched the description in Genesis 6:15, we were running around with chills down our back.[726]

Todd noticed that the Tigris River was quite high because of the extreme heat that year melting snow in the mountains. After a few years, he gave his photos to a Baptist minister who kept them and then died. Later attempts to retrieve them were unsuccessful. Todd said that the U.S. Government Archives in Washington, D.C., should have the negatives, but the archive office claims they could not be found.

> William Todd: We gave four or five copies of the pictures to all of our crew. The whole squadron of six mapping planes was abuzz about Noah's Ark and every day some plane "accidentally" went by Ararat to see it. You practically needed a control tower around Ararat that summer. After seeing the photos, we were convinced that the structure was Noah's Ark.[727]

Elfred Lee sketches other eyewitness accounts

Elfred Lee, an excellent artist/illustrator and photographer, has been interviewed many times over recent years by book authors and in television specials on the search for Noah's Ark. He is not only a personal friend of mine, but also on

---

[725] Dave Balsiger, Charles E. Sellier, Jr., "20th Century Expeditions and Sightings," *In Search of Noah's Ark, The Greatest Discovery of Our Time*, Sun Classic Pictures, Inc., Los Angeles, CA, 1976, p. 161.

[726] Bill Crouse, Rex Geissler, "Noah's Ark Sources and Alleged Sightings," *The Explorers of Ararat, and the Search for Noah's Ark*, Great Commission Illustrated Books, Long Beach, CA, 1999, pp. 327, 328.

[727] William Todd, "Noah's Ark Sources and Alleged Sightings," *The Explorers of Ararat, and the Search for Noah's Ark*, Great Commission Illustrated Books, Long Beach, CA, 1999, pp. 327-329.

this subject he is one of the most knowledgeable people alive.[728] He and his illustrations were even featured in the dramatic 1976 documentary "In Search of Noah's Ark," produced by Schick Sun Classic Pictures (and available on You Tube).

I have known Elfred since we were 8th grade classmates in 1954-55. His father, James Lee, was one of the deans in the dormitory where I went to college. Elfred and I have maintained contact over the years, including when he designed the cover of my first book. And he photographed a family wedding for me. But he's not just a friend. He's a world expert on the search for Noah's Ark, and as such, he lectures "all over the globe." And how did that happen?

In 1969, Elfred joined the SEARCH Foundation, an archaeological-research group from the Washington, D.C., area, that wanted to promote the Word of God to a skeptical world. He accepted the foundation's invitation to participate because they needed an artist and photographer and public relations person. They were working with Fernand Navarra, a French explorer and author and one of those who claimed to have found Noah's Ark.

Elfred had developed an interest in the subject when another group he knew went to Turkey in 1960; they thought they may have discovered Noah's Ark near what is now known as the Durupinar site near Greater Mount Ararat. So, he had some exposure and personal interest before his involvement with SEARCH.

Elfred had studied art in Japan, Southern California, England, and Northern California, where he earned a bachelor's degree in art at Pacific Union College. He was working toward a master's degree at the University of California, San Jose, when he was drafted into the United States Army as a motion picture photographer during the Vietnam War. Art was his main interest in life. He eventually earned a Master of Fine Arts (MFA) degree at Syracuse University, New York, after studying archaeological illustration at London University, Ankara University, and two universities in Israel. Since the MFA was at that time a terminal degree, a person could not earn a doctorate in this field. He later won an honorary doctorate in Fine Arts.

In time, Elfred was involved with several groups and expeditions, sometimes officially, sometimes unofficially, on a freelance basis. His first of four expeditions to Greater Mount Ararat was in 1969 with a team of internationally respected scientists from Washington, D.C. In Turkey, they benefited from military escorts, extra food, transportation, donkeys, horses, porters, and much scientific equipment, including coring augers and depth-sounding electronic equipment. They were led by Fernand Navarra, himself, whose son, Coco, served as his translator.

Navarra claimed to have seen a wooden structure through glacial ice 14 years earlier, in 1955. The whole glacier had since changed and he was unable to find points of interest. The group was disappointed after more than a week's work to be unable to find the Ark. Then, on July 31, 1969, while probing down in a crevasse, Navarra and Hugo Neuberg, one of the expedition's participants, started shouting and waving a piece of wood. The largest pieces found were about 17 by 4 by 5 inches in size. To find hand-tooled, very old wood on that mountain where there are no trees was very exciting. Others from the expedition joined the two men and started digging. And, sure enough, they found even more wood. "We gathered around with our arms around each other," reported Elfred, "and had a little prayer meeting." They felt sure they had located pieces of Noah's Ark.

This find and scenes recorded that day on film are included in the movie, "In Search of Noah's Ark."

On his way home, Elfred visited the homes of the Navarras where they showed him wood and photos from previous trips that convinced him that they had seen Noah's Ark. When Elfred returned home, he had the expedition's wood, and some of Navarra's wood, tested at the radio-carbon dating laboratory at the University of Pennsylvania, in Philadelphia. The samples dated between 1300 and 1900 years of age, much too young to be from Noah's Ark. He was tempo-

---

[728] "1969-1987 Elfred Lee," an oral history recorded with Dr. John Goley, Dr. Paul and Rosie Kuizinas, and Martha Lee, at Montemorelos University, Montemorelos, Mexico, and later with Rex Geissler.

This hand-tooled wood, found at an elevation of 14,000 feet on Greater Mount Ararat, grows 300 miles to the south along the Euphrates River. *(Photo Courtesy of Elfred Lee)*

rarily crushed. [As found elsewhere in this book, carbon-14 dating is often found to be unreliable for a variety of reasons, including contamination; and its dates are often considered invalid by those testing a sample if the dates are outside of what would be expected.]

The wood appeared to be the same as the wood he had received from Navarra, which Navarra had tested in Bordeaux, France, Madrid, Spain, and Cairo, Egypt.[729] Scientists had dated his wood at 5,000 years of age with well-accepted methods (lignite formation, gain in density, cell modification, growth rings, and fossilization).[730] Then why the difference in age?

Later, Elfred learned from Dr. W. F. Libby himself (the man who developed Carbon-14 dating in the late 1940s) that these particular samples were contaminated by more recent organic material. Moss, lichens, and little, low bushes grow in the area. Elfred saw goats above this very site. Goat droppings could have come down into the water. Such organic material (which contains carbon) could have contaminated the wood, resulting in the more recent Carbon-14 dates. Elfred also learned from the United States Forest Service that the wood was Quercus, a species of white oak, which (he was told by others) was a common type of wood used by seagoing vessels before the time of metal ships.

Then, through a Mrs. Mary Board of Annapolis, Maryland, Elfred learned about another man who claimed to have seen Noah's Ark. "This man has seen it," she said. "He knows where it is, but nobody will pay any attention." Elfred set up an appointment in July 1970 to meet George Hagopian. Although not in the best of health, he was very alert mentally. He had been reared near Lake Van, in eastern Turkey, and had a very strong Armenian accent. At that time, Mount Ararat was part of Armenia, which was a predominantly Christian country.

Elfred eventually conducted hours of taped interviews, wrote pages of notes, and drew sketches of the various scenes explained by Hagopian.

---

[729]Institute Forestal (Madrid), the Centre de Technique de Bois (Paris), and the Institute Preshistoire de L'Universite Bordeaux (France).
[730]Rene Noorbergen, "Chapter 6," *The Ark File*, Pacific Press Publishing Association, Mountain View, CA, 1974, p. 134.

The bottom step was hanging about 10 feet above the ground, probably because of a higher snow level at the time they were built. (*Photo Courtesy of Elfred Lee*)

As Elfred drew, the gentleman visualized his experiences, he was able to give Elfred details that cannot be reported in mere words. Elfred's artistic talents spoke volumes. When observing the finished copy of Elfred's work, Hagopian sketched in some nearby boulders that were an important part of the picture.

According to oral histories and traditions, Hagopian reported that many people from his village, including his grandfather and uncle, had seen Noah's Ark. It was common knowledge. When he was a boy, there was a three- or four-year drought in the region. Because the glacial ice had melted way back on the mountain, his uncle suggested they go toward the top of the mountain and see if the Ark was visible. Hagopian reported that he and his uncle found the fully exposed Ark resting there. Elfred reports that they were in reverence and awe. It was a very holy experience for them.

A climatology study later confirmed the dates that Hagopian said a four-year drought occurred, scientifically confirming that part of Hagopian's memory.

Hagopian remembered that the Ark was of wood and he could see the grain, color, fitted joints and wooden dowels. It was covered with something like shellac. There was green moss growing on it, and at one end was a set of stairs.

George's uncle hoisted him up to the stairs, and he walked on the top of the Ark where he saw openings all the way down the middle of the roof. He reported that the Ark was rectangular in shape (that of a barge, not a ship) and petrified. He could not give the exact measurements, but to him it was huge and fully intact. He saw the Ark again, two years later as a child and later in life.

Elfred drew pictures of the Ark based on these descriptions. The roof had a very slight pitch to it, but was almost flat. Down the middle of the roof was a small raised area running from stem to stern, and on each side of that raised area were openings. Hagopian looked inside these openings and when he shouted, his voice echoed down inside.

More recent eye witnesses claim that what they believe to be Noah's Ark has been ripped apart and moved in major sections by avalanche and glacier toward the North, above the Ahora Gorge, to a location below the Hagopian site.

He reportedly walked to the far end and looked down to see an overhanging cliff. He said the Ark was not a rock construction or an earth formation; it was obviously hand-tooled and very long and high.
(*Photo Courtesy of Elfred Lee*)

Although Elfred could not be certain that the wood he had tested was part of Noah's Ark, it was found in the Parrot Glacier, where Navarra had seen wood before and below where Hagopian claimed he saw the Ark. According to Elfred, wood fragments have been coming down both directions from that location.

At that time, there was very little international travel to that part of the world. For twelve years, there was no further research on the mountain. Then, thanks to the political connections of Colonel James Irwin, the American astronaut who walked on the moon with Apollo 15, eastern Turkey opened the area again to expeditions.

In the summer of 1985, Elfred went on his next major expedition with a very fine group of men from the Dallas, Texas, area.

On August 29, 1985, artist Elfred Lee, astronaut Colonel James Irwin, John Bradley, a businessman from Alabama and former president of SEARCH, and Ark researcher Eryl Cummings, interviewed General Ralph Havens, a two-star general and commander of United States forces in United States and NATO headquarters in Ankara, Turkey. The general said, "We've seen that. We have photos of that. Our pilots have photographed that very object. It looks just like that. It is on a ledge."

With tears in his eyes, Eryl Cummings, a pioneer Ark researcher, exclaimed, "This is the greatest day in all the years of Ark research!"

So, if true, there is confirmation of the structure on Mount Ararat in modern times by the United States military. And airmen and soldiers in the Russian military claimed to have photographed it in 1916.

According to Elfred, most of those who saw the Ark did not see it from stem to stern. It was covered with snow at one end or broken. Elfred interviewed others: "I've talked to Colonel Alexander Koor of the Russian Army who was in charge of that region at the time. Before World War I, it was still part of Russia. Mount Ararat was his command post and he was there when the Czar sent 150 troops up the mountain to photograph the Ark and measure it. As you know, in 1917, the Bolshevik Revolution took place, and we're not sure where that information

While Elfred (left) was showing his illustrations from George Hagopian's description to General Havens, the general said, "We've seen that. We have photos of that. Our pilots have photographed that very object. It looks just like that. It is on a ledge."

Elfred Lee's rendition of Noah's Ark on a ledge as described by George Hagopian before this specific detail was confirmed by two-star General Havens at NATO headquarters in the capital city of Turkey. (*Photo/drawing Courtesy of Elfred Lee*)

is. I heard one report it was somewhere in Leningrad, which is again called St. Petersburg. Maybe it will be released. Who knows? There are some Russians who know about it, but at that time they weren't willing to help us."

Because the windows of the Ark were in the middle of the top, the Ark would have to almost capsize before water could go inside. In his oral history, Elfred reported that in preparing for a Hollywood movie, a hydraulics laboratory in San Diego, made a scale model from Elfred's drawings as they studied the seaworthiness of the Ark. They found that Noah's Ark would not capsize. A Coast Guard captain told Elfred, "This is the most seaworthy design I've ever seen."[731]

In the early 17th century, Dieter Jansen, a Dutch merchant, commissioned a shipyard to build a scale model of the Ark according to Genesis 6:12-16 measurements. His model was said to be 120 feet long, 20 feet wide, and 12 feet high. The Ark was designed just to *float* and not *capsize*, but the Ark model proved more seaworthy than contemporary vessels, and its gross tonnage one-third more than ships built with more modern designs. . . . In 1844, the shipbuilder Brunel, in his quest to build the ultimate design for passenger comfort, designed an ocean liner known as the "Great Britain." Its proportions were almost identical with the dimensions of Noah's Ark. Consider the remarkable fact that Noah built the *first* ship, and Brunel relied on several thousand years of shipbuilding expertise. Interestingly, all of that accumulated knowledge could not provide a ratio more perfect than that of Noah's Ark.[732]

In a July 2017 interview, Elfred told me that, of all the people he worked with, two stood out: George Hagopian and Ed Davis. And his encounter with Mr. Davis, another key eyewitness to Noah's Ark, occurred in 1987, eleven to fifteen years *after* three of the books cited in this chapter had been published, but *before* Hagopian's story was in print.

So, who was Ed Davis? Elfred said that the hair on the back of his neck stood up as Ed started talking. He was saying the same things about the Ark that Elfred had heard from George Hagopian.

> Elfred Lee: He started describing the mountain, his experiences, how long it took to go from point A to point B, the caves, the fog, the rock formations, describing the interior of caves, and the stone steps and how they're carved. My goodness, I thought, *where did this guy hear this information? None of this information is published. I have George Hagopian on tape and in notes, but none of this*

---

[731] "1969-1987 Elfred Lee," an oral history recorded with Dr. John Goley, Dr. Paul and Rosie Kuizinas, and Martha Lee, at Montemorelos University, and later with Rex Geissler.

[732] Rene Noorbergen, "Chapter 3," *The Ark File*, Pacific Press Publishing Association, Mountain View, CA, 1974, p. 76.

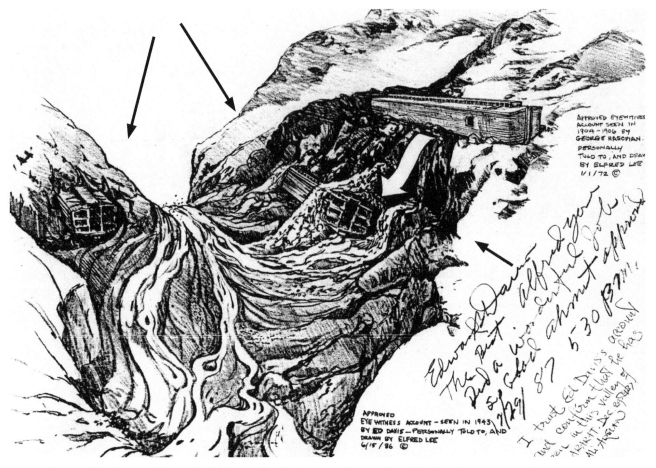

According to Davis, the Ark had slid from the Hagopian site and had broken into two major sections.

Four handwritten messages on the above art:
APPROVED EYEWITNESS ACCOUNT SEEN IN 1904-1906 BY GEORGE HAGOPIAN. PERSONALLY TOLD TO, AND DRAWN BY ELFRED LEE 1/1/72©
Edward Davis the best [to you]. [E]lfred you did a wonderful job. So glad Ahmet approved 12/29/87 530 [unclear]
I trust Ed Davis's account and confirm that he has been in this valley of Mt. Ararat. Dec. 29, 1987. Ahmet Ali Arslan
APPROVED EYE WITNESS ACCOUNT—SEEN BY ED DAVIS—PERSONALLY TOLD TO, AND DRAWN BY ELFRED LEE 6/15/86 © (*Drawing Courtesy of Elfred Lee*)

information had been published yet, so I became very interested.[733]

It turns out that Mr. Davis had been stationed with the United States Army Corps of Engineers in Hamadan, Iran, during friendlier times. They were building roads in Northern Iran to help supply the Soviet Army during World War II. During this time, Ed befriended a Muslim family named Abas. As a result of these favors, they felt indebted to him, and asked if he would like to see Noah's Ark.

The young GI responded, "Why not?"

One night the Abas family drove Davis from Iran between Greater and Lesser Ararat to the north side where they stopped near the Ahora Gorge. Because Elfred had been there, when Davis described the rock formations and a tree at the foot of Black Glacier toward Doomsday Rock, he started to document the interview as he did with George Hagopian. He turned on his tape recorder and started sketching the scene as presented. As the drawing grew larger, he asked for more paper and cellophane tape.

---
[733] "1969-1987 Elfred Lee," an oral history recorded with Dr. John Goley, Dr. Paul and Rosie Kuizinas, and Martha Lee, at Montemorelos University, and later with Rex Geissler.

He described in detail rock formations, the location of cliffs, snow banks, and moraine. He described standing in various places and what he saw from each location. He gave Elfred enough information to confirm that he had been in the same place Hagopian had been.

After Davis was discharged from the Army, he soon discovered that most people did not believe what he thought was an exciting personal story. Disappointed with this response, he decided to keep the incredible story to himself. In 1985, Dr. Don Shockey heard of the story, contacted Ed Davis, and eventually asked him to take a polygraph test. The 3.5-hour test was administered on May 1, 1988, by P. G. P. Polygraph in Albuquerque, New Mexico. The purpose of the test was to determine whether or not Ed, identified as "this subject" was truthful when he stated that he observed Noah's Ark while in the area of Mount Ararat.

> P. G. Pierangel: On the above date (5/1/88) this subject was tested utilizing the standard Backster Tri-Zone Comparison Specific Examination. Upon final analysis of all of this subject's polygrams it is the opinion of the examiner that he answered *truthfully* to the target issue.[734]

P. G. P. Polygraph is considered the local authority on test procedures and analysis. All law enforcement units abide by his conclusions.[735]

Earlier, in the early spring of 1988, Dr. Don Shockey met George Stephen II, who eventually uncovered new information concerning the remains of manmade structures on the northeast section of Mount Ararat. After explaining the challenges of obtaining any information from U-2 or satellite photos, George responded to Dr. Shockey, "Don, I have access to special technology, and can within two weeks have information on any square foot of land anywhere on the globe!"

Dr. Shockey: I challenged his statement, and he backed up his claim with information about his military background and the newer technology which he had helped develop for our [US] government in the area of infra-red analysis from satellites located 240 miles in space. He then asked me for some coordinates pertaining to my area of search. He said he would get back to me within two weeks.

George kept his word. It was two weeks to the day when he called from California. Using technology, he had analyzed the area on Mt. Ararat and found two man-made objects in the Abich II glacier. He determined these objects to be definitely man-made.

"Are you sure, George?" was my next question, and one I should not have asked. He reminded me that this is his area of expertise, then went on to explain why he was so sure. The two pieces are rectangular, and approximately one thousand feet in separation. He said that natural formations are not so specific in shape, and he emphasized his certainty that the shapes were not background rock. He then gave me the approximate elevation and also the depth which the objects are covered with ice and snow. . . . Now for the first time we had hard scientific evidence of something large and foreign on the mountain.[736]

In 1990, Dr. Shockey succeeded in getting all the permissions for an expedition to the site. But, because Kurdish rebels and Turkish military were fighting, authorities became concerned for the expedition's safety and ordered the men off the mountain.[737]

Dave Balsiger and Charles E. Sellier, Jr., authors of the book *In Search of Noah's Ark*, concluded, "Our research indicates that 200 people in 23 separate sightings since 1856 have seen Noah's Ark on Mt. Ararat."[738]

---

[734] Dr. Don Shockey, P. G. Pierangel, "Dr. Don Shockey," *The Explorers of Ararat, and the Search for Noah's Ark*, Great Commission Illustrated Books, Long Beach, CA, 1999, p. 84.
[735] Dr. Don Shockey, "Dr. Don Shockey," *The Explorers of Ararat, and the Search for Noah's Ark*, Great Commission Illustrated Books, Long Beach, CA, 1999, p. 85.
[736] Dr. Don Shockey, "Dr. Don Shockey," *The Explorers of Ararat, and the Search for Noah's Ark*, Great Commission Illustrated Books, Long Beach, CA, 1999, pp. 86, 87.
[737] Dr. Don Shockey, "Dr. Don Shockey," *The Explorers of Ararat, and the Search for Noah's Ark*, Great Commission Illustrated Books, Long Beach, CA, 1999, pp. 89-91.
[738] Dave Balsiger, Charles E. Sellier, Jr., "The Mystery of Noah's Ark Resolved," *In Search of Noah's Ark*, Sun Classic Books, Los An-

Elfred Lee believes that Noah's Ark is still in the eternal snows of "the mountains of Ararat;" and that the snow and ice very rarely melt and uncover the ark, so it is very rarely seen.

One Ark hunter summarizes his evaluation of the evidence:

> Robert Garbe: It seems logical to me that the ark exists on Ararat and that it landed high up on the mountain. It then became petrified as many reports insist and broke up due to either earthquake or ice movement. The separate portions were then carried to different parts of the northern sector of Ararat by ice flow and other natural forces. Sundry smaller pieces and timbers would then be accounted for in other locations.[739]

I have presented this chapter on Noah's Ark with the belief that the information presented here is at least fascinating. It names and quotes a number of people, with quite separate and different backgrounds, who could not possibly have conspired to report such detailed descriptions of what they saw over decades of time. Ten saw petrified wood. Three said it was covered with varnish or lacquer. Two said it was on a shelf or ledge. Many said it was locked in ice or a glacier. And four described what they saw as hand-tooled wood. The agreement between them seems to me remarkable. I believe these stories need to be shared. Perhaps some readers will wish to investigate further for themselves.

---

geles, CA, 1976, pp. 203, 204; citing Walter Lang, The Witness of Mount Ararat, Bible Science Association, Caldwell, Id, 1974, p. 2.
[739] Robert Garbe, "Robert Garbe," *The Explorers of Ararat, and the Search for Noah's Ark*, Great Commission Illustrated Books, Long Beach, CA, 1999, p. 78.

Chapter 31
Volcanoes, Their Aftereffects, and Uniformitarianism

Volcanoes provide additional evidence that gives scientific support to creation, and further calls early *strict* uniformitarianism into question. Just compare the volcanoes of today with those of the past. The scientific evidence against the once highly acclaimed uniformitarian theory is compelling, if not decisive. The eruption of Mount St. Helens on May 18, 1980, was the most destructive in recorded U.S. history. It unleashed the same energy as approximately 20,000 Hiroshima-sized atomic bombs,[740] and produced an impressive 0.25 cubic mile of volcanic ash. However, it was relatively small compared to the eruption of Taupo, New Zealand, about 1,800 years ago, which produced 8 cubic miles of ash. Mt. Tambora in the East Indies erupted in 1815, causing 36 cubic miles of rock and dust to be hurled into the upper atmosphere. It caused "the year without summer."[741] New England received six inches of snow in *June* 1816.[742] Here is a description of that event:

> Tom Canby: Frost and ice were common in every month of the year, and very little vegetation matured. The sun's rays seemed to be destitute of heat, all nature seemed to be clad in a sable hue, and men exhibited no little anxiety concerning the future of this life.
>
> All across the northern hemisphere there was misery and despair. In Canada and northern Europe dying livestock and shriveled crops brought starvation to 80,000.

What happened to bring on a cataclysm so widespread and abrupt?

The answer smoldered 10,000 miles away, in today's Indonesia. . . . The 13,000-foot volcano Tambora erupted on an island near Java. For a week thunderous explosions rocked the region and were heard 1,000 miles away. Fiery ejections of rock, flame, gas, and steam shot into the stratosphere.

Thirty-six cubic miles of earth blasted heavenward—the greatest release of energy ever known. . . . At week's end 12,000 Japanese lay dead, tsunamis had killed thousands more on distant islands, and the volcano stood a mile shorter than before.

The trillions of tons of material that Tambora shot into the atmosphere circled the earth with the winds. For more than a year they blocked sunlight from the northern hemisphere, dimming the planet with that "sable hue."[743]

Krakatoa, the volcanic island between Java and Sumatra exploded in 1883, and lowered the worldwide temperature for five years.[744]

These eruptions were dwarfed by an earlier Yellowstone eruption, perhaps soon after the Flood, which produced at least 480 cubic miles of ash.[745]

Yet these eruptions are tiny compared to a different type of volcano that we don't see happening today that deposited gargantuan stacks of thick layers known as "continental flood basalts."[746] For example, the Deccan Traps of India are over a *mile* thick and spread over nearly 200,000 square miles of the Indian subcontinent. . . . The Siberian Traps in Russia are even thicker (more than 480,000 cubic miles in

---

[740] Steven A. Austin, PhD, "Mount St. Helens and Catastrophism," in *Proceedings of the First International Conference on Creationism*, Vol. 1, Creation Science Fellowship, Pittsburgh, PA, 1986, pp. 3-9.
[741] Isaac Asimov, PhD, *Asimov's New Guide to Science*, 1984, p. 169.
[742] F. Barrows Colton, "Weather Fights and Works for Man," *National Geographic*, December 1943, p. 668.
[743] Tom Canby, "The Year Without a Summer," *Legacy*, Sandy Spring Museum, Sandy Spring, MD, 2002, Winter Edition.
[744] Isaac Asimov, PhD, *Asimov's New Guide to Science*, 1984, p. 169.
[745] P. W. Lipman, "Calderas," *Encyclopedia of Volcanoes*, ed. H. Sigurdson, Academic Press, San Diego, CA, 2000, pp. 643-662.
[746] P. R. Hooper, "Flood Basalt Provinces," *Encyclopedia of Volcanoes*, ed. H. Sigurdson, Academic Press, San Diego, California: 2000, pp. 345-359.

volume), though they cover a somewhat smaller area (130,000 square miles).[747]

> Harold G. Coffin, PhD: "The Mid-Atlantic Ridge reaches from the volcanic island of Iceland and the volcanic rim of Greenland southward through the Atlantic, forming a massive range nearly 7,000 miles long. The island mountaintops of this range, such as the Azores, are volcanic. Dredged material and magnetometer records show that much if not all of the hidden range is volcanic. Such a mountain range would require massive outpourings of lava, since it is 10,000 feet high in places and as wide as 600 miles.
>
> Formations in South Africa considered to be late Triassic or early Jurassic have been estimated to contain 50,000 to 100,000 cubic miles of igneous [volcanic] rocks. . . . Similar evidences are found around the world, and it may well be inferred that volcanic activity has been tremendous in the past in variety, extent, and effect.[748]

It's hard to imagine the scale of an event that would produce these flood basalts. Many large cracks, or fissures, had to open in the earth all at once, for so much lava to pour out over such a wide area. . . .

> Steven A. Austin, PhD: The breakup of the fountains of the great deep at [the biblical Flood's] onset and continuing for 150 days would have involved not only the bursting out of water from inside the earth, but also steam and prodigious volumes of lava. Then after the fountains were closed and plate movements slowed, volcanic activity decreased at the end of the Flood. This is also reflected in the documented declining power of post-Flood volcanoes to their relative quiescence today.[749]
>
> Andrew A. Snelling, PhD: This is another powerful example of evidence that can be explained by catastrophic plate tectonics during the biblical Flood. . . . Only the record of the cataclysmic Flood in God's Word makes sense of the evidence we see in the geologic record of God's world.[750]

Thousands of volcanoes occurring during and immediately following Noah's Flood would have caused huge amounts of volcanic aerosols to remain in the atmosphere for hundreds of years:

> Harold G. Coffin, PhD: Volcanic activity in prehistoric times was evidently so violent that in the western part of North America there must have been an almost continuous line of erupting cones. In one small area, between the Feather and Pit rivers in northern California, more than 150 cones have been counted, averaging about three miles apart. They were not necessarily all in action at once, but it is plain that there must have been a large number belching forth dust and ashes simultaneously.[751]

The aerosols would reflect much solar radiation back to space and generate a large temperature drop over the land. Once the snow cover was established, more solar radiation was reflected back into space, reinforcing the cooling over land. There seems to be evidence for one to more than 30 ice ages. The preponderance of evidence supports just one, with seasonal and climatic changes causing the supposed evidence for more than one.

> David C. Read, JD: A major seventeenth-century work on the Genesis Flood was Thomas Burnet's *A Sacred Theory of the Earth*, published in 1681. . . . Burnett argued that during the Flood the earth's axis assumed its present 23.5-degree tilt relative to the plane of its orbit around the sun. Since the axis tilt is what causes the seasons, the antediluvian [pre-flood] world would have had a constant year-round climate. . . . The theory that the earth's axis tilted either at the Fall or at the Flood pre-dated Burnet. For example, John Milton included it in [his 1667 book] *Paradise Lost*.
>
> . . . Eventually, the factors that led to the Ice Age reversed themselves. The oceans cooled to their present temperatures, reducing evaporation and precipitation to current levels. Meanwhile, the rapid plate movements of the Flood and early post-Flood years decelerated to the present slow pace. Slowed tectonic activity led to less frequent volcanic eruptions. Reduced volcanic activity allowed the skies to clear of dust and aerosols, which allowed the sun to warm up the earth. Warmer summers melted the continental

---

[747] Andrew A. Snelling, PhD, "Volcanoes—Windows into Earth's Past," *Answers*, Vol. 5, No. 3, July-September, 2010, pp. 66-69.
[748] Harold G. Coffin, PhD, "Fire in the Earth," *Creation—Accident or Design?* Review and Herald Publishing Association, Washington, DC, 1969, pp. 132-133.
[749] Steven A. Austin, PhD, "The Declining Power of Post-flood Volcanoes," *Impact #302*, Institute for Creation Research, El Cajon, CA, August 1998.
[750] Andrew A. Snelling, PhD, "Volcanoes—Windows into Earth's Past," *Answers*, Vol. 5, No. 3, July-September, 2010, pp. 66-69.

[751] Harold G. Coffin, PhD, "Climatic Conditions of the Past," *Creation—Accident or Design?* Review and Herald Publishing Association, Washington, DC, 1969, p. 238.

glaciers back to their present locations, but the much colder oceans allowed ice caps to form at the poles.[752]

Over the next few hundred years the ice caps generally retreated to their present status. And there is even more evidence against uniformitarianism's concept that the present is key to the past. A November 14, 1963, event illustrates the fact that a young geologic formation can look old, and within years, support plants and animals.

> Leonard R. Brand, PhD: In the Atlantic Ocean near Iceland, a new piece of land appeared as a volcano reached above the water and formed the island of Surtsey. A geologist visiting the island soon after it was formed commented that *processes that usually take thousands of years happened on Surtsey in days or weeks*. The reason is at least partly apparent. The island formed in the ocean with wave action constantly at work, carving cliffs and beaches and other geologic features. Surtsey shows us how quickly some geologic processes can occur when an abundance of water energy does the work and an abundant input of sediment occurs, as would be the case in a global catastrophe.[753]

According to *Wikipedia*, "The [Surtsey] eruption lasted [from 14 November, 1963] until 5 June 1967, when the island reached its maximum size of 2.7 km2 (1.0 sq mi). Since then, wind and wave erosion have caused the island to steadily diminish in size: as of 2002, its surface area was 1.4 km2 (0.54 sq mi).[754] It was intensively studied by volcanologists during its eruption, and afterwards by botanists and biologists as life forms gradually colonized the originally barren island."[755]

> *Wikipedia:* In the summer of 1965 the first vascular plant was found growing on the northern shore of Surtsey, mosses became visible in 1967 and lichens were first found on the Surtsey lava in 1970. Mosses and lichens now cover much of the island. During the island's first 20 years, 20 species of plants were observed at one time or another, but only 10 became established in the nutrient-poor sandy soil. As birds began nesting on the island, soil conditions improved, and more vascular plant species were able to survive. In 1998, the first bush was found on the island – a tea-leaved willow (Salix phylicifolia), which can grow to heights of up to 4 metres (13 ft). As of 2008, 69 species of plants have been found on Surtsey, of which about 30 have become established. . . . More species continue to arrive, at a typical rate of roughly 2-5 new species per year.
>
> Birds began nesting on Surtsey three years after the eruptions ended, with fulmar and guillemot the first species to set up home. Twelve species are now regularly found on the island. . . . As well as providing a home for some species of birds, Surtsey has also been used as a stopping-off point for migrating birds, particularly those en route between Europe and Iceland. In 2008, the 14th bird species was detected with the discovery of a Common Raven's nest.
>
> According to a 30 May 2009 report, a Golden Plover was nesting on the island with four eggs.
>
> Seals were found to be breeding on the island in 1983, and a group of up to 70 made the island their breeding spot. Grey seals are more common on the island than harbour seals, but both are now well established. The presence of seals attracts orcas, which are frequently seen in the waters around the Vestmannaeyjar archipelago and now frequent the waters around Surtsey.
>
> On the submarine portion of the island, many marine species are found. Starfish are abundant, as are sea urchins and limpets. The rocks are covered in algae, and seaweed covers much of the submarine slopes of the volcano, with its densest cover between 10 and 20 metres (33 to 66 ft) below sea level.[756]

---

[752] David C. Read, JD, "When Did the Dinosaurs Live? The 'Scientific View,'" and "Ice Age—The Flood's Aftermath," *Dinosaurs*, Clarion Call Books, Keene, TX, 2009, pp. 56, 80, 194.
[753] Leonard R. Brand, PhD, "Geologic Time," *Faith, Reason, and Earth History*, Andrews University Press, Berrien Springs, MI, 1997, p. 246, (emphasis added).
[754] Jakobsson, Sveinn P. (2007-05-06), *Surtsey—Geology*, The Surtsey Research Society, http://www.surtsey.is/pp_ens/geo_2.htm, retrieved 2008-07-08
[755] *Wikipedia*, "Surtsey."
[756] "Surtsey," *Wikipedia, the free encyclopedia*. Facts in this condensed *Wikipedia* topic are footnoted on-line.

VI
Dating Techniques

Chapter 32
Radiocarbon Dating[757]

The validity of radiocarbon dating has not proven to be as reliable as it theoretically should be. As we shall see, radiometric dating results are sometimes discarded and are used only if they are in a range that supports the expected "correct dates."

One difficulty with radiocarbon dating is that the magnetic field of the earth deflects cosmic radiation to the North and South Poles. Because this magnetic field is weakening (6 percent in the last 150 years), it is deflecting less radiation now than it once did. If so, less carbon-14 would have been formed years ago, which, in itself, also would produce carbon-14 dates older than their chronological age.

The carbon dioxide in the earth's atmosphere contains only 0.0000000001 percent radioactive carbon (carbon-14), which is produced by cosmic radiation striking the nitrogen in the atmosphere and turning it into carbon-14. Plants absorb it and animals eat the plants. When a plant or animal dies it stops taking in more carbon-14. Whatever carbon-14 it had will decay. It was decaying while the animal was alive but was replaced by new carbon-14 in the animal's food. Carbon-14 is unstable, so about half of it will break down—a statistical average—about every 5,730 ± 40 years. So, when testing for carbon-14, the amount of carbon-14 in the fossil is compared with the amount in the atmosphere, and if it is half as much, it is determined that the specimen has been dead for one half-life (5,730 years). Measurements go from a half to a fourth to an eighth to a sixteenth. Beyond about ten half-lives, there is not enough carbon-14 to measure.

Textbooks teach that some coal formed 250 million years ago during the "Carboniferous Era." So, after 10 half-lives, coal should have no carbon-14 left to measure. However, carbon-14 *has* been found in coal. How does this "anomaly" relate to the age of the earth?

The Radioisotope and the Age of The Earth (RATE) Group, sponsored by the Institute for Creation Research, asks, "Why is there any measurable radiocarbon in rocks which are supposed to be billions of years old? What's more, why is the same amount of carbon-14 found in deep-earth diamonds as is found in fossils from the top to the bottom of the fossil record, if the earth was formed billions of years before animals and organisms appeared on the scene? Is this clear evidence that the earth is thousands (not billions) of years old?"[758]

> Creation geophysicist John Baumgardner, PhD: Carbon-14 in coal simply is the carbon-14 that was in the trees and other kinds of plants that were buried in the Flood. So, the way that this carbon-14 got into the coal was simply it was in the living organisms before the catastrophe. The catastrophe ripped up the trees and plant life and buried them in thick layers that got compressed and formed into coal. So, it's relatively simple to understand how the carbon-14 got into the coal. The big issue is why is it that if the earth is hundreds of millions of years old—if it's been hundreds of millions of years since that happened, why should there be any carbon-14 still left?
>
> There's no technique on earth that can measure levels of carbon-14 in materials that are older than 100,000 years. So, the fact that we find so much carbon-14 in coal and other kinds of fossil organisms indicates that their age is much less than 100,000 years.
>
> Secular geologists have documented that very well. We had ten coal samples analyzed, all collected in a similar manner by the U.S. Department of Energy, all analyzed by what we deemed the best radiocarbon laboratory in the world, and one surprising result was that within the statistics of our samples there was no difference in the levels of carbon-14

---

[757] Paul A. L. Giem, MD, consultant to the author of *Creation*. Dr. Giem is himself the author of *Scientific Theology*, La Sierra University Press, Riverside, CA, 1997.

[758] *Thousands . . . Not Billions*, DVD, The Institute for Creation Research, an ICR Special Edition, in association with TEN31 Productions, 2005.

in the coal near the top [of the geologic column nor] the coal in the bottom, which indicates that all of this material lived at the same time and died at the same time—strong evidence that the entire fossil record has the same age.

In addition, we analyzed several diamonds for carbon-14—hoping we would find carbon-14, but unsure about it. And so, it was extremely exciting to get the first results back and to find significant levels of carbon-14 in diamond— something like a hundred times the detection threshold. This is very significant because diamonds are formed deep in the earth, on the order of a hundred miles down in the earth— generally believed to have formed early in the history of the planet. And because diamonds are the hardest substance in existence, it's essentially impossible to contaminate a diamond. So, what this means is that the earth itself must be young—only a few thousand years old—else the carbon-14 in these diamonds would long since have decayed. And so, this carbon-14 must date, we believe, to the original creation of the earth itself, just a few thousand years ago.

[Author's note: It is also possible that the catastrophic Flood and its layering of deep sediments, mountain upthrusts, etc. could account for the depth of the diamond. In either case— young earth, or recent global Flood upheaval— carbon-14 in 100-mile deep diamonds does not point to ancient age.]

> Baumgardner (cont'd): The levels of C-14 we're finding indicates that all these fossils lived at the same time and were buried at the same time, just a few thousand years ago. And this is an extremely significant result, and a strong support for the biblical account of a global Flood, just a few thousand years ago.[759]

The RATE Group's research concluded that:
1. A large amount of radioisotope decay has occurred.
2. Conventional dating methods are highly inaccurate and inconsistent.
3. There should be virtually *no* helium or carbon-14 in rocks, diamonds or coal if they're actually millions or billions of years old.
4. The fact that there *is* indicates that the earth is only thousands not billions of years old.[760]

Carbon-14 tests on oil, or hydrocarbons, in the Gulf of Mexico area from ten 1952 marine sediment sites in four areas of Texas and Louisiana, have shown that oil is found in sediments that are thousands of years old, not millions, as was held in the "most prevalent viewpoint. All Gulf of Mexico age determinations were made by J. Laurence Kulp, of the Lamont Geological Observatory of Columbia University. More recently a school of thought has developed [according to Kulp] which believes that oil formation may begin soon after deposition of the organic matter in the sediments."[761]

Carbon-dating assumptions

The amount of carbon-14 can be measured and the rate of decay can be determined. However, *no one can show how much carbon-14 any specimen had when it died, whether the rate of decay always remained the same, or whether the specimen has been contaminated.* These three unknowns become major assumptions for this dating method as well as for other radiometric dating methods (described in chapter 33).

Furthermore, after carbon-14 dating was developed in 1947 by evolutionist Willard F. Libby, PhD, he acknowledged that carbon-14 on earth had not yet reached equilibrium (the amount decaying on earth equaling the amount being created in the atmosphere). He attributed this problem to his margin of error:

---

[759] John Baumgardner, PhD, *Thousands . . . Not Billions*, DVD, The Institute for Creation Research, an ICR Special Edition, in association with TEN 31 Productions, 2005. The Dallas (Texas) Institute for Creation Research (on origins and earth history) was founded in 1979. Through its PhD specialists, it conducts laboratory, field, theoretical, and library scientific research (many, multi-year projects) at key locations (the Grand Canyon, Mount St. Helens, Yosemite Valley, Santa Cruz River Valley in Argentina, etc.). The research is on **R**adioisotopes and the **A**ge of **T**he **E**arth (the RATE Group), on Flood-Activated Sedimentation and Tectonics (the FAST group), and on the Big Bang, paleoclimatology, the ice age, geology, flood geology, zoology, dinosaurs, physics, astro/geophysics, nuclear physics, microbiology, genetics, medicine and the human body, forensic science, engineering science, etc. In addition to *research* it is a *graduate education* institution; and *produces and publishes* books, films, videos, and periodicals (such as its monthly magazine *Acts & Facts*).

[760] *Thousands . . . Not Billions*, DVD, The Institute for Creation Research, an ICR Special Edition, in association with TEN31 Productions, 2005.
[761] Paul V. Smith, Jr. "The Occurrence of Hydrocarbons in Recent Sediments from the Gulf of Mexico," Standard Oil Development Company, Linden, NJ, *Science*, October 24, 1952, pp. 437-439.

[Libby] found a considerable discrepancy in his measurements indicating that, apparently, radiocarbon was being created in the atmosphere somewhere around 25 percent faster than it was becoming extinct. Since this result was inexplicable by any conventional scientific means, Libby put the discrepancy down to experimental error.

During the 1960s, Libby's experiments were repeated by chemists who had been able to refine their techniques after a decade or so of experiments. . . . The new experiments, though, revealed that the discrepancy observed by Libby was not merely experimental error—it did exist.[762]

The *Review of Geophysics* agreed:

Richard Lingenfelter: There is strong indication, despite the large errors, that the present natural production rate exceeds the natural decay rate by as much as 25 percent.[763]

Libby determined that if a new earth were to be created, it would take about 30,000 years for carbon-14 to reach equilibrium: where the amount entering from the atmosphere matched the decay rate.[764] It has not yet reached equilibrium. And the figures are even worse than Libby acknowledged:

American Antiquity: Radiocarbon is forming 28-37 percent faster than it is decaying.[765]

If radiocarbon is still forming that much faster than it is decaying according to Libby's estimate, the earth's atmosphere is significantly less than 30,000 years old.

And there are other factors that affect the amount of carbon-14 in the atmosphere: cosmic ray penetration of the earth's atmosphere; the strength of the earth's magnetic field; carbon dioxide levels in the atmosphere; and the Genesis Flood.

American Scientist: We now know that the assumption that the biospheric inventory of C-14 has remained constant over the past 50,000 years or so is *not true.*[766]

Dating techniques were so unreliable by the 1970s and 80s that scientists could pick and choose which numbers to use.

Soderbergh and Olsson (at the *Proceedings of the Twelfth Nobel Symposium* in New York) reported: If a [carbon-14 test result] date supports our theories, we put it in the main text. If it does not entirely contradict them, we put it in a footnote. And if it is completely "out of date," we just drop it.[767]

Erich A. von Fange, PhD: In the spring of 1971 the British Museum processed palm kernels and mat reeds from the tomb of [King] Tutankhamen. Dr. Edwards, Curator of the Egyptian Department, reported to the museum at the University of Pennsylvania that the results gave a C-14 date of 899 to 846 BC. Without explanation, however, the results have never been formally published, because the dates support the "wrong" Egyptian chronology. In a follow-up of this matter, Mr. Burleigh, *director of the laboratory of the British Museum, stated* that he expected the results to be published shortly. Then he admitted that *results that deviate substantially from what is expected are often discarded and never published.*[768]

The Genesis Flood also would impact the amount of carbon-12 in the atmosphere.

The coal and oil in the ground illustrate the tremendous amount of vegetation that had to have been buried during the Flood. The carbon-12 that once existed on the earth, as evidenced by the coal and oil we see today, should cause the scientific community to question the accuracy of carbon-14 dating.

Carbon-14 can be used to date only organic material. It is not used to date sedimentary rocks, such as sandstone, shale, and limestone, which cover most of the planet.

Statements made in the last 30 to 50 years may

---

[762] Richard Milton, "The Key to the Past," *Shattering the Myths of Darwinism*, Park Street Press, Rochester, VT, 1997, p. 32; W. F. Libby, Radiocarbon Dating, University of Chicago Press, Chicago, IL, 1955, p. 7.
[763] Richard Lingenfelter, "Production of C-14 by cosmic ray neutrons," *Review of Geophysics*, 1963, p. 51.
[764] W. F. Libby, *Radiocarbon Dating*, University of Chicago Press, Chicago, IL, 1955, p. 7.
[765] R. E. Taylor, et al, "Major Revisions in the Pleistocene Age Assignments for North American Human Skeletons by C-14 Accelerator Mass Spectrometry," *American Antiquity*, Vol. 50. No. 1, 1985, pp. 136-140.
[766] Elizabeth K. Ralph and Henry M. Michael, "Twenty-five Years of Radiocarbon Dating," *American Scientist*, September/October 1974, p. 555, (emphasis added).
[767] T. Save-Soderbergh and I. U. Olsson (Institute of Egyptology and Institute of Physics respectively, University of Uppsala, Sweden); "C-14 dating and Egyptian chronology in Radiocarbon Variations and Absolute Chronology," *Proceedings of the Twelfth Nobel Symposium*, NY, 1970, p. 35.
[768] Erich A. von Fange, PhD, "Ancient Plant Oddities and Mysteries," *In Search of the Genesis World*, Concordia Publishing House, St. Louis, MO, 2006, p. 227, (emphasis added).

not illustrate today's advanced scientific methods or results. However, I share some of them for the purpose of illustrating the thinking of mainstream scientists of that time, and their sometimes-frustrating efforts to determine and portray truth.

> *Nature and New Scientist:* Thirty-eight laboratories worldwide carbon-dated samples of wood, peat and carbonate, and produced differing dates for similar objects of the same age. The overall finding of the comparative test was that radiocarbon dating was "two to three times less accurate than implied by their error terms." Ages of objects assessed by this method cannot therefore be viewed as being credible.[769]

> Sheridan Bowman, PhD, (written for the British Museum): Radiocarbon is not quite as straightforward as it may seem. The technique does not in fact provide true ages, and radiocarbon results must be adjusted—calibrated—to bring them into line with calendar ages.[770]

The major assumptions relating to carbon-14 dating (which are based on the best possible scientific methods of the time) and the resulting untrustworthy, inaccurate, and unreliable results, give the *concept of "reasonable doubt"* powerful meaning, *at least for that time.* And the facts that coal, diamonds, and dinosaurs contain appreciable amounts of carbon-14 and that carbon-14 has not yet reached equilibrium (where the amount decaying on earth would equal the amount being created in the atmosphere [implying a young earth]), provide hard-core, scientific evidence against the long ages of life on earth "resulting" from these analyses.

The presuppositions of evolutionary science have not changed. Over 40 years ago, material from supposedly 100-million-year-old Cretaceous layers (where dinosaur bones are found) gave carbon-14 dates of 34,000 years.[771] Hugh Miller, from Columbus, Ohio, had four dinosaur bone samples carbon dated at 20,000 years old. When he identified them as dinosaur bones, the lab stated that because dinosaurs lived 70,000,000 years ago, had they known, they would have never carbon dated them.[772]

> *American Antiquity:* Eleven early North American human skeletons averaged more than 28,000 years. Reinvestigation produced revised dates that averaged less than 4,000 years, but the revised dates have also been challenged.[773]

Serious problems include different dates on the same specimen. In 2005, a case of deliberate deception was exposed relating to radiometric dating. A German anthropologist misrepresented the age of Neanderthal skulls and artifacts for 30 years. A university panel exposed his frauds and he resigned in February, 2005. He had dated the "Bischof-Speyer" skeleton at 21,300 years, but testing at Oxford University showed them to be 3,300 years old. This is an 18,000-year discrepancy.[774]

This same professor of evolution had dated another skull found near Paderborn, Germany, as being 27,400 years old. It was believed to be the oldest human remain found in the region until Oxford investigations identified it as belonging to an elderly man who died in 1350.[775]

> Germany's Herne anthropological museum, which owns the Paderborn skull, was so disturbed by the findings that it did its own tests. Not only was Oxford University correct, and the skull not as old as it was claimed, but it was *not even fossilized*. Museum director Barbara Ruschoff-Thale said, "We had the skull cut open and it still smelt," and "We are naturally very disappointed."[776]

So, how does radiocarbon dating (carbon-14) compare to other kinds of radioisotope dating?

---

[769] *Nature*, September 28, 1989, p. 267; *New Scientist*, September 30, 1989, p. 10.
[770] Sheridan Bowman, PhD, *Diggings*, August, 1990, p. 8; *see also*: Sheridan Bowman, PhD, *Radiocarbon Dating—Interpreting the Past*, University of California Press/British Museum, 1990.
[771] Reginald Daly, "Origin of sedimentary mountains, opposing theories," *Earth's Most Challenging Mysteries*, The Craig Press, Nutly, NJ, 1972, p. 280.
[772] Erich A. von Fange, PhD, "The Great Prologue," *Noah to Abram: The Turbulent Years*, Living Word Services, Syracuse, IN, 1994, p. 36.
[773] R. E. Taylor; L. A. Payen; C. A. Prior; P. J. Slota, Jr.; R. Gillespie; J. A. J. Gowlett; R. E. M. Hedges; A. J. T. Jull, T. H. Zabel; D. J. Donahue; R. Berger; "Major Revisions in the Pleistocene Age Assignments for North American Human Skeletons by C-14 Accelerator Mass Spectrometry; none older than 11,000 C-14 years B. P." *American Antiquity*, 1985, 50(1):136-140.
[774] Tony Paterson, Neanderthal Man 'never walked in Northern Europe' <www.telegraph.co.uk/news/main.jhtml?xml=/news/2004/08/22/wnean22.xml&sSheet=/news/2004/08/22/ixworld.html>, 21 February 2005.
[775] www.angelfire.com/mi/dinosaurs/Murdock_TJ19_1__17_18.pdf, pp. 7-8. See also: www.worldnetdaily.com February 19, 2005.
[776] www.angelfire.com/mi/dinosaurs/Murdock_TJ19_1_17_18.pdf, pp. 7-8, (emphasis added).

Chapter 33
Radiometric Dating—Are the Dates Set in Stone?

Just how old is planet earth? and the universe? Can radiometric dating answer these questions?

Except for carbon-14 dating, radioisotope dating (also known as radiometric dating) is used to date igneous rocks (rocks that were once molten, then cooled and hardened) such as basalt, tuff, or volcanic lava. (Inorganic methods of radiometric dating cannot be used directly to date *organic* material. Organic material can be dated indirectly by dating volcanic layers above and below the fossil-bearing rocks.)

Radioisotope dating methods measure the amount of "daughter element" (such as lead) that has decayed from a "parent element" (such as uranium). Once the molten rock solidifies, the radioactive clock is thought to be "set." Potassium-argon dating is an ideal example.

> United States Geological Survey: "How do geologists date rocks? By radiometric dating!" Radiometric clocks are "set" when each rock forms. "Forms" means the moment an igneous rock solidifies from magma, . . . or a rock heated by metamorphism cools off. It's this resetting process that gives us the ability to date rocks that formed at different times in earth history.
>
> A commonly used radiometric dating technique relies on the breakdown of potassium (^{40}K) to argon (^{40}Ar). In igneous rocks, the potassium-argon "clock" is set the moment the rock first crystallizes [hardens] from magma. Precise measurements of the amount of ^{40}K relative to ^{40}Ar in an igneous rock can tell us the amount of time that has passed since rock crystallized. If an igneous or other rock is metamorphosed [transformed], its radiometric clock is reset, and the potassium-argon measurements can be used to tell the number of years that has passed since metamorphism.[776]

One of my respected consultants, Ariel A. Roth, PhD, has said that the above USGS statement is not quite correct. Competent scientists may have honest disagreements.

As Dr. Roth explained, the USGS statement is an oversimplification of the actual dates of the various constituents of the rocks.

In these chapters on radiometric dating and volcanos, I offer my understanding of the findings and reasoning of the scientists of the RATE group (**R**adioisotopes and the **A**ge of **T**he **E**arth), a seven-year research project published in 2005 by the Institute for Creation Research (ICR) and the Creation Research Society (CRS). Both organizations hold the "Young universe—recent earth—recent life" view. Online they offer many articles, books, DVDs, etc. in support of that perspective. Reflecting their origins view, the following is a summary of radiometric dating and volcano ages.

> Genesis Apologetics: Creation or evolution? Both views require faith, because no one was there thousands or millions of years ago to observe how it all got started. But can the idea of millions of years of evolution be scientifically validated? . . . Validation is the process of confirming what we believe to be true by what can be observed to be true. Has this ever been done for radiometric dating, the process that gives us the very idea of long ages?
>
> Actually, long ages and radiometric dating have not been up to the validation test. Radiometric dating has [usually not] been validated [as matching] the absolute known [eruption] ages of rocks. . . .
>
> For example, Mount St. Helens: this volcano erupted in the 1980s [catastrophically in 1980, and continuing to erupt for several years], giving scientists the opportunity to date the rocks that were formed from the eruptions.
>
> The results? Five different ages—all between 350,000 years and 2.8 million years old—from rocks that we know are [only a few decades] old.

---
[776] United States Geological Survey (USGS), "How do geologists date rocks? Radiometric dating!" U.S. Department of the Interior, April 25, 2017, (emphasis added by USGS).

This discrepancy happens all the time in radiometric dating studies.[777]

Are the radiometric dates set in stone?

As we saw, radioisotope dating methods measure the amount of "daughter element" (such as argon) that has decayed from its older "parent element" (such as potassium). Once the molten rock solidifies (enough), the radiometric clock is supposed to "begin." But it is known that argon does not always completely escape from the rock while it is still in liquid form. So, the rock starts out with an age greater than zero when it finally solidifies. That is why, depending on which of five rock fractions were tested, widely varied "ages" resulted from the tests of Mount St. Helens. With uranium, its daughter element, lead, is not driven off by melting the rock, so assumptions have to be made about how much lead was in the rock in the first place, and those assumptions can be shown to be questionable at best.

Radiometric dating—potassium-argon, uranium-lead, rubidium-strontium, samarian-neodymium, uranium 235-uranium 238—are all based on assumptions that may be flawed. The RATE group has concluded that "this straightforward dating concept is unreliable, because it's based on three unprovable and questionable assumptions." These assumptions are:

1. The amount of parent and daughter isotopes have not been altered by anything except radioactive decay.
2. When the rock was formed, it contained a known amount of the daughter isotopes, in many cases, believed to be zero.
3. The decay rate has been constant throughout history.[778]

How does radiometric dating relate to water-laid, stratified rock? Each layer in what is known as the geologic column was assigned a name (such as Cambrian), an age (in thousands or millions of years), and index fossils. Any dating technique, such as radioisotope dating of the rocks in that layer, has to match the "age" of the geologic column layer, or the age is assumed to be incorrect. So, dating of fossils is checked by their position in the geologic column, with radiometric dating only as a secondary backup—if it agrees with the previously assumed ages.

Sometimes rock specimens are radiometrically tested five or six times until the "right" date comes up—one that matches the age of the layer in the geologic column that is now considered a standard age. Question: If differing ages come up, how does one know that *any* of them are correct?

Geologic Column, the Original Foundation of Radiometric Dates

> J. E O'Rourke: Radiometric dating would not have been feasible if the geologic column had not been erected first.[779]

In 1969, James P. Dawson, Chief of Engineering and Operations for NASA's Lunar and Earth Science Division at the Manned Spacecraft Center in Houston, Texas, divided moon rock 10017 into six pieces and dated them many times. The ages ranged from 2.5 billion to 4.6 billion years. He also worked on other lunar samples, including the Genesis rock, and reported that ages in the same rock dated from 10,000 to several billion years.[780]

The fact that in the field of radiometric dating it is assumed that nothing has contaminated the specimen for eons and that the various elements have decayed at the same rate may help explain why outcomes are so varied.

> A. Hayatsu: As much as 80 percent of the potassium in a small sample of an iron meteorite can be removed by distilled water in 4.5 hours.[781]

Creationists, as well as evolutionists, don't understand everything there is to know about radiometric dating. Research is ongoing. Radiometric

---

[777] Radiometric Dating Debunked in 3 Minutes," Genesis Apologetics, You Tube, June 16, 2017, (emphasis added). See also *the reliability of radiometric dating questioned* in the online article "RATE group reveals exciting breakthroughs!" at <creation.com>, by Carl Wieland, CMI, Australia.
[778] *Thousands . . . Not Billions,* DVD, Institute for Creation Research, An ICR Special Edition, in association with TEN31 Productions, 2005.

[779] J. E. O'Rourke, "Pragmatism versus Materialism in Stratigraphy," *American Journal of Science,* Vol. 276, January 1976, p. 54.
[780] www.jpdawson.com or www.aaronc.com.
[781] L. A. Rancitelli, and D. E. Fischer, "Potassium-Argon Ages of Iron Meteorites," *Planetary Sciences Abstracts,* 48th Annual Meeting, 2016, p. 167.

dating is thought to provide reliable absolute dates for the geologic column. The science appears to be so impressive that many creationists accept it as evidence of a very Old Earth. How do evolutionary scientists cope with radiometric dating that doesn't fit the geologic column's time scale?

> *Canadian Journal of Earth Sciences:* In conventional interpretation of K-Ar [Potassium-Argon] age data, it is common to discard ages which are substantially too high or too low compared with the rest of the group or with other available data such as the geological time scale.[782]

Perhaps one of the best ways to consider "the rest of the story" in radiometric dating is to present a case study of the dating of the East African KBS Tuff strata and the famous fossil KNM-ER 1470, as recorded in the scientific journals, especially the British journal *Nature*.[783]

A major controversy evolved in 1972 when Richard Leakey, son of famous paleontologists Louis and Mary Leakey, discovered a perfectly normal, modern-looking human skull that originally was believed to be 2.9 million years old. The conflict between its modern appearance and its ancient age presented a serious challenge to all currently held theories of human evolution. The conflict lasted ten years.[784]

In 1969, before Leakey discovered the skull, he had the volcanic rock layer dated by F. J. Fitch, from Birkbeck College, University of London, and J. A. Miller, from Cambridge University, recognized authorities in potassium-argon (K-Ar) dating:

> Marvin Lubenow, PhD: Thus began the long process, based upon evolutionary and other philosophical assumptions, by which the geochronologist manipulates or "massages" the data to guarantee that he gets a "good" date. I want to stress that the geochronologist does this in absolute sincerity. He is so committed to evolution and its attendant age demands that he believes implicitly that he removes error from his data to arrive at truth. The obvious subjectivity in it escapes him. It is a perfect illustration of circular reasoning in an experimental frame of reference. The experimenter manipulates the data to guarantee that he gets the result that is "needed."[785]

Drs. Fitch and Miller first dated the tuff at 2.6 million years old, a figure that was published in both the scientific and popular press.[786] Leakey stated that [skull] 1470 was found below rock that was "accurately dated"[787] and "securely dated"[788] at 2.6 million years.

In 1974, *Nature* published a third chronology of the area by Brock (University of Nairobi, Kenya,) and Isaac (University of California, Berkeley). They based their study on the paleomagnetism of deposits below the KBS Tuff, using 247 samples, and concluded that 1470 was 2.7 to 3 million years old. This report presented a correlation of the various dating methods. The heading of the article stated that their measurements "provide a valuable check on other dating methods."[789] Later they reported that because the isotopic and paleomagnetic ages were consistent, "this independent evidence greatly strengthens our proposed chronology."[790]

> Marvin Lubenow, PhD: By late 1974, two years after skull 1470 had been presented to the world, the KBS Tuff had been dated five different times by four different dating methods. The alleged compatibility of the four different methods would seem to make all of this a geologist's dream. What better proof could one want for the reliability of the various dating methods to furnish independent confirmation of the dates for the fossil material? Because 1470 was found below rock dated to 2.61 mya [million years ago]

---

[782] A. Hayatsu, "K-Ar Isochron Age of the North Mountain Basalt, Nova Scotia," *Canadian Journal of Earth Sciences*, Vol. 16, April, 1979, pp. 973-975.
[783] F. J. Fitch, PhD, and J. A. Miller, PhD, "Radioisotopic Age Determinations of Lake Rudolf Artifact Site," *Nature*, MacMillan (Journals) LTD, London, April 18, 1970, p. 226; Marvin Lubenow, PhD, "Appendix," *Bones of Contention*, Grand Rapids: Baker, 1992, 1st ed, pp. 247-266.
[784] Marvin Lubenow, PhD, "Appendix," *Bones of Contention*, Baker Books, Grand Rapids, MI, 1992, 1st ed, pp. 247-266.
[785] Marvin Lubenow, PhD, "Appendix," *Bones of Contention*, Baker Books, Grand Rapids, MI, 1992, 1st ed, pp. 247-266.
[786] F. J. Fitch, PhD, and J. A. Miller, PhD, "Radioisotopic Age Determinations of Lake Rudolf Artifact Site," *Nature*, April 18, 1970, p. 228.
[787] *Detroit Free Press*, November 10, 1972.
[788] R. E. F. Leakey, "Evidence for an Advanced Plio-Pleistocene Hominid from East Rudolf, Kenya," *Nature*, Vol. 242, April 13, 1973, p. 447.
[789] Brock and Isaac, "Paleomagnetic stratigraphy and chronology of hominid-bearing sediments east of Lake Rudolf," Kenya, *Nature*, Vol. 247, February 8, 1974, pp. 344-348.
[790] Brock and Isaac, "Paleomagnetic stratigraphy and chronology of hominid-bearing sediments east of Lake Rudolf," Kenya, *Nature*, Vol. 247, February 8, 1974, pp. 347-348.

and above rock dated at 3.18 mya, skull 1470 was estimated to be an incredible 2.9 million years old. Richard Leakey had found the world's oldest fossil belonging to the genus *Homo*. On the surface all seemed serene.

However, under the surface paleoanthropology was seething in ferment. Skull 1470 with its estimated date of 2.9 mya presented the evolutionary world with an intolerable situation. Richard Leakey did not exaggerate when he declared: "Either we toss out this skull or we toss out our theories of early man."[791] The problem was quite simple. The theory of human evolution did not allow for a skull so modern in morphology [form and structure] to be that old."[792]

J. B. Birdsell, PhD: The real question in the case of 1470 is whether the discovery should be considered as representative of its population or as an extreme variant within it.... The estimates do not indicate that 1470 should be classed with any other contemporary [2.9 million years ago] group.

It is worth asking what the sex of cranium 1470 might be and what effect differences in sex would have on the problem. The skull is so modern in its general form that it is very tempting to consider that it is indeed female. In this case it can be estimated that an average male from the same population would be about 55 centimeters larger, or have a cranial capacity of 835 cc. This places the population from which we have this interesting sample even higher on the evolutionary scale and so creates a greater problem because of the very early date at which it lived. It cannot be made to disappear within the normal range of any other known early human population.[793]

How to solve the problem? *Ten more samples of the KBS tuff then dated at 0.52 to 2.64 million years old.* Problem solved? In some cases, "naughty" crystals were removed to give results more appropriate to the overriding principle behind it all—human evolution.[794]

Dennis R. Petersen: The anthropologist said this fossil was 2.8 million years old, yet it belongs to man's genus. In other words, Leakey claimed it was more man-like than any of the other near-man relics on the chart. The problem was that the skull was found beneath volcanic ash that had been acceptably dated for years by evolutionist's reckoning as 2.6 million years old. *That would make a human-like ancestor over a million years older than our nearest ape-like ancestor.*[795]

In 1973, Richard Leaky had said: *It simply fits no previous models of human beginnings.... It leaves in ruins the notion that all early fossils can be arranged in an orderly sequence of evolutionary change.*[796]

Princeton University physical anthropologist Alan Mann, PhD, spent four weeks with Leakey in Kenya. He reportedly had been very skeptical of Leakey's report concerning 1470, but became convinced that Leakey had revolutionized anthropology. He, like most other anthropologists, was left thoroughly confused by the astounding implications of Leakey's discovery, saying, "We just don't know what happened. There's no real theories. Everybody's sort of astounded.... It just throws us back to 'go.'"[797]

Duane Gish, PhD: Evidence produced by Richard Leakey in the past two or three years has now established strong support for the fact that the *australopithecines* did not walk upright, but were long-armed, short-legged knucklewalkers, similar to the extant [present] African apes.

Leakey's latest find may now have delivered the final shattering blow to the *australopithecines* as candidates for man's ancestor; in fact, if accepted, it will destroy all presently held theories on man's evolutionary ancestry. In his lecture last year in San Diego (which the author [Gish] attended) Leakey reported that what he has found destroys all that we have ever been taught about human evolution, and, he said, "is thought to have nothing to offer in its place!"[798]

---

[791] Richard E. Leakey, "Skull 1470," *National Geographic*, June 1973, p. 819.
[792] Marvin Lubenow, "Appendix," *Bones of Contention*, Baker Books, Grand Rapids, MI, 1992, 1st ed, pp. 247-266.
[793] J. B. Birdsell, PhD, "Men of the Lower and Middle Pleistocene," *Human Evolution: An Introduction to the New Physical Anthropology*, Rand McNally College Publishing Company, Chicago, 1975, pp. 288-289.
[794] Marvin Lubenow, "Appendix," *Bones of Contention*, Baker Books, Grand Rapids, MI, 1992, 1st ed, pp. 247-266.
[795] Dennis R. Petersen, "The Discovery That Rattled All the Other Bones," *Unlocking the Mysteries of Creation: the Explorer's Guide to the Awesome Works of God*, Bridge-Logos Publishers, Alachua, FL, 2002, p. 130, (emphasis added).
[796] Richard Leakey, quoted by Dennis R. Petersen, "The Discovery That Rattled All the Other Bones," *Unlocking the Mysteries of Creation: the Explorer's Guide to the Awesome Works of God*, Bridge-Logos Publishers, Alachua, FL, 2002, p. 130.
[797] Alan Mann, PhD, reported by J. N. Shurkin (Knight Newspapers writer) *The Cincinnati Engineer*, October 10, 1973, p. 6.
[798] Duane T. Gish, PhD, "Brainwashed," Lecture given at the University of California at Davis, http://www.skepticfiles.org/evolut/evolve7i.htm, no date recorded.

Dating volcanoes

We turn now to the radioisotope dating methods used to date once-molten igneous and metamorphic rocks (basalt, or lava) after it has hardened. They cannot be used to date most sedimentary rocks, such as limestone, sandstone, or shale.

When a volcano erupts and its lava hardens, its age clock is thought to restart. Another theory is that this material is really old. But serious questions arise when dates from the same rocks vary as much as they do. So how does radiometric dating compare with historically-known dates? Basalt (a lava flow) from Mt. Etna, Sicily (which erupted in 122 BC) gave a potassium-argon age of 250,000 years.[799] Wood buried beneath New Zealand's volcanic island of Rangitoto dated at less than 350 years. Yet the lava itself has been potassium-argon dated at 145,000 to 465,000 years old.[800]

Lava from a Hawaiian volcano that erupted in 1801 gave a potassium-argon date of 1.6 million years.[801] Basalt from Hawaii's Mt. Kilauea Iki, which erupted in 1959 was dated with potassium-argon to 8.5 million years.[802] A 1964 basalt from Mt. Etna, Sicily, dated at 700,000 years old. A 1972 eruption gave an age of 350,000 years.[803]

Recent material from the lava dome at Mt. Saint Helens, Washington, which erupted in 1980, dated at 350,000; 900,000; and 2.8 million years.[804]

> Derek Isaacs: At [New Zealand's volcanic Mount Ngauruhoe], eleven samples were collected from eruptions in 1949, 1954, and 1975. Geochron Laboratories of Cambridge, Massachusetts, then dated these samples. . . . The laboratory tests produced ages that range from 270,000 years up to 3.5 million years.[805]

> John D. Morris, PhD: When the same rock is dated by more than one method, it will often yield different "ages." And when the rock is dated more than one time by the same method, it will often give different results.[806]

Scientists were sent to the bottom of the Grand Canyon to sample the Cardenas Basalt, in the Precambrian area. These scientists delivered the rocks to a dating laboratory without providing any range of expected ages or reporting the location from which they came. Their only instructions were, "Date these rocks." The lab dated the rocks 13 different ways: five potassium-argon; six rubidium-strontium; one potassium-argon isochron, and one rubidium-strontium isochron. The rocks from this oldest layer were dated from 715 million to 1.07 billion years of age. The lab chose to report the oldest results (from the rubidium-strontium isochron method) as 1.07 billion years.

Scientists then took rocks from the top of the Grand Canyon, the Unikaret Plateau (supposed to be the youngest rocks), to the same dating laboratory. Again, these scientists did not provide a range of expected ages nor did they identify the location from which the rocks came. They just asked the lab to "Date these rocks." Again, the lab used a variety of dating methods and came up with just as many different ages (which in itself should raise serious questions). To be consistent, because the lab had reported on the rubidium-strontium isochron age for the first rocks, they decided to report results from the same dating method for the second series. The result was 1.34 billion years old. The youngest rocks dated 330 million years older than the oldest rocks.[807]

> Andrew A. Snelling, PhD: The radioisotope methods, long touted as irrefutable dating the earth's rocks as countless millions of years old, have repeatedly failed to provide reliable and meaningful absolute ages for Grand Canyon

---

[799] G. B. Dalrymple, "40 Ar/36 Ar Analyses of Historical Lava Flows," *Earth and Planetary Science Letters 6*, 1969, pp. 47-55.
[800] Ian McDougall, H. A. Polach, and J. J. Stipp, "Excess Radiogenic Argon in Young Subaerial Basalts from Auckland Volcanic Field, New Zealand," *Geochimica et Cosmochimica Acta*, Vol. 33, December 1969, pp. 1485, 1499.
[801] Funkhouser and Naughton, *Journal of Geophysical Research*, Vol. 73, July 15, 1968, p. 4601; G. Brent Dalrymple, Earth and Planetary Science Letters, 1969, p. 6-55.
[802] *Creation Ex Nihilo*, December 1999, p. 18.
[803] Andrew A. Snelling, PhD, "Excess Argon": The "Achilles' Heel" of Potassium-Argon and Argon-Argon "Dating" of Volcanic Rocks," *Impact #307*, Institute for Creation Research, El Cajon, CA, January 1999.
[804] S. A. Austin, "Excess Argon Within Mineral Concentrates from the New Dacite Lava Dome at Mount St. Helens Volcano," *CEN Technical Journal*, 1996, 10(3):335-343.
[805] Derek Isaacs, "Upon This Rock," *Dragons or Dinosaurs? Creation or Evolution?* Bridge-Logos, Alachua, FL, 2010, p. 149.
[806] John D. Morris, PhD, *The Geology Book*, Master Books, Inc., Green Forest, AR, 2000, p. 52.
[807] *Dating Fossils and the Rocks: Scientific Evidence and the Age of the Earth*, DVD, Answers in Genesis-USA, P. O. Box 510, Hebron, KY, 2004.

Rocks. Irreconcilable disagreement within and between the methods is the norm, even at the outcrop scale. This is a devastating "blow" to the long ages that are foundational to uniformitarian geology and evolutionary biology. Yet the discordance patterns are consistent with past accelerated radioisotope decay, which would also render these "clocks" useless. Thus, there is no reliable evidence to dispute that these metamorphosed basalt lava flows deep in Grand Canyon date back to the Creation Week only thousands of years ago.[808]

In 1993, scientists found trees buried in a 69-foot deep lava flow. Interesting. . . . The trees were subjected to carbon-14 dating methods and the lava was subjected to potassium-argon dating. Because the trees were caught up in the lava flow when it was molten, the trees and the basalt should all date the same. The trees dated at 44,000 years old and the lava flow dated at 45 *million* years old.[809]

Despite these inconsistencies, radiometric dating is still considered by secular scientists to be accurate enough to publish in peer-reviewed scientific journals, and text books. The data is sometimes varied enough to offer support for their own worldview. Even though the testing yields a significant number of erroneous, not always reproducible results, it still is a "scientific" standard believed by most scientists and the lay public to be irrefutable truth.

The concept of an "absolute date" reached its ultimate scientific subjectivity in 1975. Notice the word *absolute* in the quotation. "In the last two years an *absolute* date has been obtained for them [the Ngandong in Java, Indonesia, above the Trinil beds], and it has the very interesting value of 300,000 years, *plus or minus 300,000 years.*"[810]

---

[808] Andrew A. Snelling, PhD, "Radioisotope Dating of the Grand Canyon Rocks: Another Devastating Failure For Long-Age Geology," *Impact #376*, Institute for Creation Research, El Cajon, CA, October 2004, pp. iii-iv.

[809] *Dating Fossils and Rocks: Scientific Evidence and the Age of the Earth*, DVD, Answers in Genesis-USA, P. O. Box 510, Hebron, KY, 2004.

[810] J. B. Birdsell, PhD, "Men of the Lower and Middle Pleistocene," *Human Evolution: An Introduction to the New Physical Anthropology*, Rand McNally College Publishing Company, Chicago, IL, 1975, p. 295, (emphasis added).

VII
Dinosaurs and Humans

Chapter 34
Did Early Humans See Dinosaurs?

National Geographic says, "No human being has ever seen a live dinosaur."[811] Evolutionists now believe that dinosaurs lived 65 to 70 million years before man evolved. As already noted, unfossilized, soft microscopic portions of dinosaur tissue were recently found.[812]

Now the question is: How could it stay soft for 70 million years? Could it be that it is not 70 million years old?

If evidence were ever found that dinosaurs and man lived together, it would further prove false the theory of evolution. Is there such evidence?

The word "dinosaur" wasn't created until 1842. Sir Richard Owen coined the name from the Greek *deinos*, meaning "terrible" or "fearfully great," and *sauros*, meaning "lizard" or "reptile."

No scientist today can *prove* or *disprove* whether man ever saw dinosaurs. However, there is an underlying foundation for the dragon stories that have been reported throughout human history.

The following accounts in this chapter would, of course, be dismissed by scientists who believe that the earth (and dinosaurs) must be billions or millions of years old (to allow time for evolution). Yet the large number of such accounts across many cultures does clarify an anthropological and cultural perspective.

> *World Book Encyclopedia:* The dragons of legend are strangely like actual creatures that have lived in the past. They are much like the great reptiles, which inhabited the earth long before man is supposed to have appeared on earth.[813]

> Derek Isaacs: The very fact of the matter is this—dragons were described in vivid detail as large fierce reptiles that match . . . the physical specimen that modern science now calls the dinosaur.[814]

> At their base, we have a scholastically recognizable consistency—dragons are giant reptile-like creatures that live in remote areas, eat livestock, have scales and dermal spines along serpentine, spiked tails; some, but not all, have bat-like wings, and they lay eggs. . . . The claim that dragons are completely mythological and the subject of fairy-tales begins to look a bit loose when cultures from literally the entire ancient world created fabled creatures with very similar appearance, habits, ferocity, and diet. . . .

> In fact, dragons appear to be the most documented creature in all of ancient history—only the chronicling of humanity surpasses them. . . . Many have recognized that the sheer number of historical dragon accounts, coupled with their remarkable similarity, present a conundrum for those who believe they are a product of simple myth.[815]

One of the oldest pieces of pottery on earth, a piece of slate from Hierakonpolis, Egypt, shows two long-necked dragons 3,800 years ago.[816] The saga of Beowolf, set in Denmark and Sweden from the 8th to the early 11th century and cited as one of the most important works of Anglo-Saxon literature, is a description of violence between humans and a dragon.[817] And from Africa and Europe to Asia.

> Derek Isaacs: In the thick jungles of Cambodia, the temples of Angkor were built between the ninth and twelfth centuries. The antiquity and the authenticity of this archaeological site

---

[811] Geoguide, "Age of Dinosaurs," *National Geographic*, January 1993, p. 142.
[812] John N. Wilford, "Tissue Find Offers New Look Into Dinosaurs' Lives," *New York Times*, March 24, 2005; *see also: Science*, March 25, 2005; *see also:* M. Schweitzer and T. Staedter, "The Real Jurassic Park," *Earth*, June 1997, pp. 55-57.
[813] Knox Wilson, "Dragon," *The World Book Encyclopedia*, Vol. 5, 1973, p. 265.
[814] Derek Isaacs, "Eighteen Hundred," *Dragons or Dinosaurs? Creation or Evolution?* Bridge-Logos, Alachua, FL, 2010, p. 16.
[815] Derek Isaacs, "Magic Memories," *Dragons or Dinosaurs? Creation or Evolution?* Bridge-Logos, Alachua, FL, 2010, pp. 19-20.
[816] James B. Pritchard, PhD, "Scenes from History and Monuments," *The Ancient Near East in Pictures, Relating to the Old Testament*, Princeton University Press, Princeton, NJ, p. 93; *see also:* David Hatcher Childress, *Technology of the Gods*, p. 155, Original in the Cairo Museum.
[817] Beowolf, Wikipedia, see also: Bill Cooper, "Beowolf and the Creatures of Denmark," *After the Flood: The Early Post-flood History of Europe*, New Wine Press, Chichester, West Sussex, England, 1995, pp. 155-157.

is beyond any dispute. It is truly ancient. The Khmer people—the inhabitants of the region who lived from A.D. 50 to A.D. 1400—proved to be very skilled builders and carvers. On one of the dramatic pillars within the temple grounds a *Stegosaurus* is carved in stone.[818]

The shape and size of the plates that run the entire length of the creature's back and tail are perfectly in proportion to what a *Stegosaurus* would have had. The Khmer people seem to have had an intimate knowledge of this dinosaur.[819]

Dennis Swift, PhD: The impressive glyph is showcased in a round circle. The *Stegosaurus* has diamond-shaped plates sticking upright on its humped back and along the first part of its tail. The tail is tapered and has sharp spikes sticking out near the tip. The creature has a small neck, small head, and long snout. It has short stubby front legs and longer stubby legs in the back. The stone sculpture shouts, "Look at me! I'm a *Stegosaurus*—alive just 800 years ago!" . . . The *Stegosaurus* glyph's authenticity [as a temple carving] is beyond dispute and dated at approximately 1200 A.D.

The archaeologists and paleontologists are deeply troubled by its existence and offer ever expanding elastic explanations as to how the *Stegosaurus* got there. They claim a Khmer saw the fossilized remains of a *Stegosaurus*. He was so impressed at the sight of a disarticulated bone pile puzzle that he was able to artistically and accurately sculpt the *Stegosaurus* in stone. The Khmer even have the plates sticking up in proper arrangement: . . .

Paleontologists will tell you that putting a dinosaur skeleton together is an exceedingly complicated task. It is like putting together an enormous jigsaw puzzle with many of the pieces missing or destroyed. It is trial and error, much guesswork, and years of sorting through fossil fragments, to fill in the gaps: to ascertain what the dinosaur might have looked like. Paleontologists do not find dinosaur fossils with labels, diagrams, or photographs showing what the animal looked like.[820]

Petroglyphs on a cliff at Natural Bridges State Park, Blanding, Utah, by the Anasazi Indians from 400 A.D. to 1300 A.D., show animals that look like they may be dinosaurs:

F. A. Barnes and M. Pendleton: There is a petroglyph in Natural Bridges National Monument that bears a startling resemblance to a dinosaur, specifically a *Brontosaurus* with long tail and neck, small head and all. In the San Rafael Swell there is a pictograph that looks very much like a pterosaur, a Cretaceous flying reptile. . . . So far, archaeologists have chosen barely to mention such oddities, then ignore them, but sooner or later the problem of extinct or anachronistic animal rock art must be scientifically studied and resolved.[821]

What is the significance of the small head? *Brontosaurus* was the most popular dinosaur of all times. In 1883, O. C. Marsh, a famous paleontologist at Yale University, described a headless skeleton excavated in Colorado in 1879. To make the skeleton complete, Marsh added a skull he found three or four miles away and never told his secret. Many years later, Earl Douglas, a paleontologist with the Carnegie Museum in Pittsburgh, found a *Brontosaurus* skeleton with a skull right beneath it that looked less like the *Camarasaurus* skull Marsh had used and more like a *Diplodocus* skull. This was the first indication that something was wrong, but it was not until 1975, when two scientists published an article on the subject, that the scientific community finally admitted Marsh's error. Major museums had depicted the dinosaur with the wrong head for almost 100 years.[822] (They did make the correction.)

But that's not all.

Los Angeles Herald Examiner: A fantastic mystery has developed over a set of cave paintings found in the Gorozomzi Hills, 25 miles from Salisbury, Rhodesia (now Zimbabwe), for the paintings include a *Brontosaurus*, the 67-foot, 30-ton creature scientists believed became extinct millions of years before man appeared on earth. Yet the Bushmen who did the paintings ruled Rhodesia from only 1500 BC until a couple of hundred years ago. And

---

[818] *The Fossil Record*, DVD, Answers in Genesis-USA, PO Box 510, Hebron, KY, 2004.
[819] Dennis Swift, PhD, "Stumbling Upon History," *Secrets of the ICA Stones and Nazca Lines*, (there is no publisher or date listed for this book), p. 81; quoted by Derek Isaacs, "Dragon History," *Dragons or Dinosaurs? Creation or Evolution?* Bridge-Logos, Alachua, FL, 2010, pp.58-59, (bracketed supplied).
[820] Dennis Swift, PhD, "Dinosaurs Among the Moche," *Secrets of the ICA Stones and Nazca Lines*, (there is no publisher or date listed for this book), pp. 81-82.

[821] F. A. Barnes and Michaelene Pendleton, *Canyon Country: Prehistoric Indians—Their Cultures, Ruins, Artifacts, and Rock Art*, Wasatch Publishing Company: Salt Lake City, UT, 1979, pp. 201-203.
[822] Dennis Swift, PhD, "The Quest for Discovery," *Secrets of the ICA Stones and Nazca Lines*, p. 49; David C. Read, JD, "Bones of Contention," *Dinosaurs*, Clarion Call Books, Keene, TX, 76059, 2009, p. 30.

the experts agree that the Bushmen always painted from life. The belief is borne out by Gorozomzi Hills cave paintings—accurate representatives of the elephant, hippo, buck and giraffe.[823]

Even a science journal admitted 20 years later the possibility that Bushmen had firsthand knowledge of dinosaurs:

> David J. Mossman in *Ichnos:* In Lesotho, the Bushmen left cave paintings. . . . One of the creatures portrayed by those most distinct[ly] superb of all trackers is the distinct outline of an iguanodontid, in bipedal stance, with reduced forelimbs and other details remarkably compatible with our present-day enlightened understanding of the ancient creatures. . . . Ammunition here for creationists.[824]

Dinosaur-like animals are depicted on a cliff in Utah at a place called Black Dragon Wash.

Even the walls of the Havasupai Canyon in the Grand Canyon display an Indian pictograph of a dinosaur-like animal. One must ask why would prehistoric men illustrate what a scientific expedition called "a dinosaur"? Some scientists ask the same question. In 1925, men who had just traversed one of these canyons reported—

> Doheny Scientific Expedition: The fact that some prehistoric man made a pictograph of a dinosaur on the walls of this canyon upsets completely all of our theories regarding the antiquity of man. Facts are stubborn and immutable things. If theories do not square with the facts, then the theories must change; the facts remain.[825]

> Google: The Doheny Expedition to the Hava Supai [sic] Canyon: In October and November 1924, a scientific expedition led by Samuel Hubbard, curator of archaeology at the Oakland Museum, Charles W. Gilmore, curator of vertebrate paleontology at the United States National Museum, and funded by the oil magnate . . . E. L. Doheny, went to Havasuapi Canyon in northern Arizona to search for evidence of prehistoric man. Hubbard and Doheny had visited this area before, Doheny as a young prospector and Hubbard as a scientist.

One notes immediately that the people involved in this expedition have serious credentials. Similarly, many people hearing of a scientific report reputedly documenting a sauropod image petroglyph will expect the material to come from some little-known fundamentalist religious college; that does not appear to be the case in this instance:[826]

> "About a year ago [written in 1925] a photograph of the 'dinosaur' was shown to a scientist of national repute, who was then specializing in dinosaurs. He said, 'It is not a dinosaur, it is impossible, because we know that dinosaurs were extinct 12 million years before man appeared on earth.'"[827]

Dinosaurs are now believed to have become extinct 65 to 70 million years ago. However, in this 1925 quotation they were said to have become extinct ("only") 12 million years ago.

Vine V. Deloria, Jr., noted author of *Custer Died for Your Sins* and other books, and past executive director of the National Congress of American Indians, writes in *Red Earth, White Lies:*

> Following the "Tobocobe Trail" to where it intersects with Lee Canyon, the party soon discovered what they described as "wall pictures," figures scratched long, long ago depicting the local fauna. The most spectacular of these pictures was one of a dinosaur, identified by them as *Diplodocus,* standing upright. . . .

> Just as spectacular, however, were other discoveries in the canyon. In [Havasupai-Canyon Doheny-expedition scientist] Hubbard's words: "On the same wall with the dinosaur pictograph, and about 16 feet from it, we found a pictograph representing an animal which was evidently intended for an elephant, attacking a large man." . . . This pictograph scene accurately depicts the manner in which scholars believe that man hunted the mammoth—an ambush at a waterhole; and in southern Arizona there are several sites which have man and mammoth remains together in an obvious hunting format, with butchering marks on the mammoth's bones.[828]

The evolution perspective does not allow the possibility of humans and dinosaurs co-existing (and, so, would discount such findings and re-

---

[823]*Los Angeles Herald Examiner,* "Bushman's Painting Baffling to Scientists," January 7, 1970.
[824]David J. Mossman, "Dinosaur Tracks and Traces," *Ichnos,* Vol. VI, 1990, p. 151-153.
[825]*Discoveries Relating to Prehistoric Man, by the Doheny Scientific Expedition in the Hava Supai* [i.e., Havasupai] *Canyon, Northern Arizona,* Oakland Museum [publisher], October and November 1924, pp. 5-10.

[826]Google: The Doheny Expedition to the Hava Supai Canyon.
[827]*Discoveries Relating to Prehistoric Man, by the Doheny Scientific Expedition in the Hava Supai Canyon, Northern Arizona,* Oakland Museum [publisher], October and November 1924, pp. 5-10.
[828]Vine Deloria, Jr., "At the Beginning," *Red Earth, White Lies,* Fulcrum Publishing, Golden, CO, 1997, pp. 224-225.

ports as "fraud.") Yet many may be surprised at the volume of ancient legends and cave paintings and other depictions apparently made in recent millennia. A Viking woodcut from the eleventh century shows a dragon swallowing a man. Vikings put dragon heads on their ships a thousand years ago.[829]

> Derek Isaacs: [February 5, 2000] on the Chinese calendar . . . was the beginning of the year 4698, and the Year of the Dragon. . . . A series of twelve animal names are used to denote a sequence in the Chinese calendar. In order, these names are rat, ox, tiger, hare, dragon, snake, horse, sheep, monkey, fowl, dog, and pig.[830]

> Amongst animals that are familiar to people around the world is the dragon—a creature that has been thought to be only a myth. Yet, one of the oldest societies in the world has used the dragon as a cultural icon and for centuries the Chinese have included the dragon in a list of common animals like the sheep, rabbit, and dog.[831]

The Roman historian, Pliny the Elder, wrote an encyclopedia entitled *Natural History*. It was relied upon for scientific matters up to the Middle Ages. In it, he spoke freely of dragons:[832]

> Pliny the Elder: It is India that produces the largest (elephant) as well as the dragon, who is perpetually at war with the elephant, and is itself of so enormous a size as to envelop the elephant with its folds, and encircle them in its coils. The contest is equally fatal to both.[833]

There is not a known creature today that treats the elephant like common prey.[834]

Even the Greek historian Herodotus provided us with an account of winged serpents in Arabia: "The form of the serpent is like that of the water-snake; but he has wings without feathers, and as like as possible to the wings of a bat."[835]

Saint George, the patron saint of England and Portugal, is famous for slaying a dragon in 275 A.D.[836]

Bricks from Burg Nanstein, a 12th century Castle in Landstuhl, Germany, show pictures of dragons.

The tomb of a bishop in the 15th century cathedral in Cumbria, England, displays a brass cast of two long-necked dinosaurs.

A cave painting in Australia shows a man running away from what appears to be a dinosaur. Another painting shows aboriginals in the far north Queensland, Australia, dancing around what appears to be a long-necked dinosaur.

Silver dollars from the 1500s to the 1600s show people slaying dragons. A Russian medallion shows a man on horseback killing a dragon. A Bulgarian postage stamp shows a man aiming a bow and arrow at a dragon. The crest of Lithuania shows a man on horseback killing a dragon. The city of Nerluc, France, was renamed in honor of the "dragon" slain there. It was described as being bigger than an ox and having long, sharp, pointed horns on its head.[837]

> Derek Isaacs: Herodotus, just a few thousand years ago, saw some sort of flying reptile, which, according to evolutionary ideals, should have been extinct by millions of years.[838]

A cylinder seal from the fourth millennium B.C. on display at the Louvre in Paris shows what appear to be long-necked dinosaurs.[839]

In a completely different setting, 33,500 ceramic figurines,[840] including dinosaurs, were

---

[829] *The Vikings*, by Tony Allan Duncan, Baird Publishers, London, 2002, p. 64.
[830] Chinese calendar, *Encyclopædia Britannica*, 2009; *Encyclopædia Britannica* online, 18 May, 2009; quoted by Derek Isaacs, "Dragon History," *Dragons or Dinosaurs? Creation or Evolution?* Bridge-Logos, Alachua, FL, 2010, p. 29.
[831] Derek Isaacs, "Dragon History," *Dragons or Dinosaurs? Creation or Evolution?* Bridge-Logos, Alachua, FL, 2010, p. 29.
[832] Derek Isaacs, "Dragon History," *Dragons or Dinosaurs? Creation or Evolution?* Bridge-Logos, Alachua, FL, 2010, p. 31.
[833] Plinius Secundus, *The Natural History of Pliny*, Translated by John Bostock, MD, FRS, and H. T. Riley, Esq., BA, Vol. II, Book VIII, Chapter 11; quoted by Derek Isaacs, "Dragon History," *Dragons or Dinosaurs? Creation or Evolution?* Bridge-Logos, Alachua, FL, 2010, pp. 31-32.
[834] Derek Isaacs, "Dragon History," *Dragons or Dinosaurs? Creation or Evolution?* Bridge-Logos, Alachua, FL, 2010, p. 32.
[835] "Herodotus," *Encyclopædia Britannica*, 2009, *Encyclopædia Britannica* Online, 18, May, 2009; quoted by Derek Isaacs, "Dragon History," *Dragons or Dinosaurs? Creation or Evolution?* Bridge-Logos, Alachua, FL, 2010, p. 33.
[836] Duane Gish, "Dinosaurs, Dragons, and Beetles," *Dinosaurs by Design*, Master Books, Green Forest, AR, 1992, p. 81.
[837] Paul Stanley Taylor, "After the Flood, What Happened to Dinosaurs?" *The Great Dinosaur Mystery, and the Bible*, Chariot Family Publishing, Elgin, IL, 1992, p. 43.
[838] Derek Isaacs, "Dragon History," *Dragons or Dinosaurs? Creation or Evolution?* Bridge-Logos, Alachua, FL, 2010, p. 35.
[839] "The Story of Man," *Bible Times*, Vol. 1, p. 4
[840] David Hatcher Childress, "Foreword," *Mystery in Acámbaro*, Adventures Unlimited Press, One Adventure Place, Kempton, IL, 2000, p. 13.

excavated from six to ten acres on the side of a hill on the outskirts of Acámbaro, Mexico. The site was called Toro or Bull Mountain. These figurines range in size from a few inches long to statues three feet high and five feet long and made of ceramic, jade, stone, and obsidian. It is thought that they were buried 20 to 40 in a pit, apparently in haste, to avoid capture by the Spanish in the 1600s. The collection has been branded as "fake" by archaeologists whose worldview does not allow them to consider the possibility that man has seen dinosaurs.[841]

Charles H. Hapgood, professor of history and anthropology at Keene State College of the University of New Hampshire, initially an open-minded skeptic, became a believer in 1955 after his first visit while witnessing some of the figures being excavated. He even dictated to the excavators where he wanted them to dig. He returned in 1968 and became further convinced of the authenticity of the artifacts. After 18 years of research, he published the book, *Mystery in Acámbaro*.[842]

Human Evidence Found in Dinosaur Rock Layers

Dennis R. Petersen: In 1971, a rock collector spotted some bones recently exposed by a quarry bulldozer in hard sandstone near Moab, Utah. Requesting the excavators to pause their earthmoving operation, he brought a university anthropologist, a journalist and photographer to the site. The lower halves of two human skeletons were removed and taken to the university for further study. The rock formation was confirmed to be the same "100 million-year-old sandstone" [layer] containing dinosaur bones not far away in the famous Dinosaur National Monument near Vernal, Utah.

Strangely, the bones were never subjected to the technical analysis expected. No scientific report was released to the press, and the discoverer had to reclaim his fossils.[843]

Malachite Man

Dennis R. Petersen: In 1990, an independent team of researchers, including Dr. Don Patton, excavated further and found more. Thanks to their work, it's now clear that skeletons of 10 modern humans were buried under fifty-eight feet of Dakota Sandstone, in an area spanning about 50 by 100 feet. This rock formation is called Lower Cretaceous and is supposedly 140 million years old. At least four of the ten bodies are female. One is an infant. Some of the bones are articulated [connected in the rock as they were when alive]. Some are not, appearing to have been washed into place. No obvious tools or artifacts were found associated with the bones.[844]

The bones are partially replaced with malachite (a green material) and turquoise. The name "malachite man" is appropriate. . . . Some insist this is a mass grave. Think about that! Who would dig a grave up to 54 feet deep through extremely hard sandstone layers? The modern mining operation was halted in the 1970s because the sandstone was so hard it was wearing out the bulldozers.

It seemed obvious to Peterson that "these 10 men, women and children, were buried rapidly by some catastrophe, like a flood. Articulated skeletons indicate rapid burial. Some argue that these people were mining in a cave, when the ceiling collapsed on them. However, there are no signs of tunnels. Women and small children would not likely be included in a mining operation. No tools have been found and there are no crushed bones as you would expect if a mine caved in."[845]

And in South America, Dr. Jamie Gutierrez found fossil human hands in Cretaceous rock near a large ichthyosaurus (marine dinosaur).[846]

In 1979, in Laetoli, Northern Tanzania, Africa, scientists found what appeared to be perfectly formed human footprints in a layer of volcanic ash that had turned into stone. Long-age scientists have assumed that this provided evidence that two

---

[841] Charles H. Hapgood, "The Investigation of a Discovery," *Mystery in Acámbaro*, Adventures Unlimited Press, One Adventure Place, Kempton, IL, 2000, pp. 74-80.
[842] David Hatcher Childress, "Foreword," *Mystery in Acámbaro*, Adventures Unlimited Press, One Adventure Place, Kempton, IL, 2000, pp. 15-16.
[843] Dennis R. Petersen, "Is There Evidence of Humans Buried by the Great Flood?" *Unlocking the Mysteries of Creation: The Explorer's Guide to the Awesome Works of God*, Bridge-Logos Publishers, Alachua, FL, 2002, p. 144.

[844] Dennis R. Petersen, "Is There Evidence of Humans Buried by the Great Flood?" *Unlocking the Mysteries of Creation: the Explorer's Guide to the Awesome Works of God*, Bridge-Logos Publishers, FL, Florida, 2002, p. 145. Don R. Patton offers further information at the website www.bible.ca/tracks.
[845] Dennis R. Petersen, "Is There Evidence of Humans Buried by the Great Flood?," *Unlocking the Mysteries of Creation: the Explorer's Guide to the Awesome Works of God*, Bridge-Logos Publishers, Alachua, FL, 2002, p. 146.
[846] www.creationevidence.org.

adults and a child walked upright some 3.6 million years ago. (It had been said that humans didn't evolve until three million years ago.) According to Mary Leakey, the footprints were described as "remarkably similar to those of modern man. . . . The form of his foot was exactly the same as ours. . . . Weight-bearing pressure patterns in the prints resemble human ones. . . . Footprints, so very much like our own."[847]

Russell H. Tuttle, PhD, from the University of Chicago went to a place where people never wear shoes and studied their footprints in mud. He found that "the 3.5-million-year-old footprint trails at Laetoli Site G. resemble those of habitually unshod modern humans. None of their features suggest that the Laetoli hominids were less capable bipeds than we are. If the [Laetoli] footprints were not known [from evolution's perspective] to be so old, we would readily conclude that they were made by a member of our own genus, *Homo*."[848]

Evolutionist Tim White, probably recognized as the leading authority on the Laetoli footprints, could not believe his own eyes.

Timothy D. White, PhD: Make no mistake about it, they are like modern human footprints. If one were left in the sand of a California beach today, and a four-year-old were asked what it was, he would instantly say that someone had walked there. He wouldn't be able to tell it from a hundred other prints on the beach, nor would you. The external morphology [structure and form] is the same. There is a well-shaped modern heel with a strong arch and a good ball of the foot in front of it. The big toe is straight in line. It doesn't stick out to the side like an ape toe, or like the big toe in so many drawings you see of *australopithecines* in books.[849]

In discernible features, the Laetoli [Site] G prints are indistinguishable from those of habitually barefoot Homo sapiens.[850]

Yet, *National Geographic* illustrated the story by representing the beings who made the footprints as dark-skinned, ape-like creatures when there were clearly only modern footprints in the ash.[851]

---

[847] Mary Leakey, "Ashes of Time," *National Geographic*, April 1979, pp. 446-457.
[848] Russell H. Tuttle, "The Pattern of Laetoli Feet," *Natural History*, March 1990, p. 64.
[849] Timothy D. White, PhD, (probably recognized as the leading authority on the Laetoli footprints), *The Beginnings of Humankind*, Donald C. Johanson and Maitland A. Edey, NY, Simon & Schuster, 1981, p. 250.
[850] Russell H. Tuttle, "The Pattern of Laetoli Feet," *American Journal of Physical Anthropology*, Vol. 78, No. 2, February 1989, p. 316.
[851] Mary D. Leakey, "Footprints in the Ashes of Time," *National Geographic*, April 1979, pp. 446-457.

Chapter 35
Are Dinosaurs Mentioned in the Bible?

Evolution says dinosaurs are millions of years old. The following does not support that view and adds another perspective: Dragons are mentioned in the Bible 34 times. And there are two other words that may possibly describe dinosaurs. In the book of Job, God said, "Behold now behemoth, which I made with thee; he eateth grass as an ox" (Job 40:1). The authors of some Bible commentaries state that the behemoth is either an elephant or hippopotamus.

Don't elephants eat grass? Yes. "Lo now, his strength is in his loins, and the force is in the navel of his belly" (Job 40:16).

Do elephants or hippos have a big belly? Yes.

But there is a third description. "He moveth his tail like a cedar" (Job 40:17). Do either elephants or hippos move their tails "like a cedar"? No.

They have tails like ropes. The Bible states that "behemoth" moved its "tail like a cedar." Some dinosaurs had tails as huge and as strong as cedar trees. If the writer of Job described a dinosaur, then the evolution position that man never saw dinosaurs should be questioned.

The 1980 Abingdon edition of *Abingdon's Strong's Concordance* defines the word "behemoth" as "a water ox, i.e. the *hippopotamus* or Nile-horse." Thomas Nelson's 2010 edition of *The New Strong's Expanded Exhaustive Concordance of the Bible* [expanded to include several standard theological dictionaries] includes the above definition and adds "could possibly be an extinct dinosaur."

> Walt Brown, PhD: The next chapter of Job describes another huge, fierce animal, a sea monster named leviathan. Leviathan is also mentioned in Psalm 74:14; Psalm 104:26; and Isaiah 27:1. It was not a whale or a crocodile, because the Hebrew language had other words to describe such animals.[852]

Thou didst divide the sea by thy strength: thou brakest the heads of the dragons in the waters. Thou brakest the heads of leviathan in pieces." (Psalm 74:13-14).

Canst thou draw out leviathan with a hook? (Job 41:1).

In that day the LORD with his sore and great and strong sword shall punish leviathan the piercing serpent, even leviathan that crooked serpent; and he shall slay the dragon that is in the sea (Isaiah 27:1-2).

Did dinosaurs live recently?

In 1883, *Scientific American* published the following:

> The Brazilian Minister at La Paz, Bolivia, had remitted to the Minister of Foreign Affairs in Rio photographs of drawings of an extraordinary saurian killed on the Beni after receiving thirty-six balls. By order of the President of Bolivia the dried body, which had been preserved in Asunción, was sent to La Paz. It is 12 meters long (39 ft) from snout to point of the tail, which latter is flattened. . . . The legs, belly, and lower part of the throat appear defended by a kind of scale armor, and all the back is protected by a still thicker and double cuirass [an armor of bony plates], starting from behind the ears of the anterior head, and continuing to the tail. The neck is long, and the belly large and almost dragging on the ground. Professor Gilveti, who examined the beast, thinks it is not a monster, but a member of a rare or almost lost species, as the Indians in some parts of Bolivia use small earthen vases of identical shape, and probably copied from nature.[853]

---

[852] Walt Brown, Jr., PhD, "What About Dinosaurs," *In the Beginning: Compelling Evidence for Creation and the Flood,* Center for Scientific Creation, Phoenix, AZ, 2001, p. 249.

[853] *Scientific American,* 1883, Vol. 49, No. 3; William Corliss, *Incredible Life: A Handbook of Biological Mysteries,* p. 531.

Chapter 36
Did Dinosaurs Evolve into Birds?

A "missing link," the *Archaeopteryx*, dated to 150 million years in the evolution paradigm and discovered in Germany immediately after the publication of Charles Darwin's book, *On the Origin of Species*, helped to establish the credibility of Darwinian evolution. Other examples can be given:

> *Miller and Levine Biology:* Ask many paleontologists what a bird is and they'll reply with a grin, "a hot-blooded dinosaur with feathers." Although that answer may sound odd, there really is reason for it. The first fossil ever found of an early bird-like animal is called *Archaeopteryx* and dates from late in the Jurassic Period.[854]

Not everyone agrees:

> *Geotimes:* But there are plenty of other reasons to refute the dinosaur-bird connection, says [evolutionist Alan] Feduccia. "How do you derive birds from a heavy, earthbound, bipedal reptile that has a deep body, a heavy balancing tail, and foreshortened forelimbs?" he asks. "Biophysically, it's impossible."[855]

How do we decide what to believe? Evolutionists claim that *Archaeopteryx* fossils are the most famous fossils in the world.

If *Archaeopteryx* did not have a few perfectly formed, modern feathers, clearly visible on only two of the six [now at least ten] known specimens,[856] *Archaeopteryx* would be considered Compsognathus.[857] But there is a major problem.

> Phillip E. Johnson, JD: If we are testing Darwinism rather than merely looking for a confirming example or two, then a single good candidate for ancestor status is not enough to save a theory that posits a *worldwide* history of continual evolutionary transformation.[858]

Recent molecular studies cast doubt on the assumption that feathers evolved from scales. The new data suggests to Darwinists that feathers may be a completely novel structure.[859] Two researchers wrote that, "progress in solving the particularly puzzling origin of feathers has also been hampered by what now appear to be false leads, such as the assumption that the primitive feather evolved by elongation and division of the reptilian scale. . . . The new evidence from developmental biology is particularly damaging to the classical theory that feathers evolved from the elongated scales."[860]

Feathers are a problem for Darwinists. One can appreciate why Darwin wrote that, "the sight of a feather in a peacock's tail, whenever I gaze at it, makes me sick!"[861]

For a reptile to evolve into a bird, it would have to develop primary and secondary feathers, reform its respiratory, digestive, circulatory, and nervous systems, develop a skeletal system with hollow bones, construct a beak, master nest building, acquire flight, and develop a sound producing organ. *There is no scientific evidence that it ever happened.*[862]

Another "missing link" between dinosaurs and birds was first presented on October 15, 1999, at a major press conference scheduled by the National Geographic Society in its Explorer's Hall in Washington, DC. The fossil, billed as a

---

[854] Kenneth R. Miller, PhD, Joseph Levine, PhD, "Biology Update, The Evolution of Flight," *Biology*, Prentice-Hall, Inc., Upper Saddle River, NJ, 2000, p. 724.
[855] Alan Feduccia, PhD, "Jurassic Bird Challenges Origin Theories," *Geotimes*, Vol. 41, January 1990, p. 7.
[856] Ian Taylor, "The Ultimate Hoax: Archaeopteryx Lithographica," *Proceedings of the Second International Conference on Creationism*, Vol. 2, 1990, p. 280.
[857] F. Hoyle, N. C. Wickramasinghe, and R. S. Watkins, "Archaeopteryx," *The British Journal of Photography*, June 21, 1985, p. 694.
[858] Phillip E. Johnson, JD, "The Vertebrate Sequence," *Darwin on Trial*, Regnery Gateway, Washington, DC, 1991, p. 79, (emphasis added).
[859] Thomas Quinn, transcript of a paper presented April 18, 1998, at Dinofest. Sponsored by the National Academy of Sciences, Philadelphia, PA.
[860] Richard O. Prum and Alan H. Brush, "Which came first, the feather or the bird?" *Scientific American*, March 2003, pp. 84-93.
[861] Charles Darwin, in an April 3, 1860, letter to American botanist Asa Gray, The TalkOrigins Archive, Creation or Evolution #3, Darwin Quotations on Complexity.
[862] Mike Riddle, *The Fossil Record*, DVD, Answers in Genesis-USA, P. O. Box 510, Hebron, KY, 2004.

missing link between dinosaurs and birds, had a bird's body, and a dinosaur-like tail. Some of the scientists present, who had studied the fossil, commented: "We're looking at the first dinosaur capable of flying. . . . It's kind of overwhelming." The announcement preceded the publication of the November 1999 issue of *National Geographic* that featured the fossil under the title of "Feathers for *T. rex?*" The subtitle read, "New Birdlike Fossils Are Missing Links in Dinosaur Evolution." The article, by Christopher P. Sloan, characterized *Archaeoraptor* as "a missing link between terrestrial dinosaurs and birds that could actually fly."[863]

According to Dr. Ariel Roth, "It was just the kind of find the paleontologists' camp needed to support their case that birds evolved from dinosaurs."[864]

> Christopher P. Sloan, quoting Stephen Czerekas: It's a missing link between terrestrial dinosaurs and birds that could actually fly. With arms of a primitive bird and the tail of a dinosaur, this creature found in Liaoning Province, China, is a true missing link in the complex chain that connects dinosaurs to birds. Scientists funded by *National Geographic* studied the animal, named *Archaeoraptor liaoningensis*, under ultraviolet light and used CT scans to view parts of the animal obscured by rock.[865]

The National Geographic Society imposed absolute secrecy on the study in order to enhance the effectiveness of the publicity regarding the amazing "missing link." Eventually it was exposed that *Archaeoraptor* was a fake fossil consisting of many parts carefully glued together. It is now known as "the Piltdown Bird," named after the famous Piltdown hoax in which someone during the early part of the past century crudely fitted an apelike jaw to a human skull.[866]

Dr. Storrs L. Olson is one of the scientists critical of how popular magazines and news organizations have reported the theory of bird evolution to the public. One of the world's foremost avian paleontologists, he is currently [2017] curator Emeritus and Senior Scientist at the Smithsonian's National Museum of Natural History, Division of Birds, Department of Vertebrate Zoology. In 1998, he sent an open letter to Peter H. Raven, PhD, Secretary, National Geographic Committee for Research and Exploration, chiding the National Geographic Society for articles that were excessively supportive of the dinosaur-to-bird evolutionary theory—

> Storrs L. Olson, ScD: With the publication of "Feathers for *T. rex?*" by Christopher P. Sloan in its November issue, *National Geographic* has reached an all-time low for engaging in sensationalistic, unsubstantiated, tabloid journalism. . . . Prior to the publication of the article "Dinosaurs Take Wing" in the July 1998 *National Geographic*, Lou Mazzatenta, the photographer for Sloan's article, invited me [Olson] to the National Geographic Society to review his photographs of Chinese fossils and to comment on the slant being given to the story. At the time, I tried to interject the fact that strongly supported alternative viewpoints existed to what *National Geographic* intended to present, but it eventually became clear to me that *National Geographic* was not interested in anything other than the prevailing dogma that birds evolved from dinosaurs.
>
> The idea of feathered dinosaurs and the theropod origin of birds is being actively promulgated by a cadre of zealous scientists acting in concert with certain editors at *Nature* and *National Geographic* who themselves have become outspoken and highly biased proselytizers of the faith. Truth and careful scientific weighing of evidence have been among the first casualties in their program, which is now fast becoming one of the grander scientific hoaxes of our age.[867]

The media seldom retract evolutionist claims after they are shown to be false. *National Geographic* has provided one refreshing exception. Details of this fiasco were explained on a few back pages of *National Geographic* by Lewis M.

---

[863] Christopher P. Sloan, quoting Stephen Czerkas, "Feathers for T. Rex?" *National Geographic*, November 1999, p. 98-103.
[864] Ariel A. Roth, PhD, "Fashions in Science," *Science Discovers God: Seven Convincing Lines of Evidence for His Existence*, Autumn House Publishing, a division of Review and Herald Publishing, Hagerstown, MD, 2008, p. 173.
[865] Christopher P. Sloan, quoting Stephen Czerkas, "Feathers for T. Rex?" *National Geographic*, November 1999, p. 100.
[866] Ariel A. Roth, PhD, "Fashions in Science," *Science Discovers God: Seven Convincing Lines of Evidence for His Existence*, Autumn House Publishing, a division of Review and Herald Publishing, Hagerstown, MD, 2008, p.174.
[867] Storrs L. Olson, quoted by Carl Werner, [re Olson's open letter to Dr. Peter Raven] in "Evolution's 'Best Proof' Is Under Attack from Top Scientists!" *Evolution, the Grand Experiment: The Quest for an Answer*, New Leaf Press, Green Forest, AR, 2007, pp. 182-183.

Simons, an independent investigative reporter and winner of the Pulitzer Prize who conducted some of his research in China at the request of Bill Allen, *National Geographic's* editor. The report was summarized as follows:

> Lewis M. Simons: In his one-room house, the farmer laid the counter slab of the tail aside. Using a homemade paste, he glued the slab of the tail to the lower portion of the birdlike body. With counter slab pieces from the body itself—and possibly other scraps he'd kept over time—he glued in missing legs and feet. Aware that fossil fanciers, unlike paleontologists, prefer specimens assembled and suitable for display, the farmer was following basic market economics.
>
> The result was the "missing link"—the body of a primitive bird with teeth and the tail of a land-based little dinosaur, or dromaeosaur.[868]

When the story started to unravel, Xu Xing, from the Institute of Vertebrate Paleontology and Paleoanthropology, Chinese Academy of Sciences in Beijing, China, one who had investigated the "missing link," wrote "I am 100 percent sure we have to admit that *Archaeoraptor* is a faked specimen."

National Geographic published a cleaned-up version of Xu's letter in its March issue, at his request changing "faked" to "composite."

> Lewis M. Simons: It's a tale of misguided secrecy and misplaced confidence, of rampant egos clashing, self-aggrandizement, wishful thinking, naïve assumptions, human error, stubbornness, manipulation, backbiting, lying, corruption, and most of all, abysmal communication.[869]

This investigative reporter made these strongly-worded declarations and *National Geographic*, to its credit, published it. *Science News* commented:

> Red-faced and downhearted, paleontologists are growing convinced that they have been snookered by a bit of fossil fakery from China. The "feathered dinosaur" specimen that they recently unveiled to much fanfare apparently combines the tail of a dinosaur with the body of a bird.[870]

Discover magazine addressed the fraud issue:

> Evolutionist Alan Feduccia, PhD: *Archeoraptor* is just the tip of the iceberg. There are scores of fake fossils out there, and they have cast a dark shadow over the whole field. When you go to these fossil shows, it's difficult to tell which ones are faked and which ones are not. I have heard there is a fake-fossil factory in northeast China, in Liaoning Province, near the deposits where many of these recent alleged feathered dinosaurs were found.[871]

In an interview with Lee Strobel, Dr. Jonathan Wells reported another incident he witnessed at a conference in Florida. The star of the show was a fossil called *Bambiraptor*, a chicken-sized dinosaur with purportedly bird-like characteristics paleontologists identified as "the missing link." "And, sure enough, the reconstructed animal on display had feathers or feather-like structures on it. The problem was that no feathers were ever found with the fossil! But because scientists said they should be there, they were added."[872]

Science referred to the *Bambiraptor feinbergi* as the *"Archaeoraptor* fiasco."[873]

At the same conference a group of molecular biologists reported finding bird DNA in *Triceratops* dinosaur bones that were 65 million years old. This, they declared, was genetic evidence that birds are closely related to dinosaurs.[874]

According to Dr. Wells, "The problem is that the bones from which the DNA was supposedly extracted are from a branch of dinosaurs that [according to evolution's position] had nothing to do with bird ancestry. Furthermore, the DNA they found was not *ninety* or *ninety-nine percent similar* to birds—it was *one-hundred-percent turkey* DNA! Even chickens don't have DNA that is one-hundred-percent similar to turkey DNA. Only turkeys have one-hundred-percent turkey DNA. So, these people said they found turkey DNA in a dinosaur bone—and it actually got published in *Science* magazine. This is just

---

[868] Lewis M. Simons, *"Archaeoraptor* Fossil Trail," *National Geographic*, Vol. 198, No. 4, October 2000, pp. 128-131.
[869] Lewis M. Simons, *"Archaeoraptor* Fossil Trail," *National Geographic*, Vol. 198, No. 4, October 2000, pp. 128-131.
[870] R. Monastersky, "All mixed up over birds and dinosaurs," *Science News*, January 15, 2000.
[871] Alan Feduccia, PhD, "Plucking Apart the Dino-Birds," *Discover*, February 2003, p. 16.
[872] Jonathan Wells, PhD, quoted by Lee Strobel, "Doubts About Darwinism," *The Case for a Creator*, Zondervan, Grand Rapids, MI, 2004, p. 59.
[873] Constance Holden, "Florida Meeting Shows Perils, Promise of Dealing for Dinos," *Science*, April 14, 2000, pp. 238-239.
[874] Constance Holden, "Florida Meeting Shows Perils, Promise of Dealing for Dinos," *Science*, April 14, 2000, pp. 238-239.

incredible to me. The headline in the magazine said with a straight face: 'Dinos and Turkeys: Connected by DNA?'"[875]

"Yet people were willing to seize on it to support their belief in Darwinian theory."[876]

Problems with reptile-to-bird evolution include:
- Modern birds are found in sediments with dinosaurs;
- Birds have a four-chambered heart, but most reptiles have a three-chambered heart;
- There is no similarity between reptile and bird wings;
- The way dinosaur scales and bird feathers attach is not similar;
- Scales and feathers develop from different genes on the chromosomes; and
- Reptiles lay leathery eggs.

In 2009, new research supported the creationist position that birds did not evolve from dinosaurs:

A. P. Galling: The notion that theropod dinosaurs evolved into birds has almost certainly become one of the most widely accepted "facts" of evolution. . . . Except for a few notable critics, such as University of North Carolina paleobiologist Alan Feduccia, evolutionists seem to have all but agreed on birds' dinosaurian origins.[877]

Now, a new paper in the *Journal of Morphology* presents the research of two Oregon State University scientists who don't agree with the evolutionary dogma on bird origins.[878]

According to Devon Quick, a doctoral student coauthor of the paper, "The position of the thigh bone and muscles in birds is critical to their lung function, which in turn is what gives them enough lung capacity for flight." Dinosaurs, including the theropod dinosaurs from which birds supposedly evolved, lacked a fixed femur.[879]

According to John A. Ruben, PhD, Oregon State University professor of zoology and coauthor of the paper, "Theropod dinosaurs had a moving femur and therefore could not have had a lung that worked like that in birds. Their abdominal air sac, if they had one, would have collapsed. That undercuts a critical piece of supporting evidence for the dinosaurs-bird link. It's really kind of amazing that after centuries of studying birds and flight we still didn't understand a basic aspect of bird biology."

Then Ruben outlined an equally difficult dilemma. *"The appearance of birds **before** dinosaurs in the fossil record is a 'serious problem' that is ignored by those who advocate dinosaur-to-bird evolution."*[880]

An Oregon State University press release quoted Dr. Ruben with additional historical perspective: "Frankly, there's a lot of museum politics involved in this, a lot of careers committed to a particular point of view even if new scientific evidence raises questions." In some museum displays, he said, the birds-descended-from-dinosaurs evolutionary theory has been portrayed as a largely accepted fact, with an asterisk pointing out in small type that 'some scientists disagree.' . . . Our work at OSU used to be pretty much the only asterisk they were talking about," he said. "But now there are more asterisks all the time. That's part of the process of science."[881]

. . . A polite way of saying "more scientists disagree with evolution all the time."

---

[875]Constance Holden, "Dinos and Turkeys: Connected by DNA?" *Science*, April 14, 2000, pp. 238-239, (emphasis added).
[876]Jonathan Wells, PhD, quoted by Lee Strobel, "Doubts About Darwinism," *The Case for a Creator*, Zondervan, Grand Rapids, MI, 2004, p. 60
[877]A. P. Galling, "Birds Did Not Evolve from Dinosaurs, Say Evolutionists, Stunning New Research Overturns Widely Held Evolutionary Idea," *AIG-US*, June 12, 2009, p. 1.
[878]D. E. Quick and J. A. Ruben, "Cardio-pulmonary anatomy in theropod dinosaurs: Implications from extant archosaurs," *Journal of Morphology*, 2009.
[879]D. E. Quick, Quoted by A. P. Galling, "Birds Did Not Evolve from Dinosaurs, Say Evolutionists, Stunning New Research Overturns Widely Held Evolutionary Idea," *AIG-US*, June 12, 2009, pp. 1-2.
[880]J. A. Ruben, quoted by A. P. Galling, "Birds Did Not Evolve from Dinosaurs, Say Evolutionists, Stunning New Research Overturns Widely Held Evolutionary Idea," *AIG-US*, June 12, 2009, p. 2, (emphasis added).
[881]J. A. Ruben, "Discovery Raises New Doubts About Dinosaur-Bird Links," OSU Press Release, quoted by A. P. Galling, "Birds Did Not Evolve from Dinosaurs, Say Evolutionists, Stunning New Research Overturns Widely Held Evolutionary Idea," *AIG-US*, June 12, 2009, p. 2.

VIII
Prophecy and Astronomical Phenomena

Chapter 37
Evidences for Scripture's Reliability

Those who accept the claims of the Creator and the Bible's account of creation discover in this spiritual world, and in science, evidences that support the reliability of the Bible as history. Authenticated miraculous answers to prayer are one of those evidences. Another is fulfilled Bible prophecies. A major reason people of faith believe in the Holy Bible is that its predictions about the future have been fulfilled and are being fulfilled. A good example is a series of prophecies regarding the Messiah, documented in the Old Testament, that have been fulfilled by Jesus, as recorded in the New Testament. These prophecies were written hundreds of years before Jesus was born in Bethlehem:

Prophecies Concerning Jesus' Birth	**Prediction in Advance**	**Historical Fulfillment**
• Born of the Seed of Woman	Genesis 3:15	Galatians 4:4
• Born of a virgin	Isaiah 7:14	Matthew 1:18, 24-25
• Son of God—Divine	Psalm 2:7	Matthew 3:17
• Seed of Abraham—human	Genesis 22:18	Galatians 3:16
• Son of Isaac	Genesis 21:12	Luke 3:34
• Son of Jacob	Numbers 24:17	Luke 3:34
• Born into the Tribe of Judah	Genesis 49:10	Luke 3:23, 33
• Born into the Family line of Jesse	Isaiah 11:1	Luke 3:23, 32
• Born into the House of King David	Jeremiah 23:5	Luke 2:4
• Born in Bethlehem	Micah 5:2	Matthew 2:1
• Presented with gifts	Psalm 72:10	Matthew 2:1, 11
• King Herod kills potential rivals	Jeremiah 31:15	Matthew 2:16

Prophecies Concerning Jesus' Ministry		
• Preceded by a messenger	Isaiah 40:3	Matthew 3:1-2
• Year Christ begins ministry	Daniel 9:25 Isaiah 61:1-2	Matthew 4:16-20
• Ministry to begin in Galilee	Isaiah 9:1	Matthew 4:12-17
• Will perform miracles	Isaiah 35:5-6	Matthew 9:35
• Will teach in parables	Psalm 78:2	Matthew 13:34
• Will enter Jerusalem on a donkey	Zechariah 9:9	Matthew 21:7-10

Prophecies of Jesus' Last Days		
• Betrayed by a friend	Psalm 41:9	Matthew 10:4
• Sold for 30 pieces of silver,	Zechariah 11:12-13	Matthew 27:3-7
• Forsaken by His disciples	Zechariah 13:7	Mark 14:50
• Accused by false witnesses	Psalm 35:11	Matthew 16:59-60
• Did not answer His accusers	Isaiah 53:7	Matthew 27:12
• Wounded and bruised	Isaiah 53:5	Matthew 27:26
• Beaten and spit upon	Isaiah 50:6	Matthew 26:67

• Mocked	Psalm 22:7-8	Matthew 27:31
• Hands and feet pierced	Psalm 22:16	John 20:25-27
• Crucified with thieves	Isaiah 53:12	Matthew 27:38
• Interceded for persecutors	Isaiah 53:12	Luke 23:34
• Rejected by His own people	Isaiah 53:3	John 1:10-11; 7:5, 48
• His friends stood afar off	Psalm 38:11	Luke 23:49
• His reproachers shook their heads	Psalm 109:25	Matthew 27:39
• They stared at Him on the cross	Psalm 22:16-17	Luke 23:35
• Clothes divided and gambled for	Psalm 22:18	John 19:23-24
• He suffered thirst	Psalm 69:21	John 19:28
• Gall and vinegar offered to Him	Psalm 69:21	Matthew 27:34
• Felt forsaken by God His Father	Psalm 22:1	Matthew 27:46
• Committed Himself to God	Psalm 31:5	Luke 23:46
• His bones were not broken	Psalm 34:20	John 19:33
• His heart was broken	Psalm 22:14	John 19:34
• His side was pierced	Zechariah 12:10	John 19:34
• Darkness was over the land	Amos 8:9	Matthew 27:45
• He was buried in a rich man's tomb	Isaiah 53:9	Matthew 27:57-60

Prophecies of Jesus' Life After His Death

• Resurrection	Psalm 16:10	Acts 2:31
• Ascension	Psalm 68:18	Acts 1:9, Ephesians 1:20
• Jesus intercedes in heaven's Sanctuary/Temple. The Old Testament Sanctuary worship service prophesied Jesus' work on earth, described in Hebrews. The New Testament shows the Sanctuary symbols predicted Jesus' priest-work now for His earth family.	Exodus, Leviticus	Hebrews (especially 7:25-10:22)

Chapter 38
The Cosmos

The Old Testament book of Jeremiah refers to the stars as "the host of heaven," and says they, along with the sand of the sea, "cannot be numbered" (Jeremiah 33:22). The New Testament book of Hebrews agrees, and specifically mentions "the stars of the sky in multitude, and as the sand which is by the sea shore, innumerable" (Hebrews 11:13).

These two statements were written before the invention of the telescope, when the sky could be seen only by "the naked eye." So how many stars can be seen without a telescope? Figures range from 3,000 to 6,000.[882]

No one looking at the sky would have thought then that there were as many stars as there are grains of sand by the sea. However, today, scientists are acknowledging this.[883]

It has been estimated from evidence provided by the Hubble Space Telescope in 2003 that there are up to 70 sextillion stars (that's 70 with 21 zeroes after it), enough for every individual on earth (six billion at the time) to own 11 trillion of them.[884]

The Hubble Ultra-Deep Field image is a composite of pictures taken by the Hubble Space Telescope accumulated from September 24, 2003 to January 16, 2004. It was taken from a part of the sky with a low density of visible stars within our own galaxy, which allowed a much better view of dimmer, more distant objects beyond. In August and September of 2009, the Hubble's Deep Field was expanded by using the infrared channel of the recently attached Wide-Field Camera 3. Located southwest of the constellation Orion, in the southern hemisphere constellation Fornax, the tiny image (smaller than a 1 millimeter [.03937 inches] by 1-millimeter square of paper held 1 meter [39.37 inches] away) is equal to approximately *"one thirteen-millionth of the total area of the sky."* And just that square millimeter of sky contains an estimated *10,000 galaxies.*[885]

From this study, we cannot say that the stars have increased in number, only that our ability to see them has improved.

According to NASA's online site, October 13, 2016: "The landmark Hubble Deep Field, taken in the mid-1990s, gave the first real insight into the universe's galaxy population. Subsequent sensitive observations such as Hubble's Ultra Deep Field revealed a myriad of faint galaxies. This led to an estimate that the observable universe contained about 200 billion galaxies. The new research shows that this estimate is at least 10 times too low."[886]

So, *two hundred billion* galaxies was *ten times too low* an estimate? If we multiply ten by two hundred, that is two thousand—so it's not just two *hundred* billion galaxies can be seen—now the estimate is two *thousand* billion galaxies. And that's *2 trillion* galaxies (2,000 x 1,000,000,000 =2,000,000,000,000) astronomers can see!

Now that we can have a greater appreciation for the word "astronomical," let's look with greater detail at the concept of the Big Bang. It began with a Belgian astronomer, Georges Edward Lemaitre (1894 to 1996). According to Isaac Asimov, Lemaitre conceived this exploding mass to be "no more than a few light-years in diameter." At the very least, that would be two light-years or about 12 trillion miles. By 1965, that figure was reduced to 275 million miles; by 1972, to 71 million miles; by 1974, to 54 thousand miles; by 1983 "a trillionth the diameter of

---

[882]"How Many Stars Can You See At Night?" *About.com.Space/Astronomy*; Google Answers: *Number of visible stars, visible with the unaided eye*, www.alcyone.de/SIT/bsc/, see also: *The Yale Bright Star Catalog.*
[883]"The Universe," Direct TV, The History Channel, 9 AM, December 8, 2015.
[884]"Astronomers count the stars," BBC News. July 22, 2003. http://news.bbc.co.uk/2/hi/science/nature/3085885.stm. Retrieved July 18, 2006.
[885]"Hubble Ultra-Deep Field," *Wikipedia, the free encyclopedia*, January 3, 2012, p. 1.
[886]www.nasa.gov/feature/goddard/2016/hubble-reveals-observable-universe-contains-10-times-more-galaxies-than-previously-thought

a proton"; and now, to . . . a singularity.[887]

This singularity was at first called "a cosmic egg," "a nugget of mass energy," and "a primeval atom." The Big Bang has also been called "a primordial fireball," "a flash of lawlessness," "a quantum fluctuation," and "a tremor of uncertainty."

A problem with the Big Bang is that it appears to have happened without an explainable *natural* cause. There are different ideas about the cause of the Big Bang, but no consensus. It might be that the researchers describing the Big Bang are actually seeing and describing evidence of the start of our universe at the time God created it. Their observations about the Big Bang may be correct, though they have no explanation for the *origin* of the singularity and its Big Bang!

Then what is a key benefit of creation science?

When we put statements by secular scientists and creationists side by side, secular scientists are basing their statements on the philosophy of materialism. According to the *American Heritage Dictionary*, materialism is "the theory that physical matter is the only reality and that everything, including thought, feeling, mind, and will, can be explained in terms of matter and physical phenomena."

In other words, from a materialism/scientism perspective all answers to questions are looked for within nature, within a cause-and-effect sequence that must be coherent and *totally materialistic*. This places the evolution model in a box that will not include anything outside of nature, anything *Super*-natural.

However, the creation model comes from the position of "theistic" science—science willing to include God when studying nature and the natural universe in search of answers. This allows the creationist who is a scientist to consider all knowledge about reality from the *natural* sciences and also from the Supernatural—*above and outside* the box of nature. They believe that nature is not "all there is," that there is a Creator.

The laws of physics do not explain the Big Bang. Here's an alternative: "In the beginning God . . ."

Frank Lewis Marsh, PhD: The material substances of our earth are here, and their presence can be demonstrated. But today man quite universally recognizes that these substances have not always existed. By faith it is assumed that they originated sometime, somewhere. To believe that they brought themselves into existence from nowhere is naturally preposterous.[888]

The Bible indicates that *life* was created on this earth thousands of years ago, not millions. However, the *universe*, the *cosmos*, may be very old. Genesis 1:14-19 states that on the fourth day God not only created two lights—*sun and moon*—but "he made the stars also." The phrase about Him making the stars also has been seen by many creationists and Hebrew linguists to be a parenthetical phrase. In other words: *By the way, He made the stars also*. And they could have been created eons ago.

Genesis 1:2 also states that in the beginning "the earth was without form, and void." That may mean that the earth existed but in an uninhabitable state, for possibly eons—the "old earth" model). Different interpretations are possible.

For many creationists, the Genesis model includes both a six-day creation of habitable earth, and the Flood. What does Genesis say about earth's climate at the close of the first week—after earth had been made into a fit habitat for plants, animals, and humans? In nature itself, there are several good indicators of a warmer climate in the past, i.e., in the time before the Flood.

And the *warm climate* is explained as follows: Much earlier, on day two of creation week "God separated the waters below the firmament [earth's atmosphere where we see the clouds] from those above the firmament." The former formed lakes and seas, the latter formed a water vapor canopy around the earth that caused a greenhouse effect with warm temperatures in which vegetation grew rapidly—a pre-Flood warm climate.

Then at the time of the Flood, "the windows of heaven were opened" (Gen. 7:11) and most or all of the waters above the firmament rained down. As a result, the warming effect of the water-canopy was eliminated, and Earth's climate

---

[887] Bolton Davidheiser, PhD, "A statement concerning the ministry of Dr. Hugh Ross," Box 22, La Mirada, CA.

[888] Frank Lewis Marsh, PhD, "Origin of Earth," *Studies in Creationism*, Review and Herald Publishing Association, Washington, DC, 1950, p. 125.

became more temperate, largely as we have known it for thousands of years. This seems to be confirmed in Genesis 8:22 KJV—"While the earth remains, seed time [springtime] and harvest, and cold and heat, and summer and winter shall not cease."

The presence of extensive layers of coal in many places point to trees and other vegetation that grew profusely and were buried rapidly. Laboratory experiments have shown that in as short a time as one to five years trees can form coal. All it requires is a *warm climate* (to make the trees grow rapidly and to great height) and high pressure to compact the trees after burial.

The burial under high pressure was provided in the calamitous events of the Flood—considering all the material that was floating in the hurricane-wind-tossed waters, and then settled back on earth as the waters dried.

The Biblical model is elegant in its simplicity and in creation's evident custom-design. It points with clarity to just about everything we find in nature today. For example, the geologist open to the creation model finds in research that the Bible record fits the findings exactly, like a key in a lock. That is, the geologists' findings in nature are what would be expected from the Bible's description of the event, circumstance, location, etc.

The Bible tells us there was once a man named Tubal-cain who was "an artificer in brass and iron . . . " Genesis 4:22. Is there any evidence for a pre-Flood civilization? Is it possible that a global flood could not only have created coal, oil, but also buried human artifacts?

A 4.5-inch by 6.5-inch zinc and silver, 1/8th inch thick, bell-shaped vessel was found in Dorchester, Massachusetts, in 1851, in solid rock supposedly over 600 million years old. On the side were six figures beautifully inlaid with pure silver, and around the lower part of the vessel a vine, or wreath, also inlaid with silver. The carving and inlaying were exquisitely done.[889]

In 1865, workmen found human bones and a well-tempered copper arrowhead in a vein of silver at the Rocky Point Mine in Gilman, Colorado.[890]

Mind-expending possibilities in an ancient book

What about the *redshift?* The recession of distant galaxies implies that the universe is expanding. The expansion is isotropic—that is, the same in all directions—somewhat like the dots on a balloon would expand away from each other as the balloon is blown up (although that illustration is only two dimensions). "Redshifts help scientists measure the expansion of the universe—the distance of galaxies from earth.[891]

Was the "redshift" astronomers now see in the expanding universe written about by ancient prophets? The Bible tells us in 11 places that God *stretched out* or spread out the heavens (Psalm 104:2; Job 9:8; Job 26:7; Job 37:18; Isaiah 40:22; Isaiah 42:5; Isaiah 44:24; Isaiah 45:12; Isaiah 48:13; Isaiah 51:12-13; Jeremiah 10:10-12).

In one of those verses—Isaiah 48:13—the Cosmos-maker says, "My own hand laid the foundations of the earth, and my right hand spread out the heavens" (New International Version).

Are these verses more than poetry? Is it coincidence that they do not contradict but align with what modern science says about the redshift? Those who accept the rational and spiritual claims of the Creator and the Bible's account of creation discover in this spiritual world, and in science, many such alignments and evidences that give satisfying answers to our basic questions—questions such as what is the origin of life? Of the two sexes? speech? consciousness? creativity? desire? love? hope? belief? morality? intelligence? intuition? beauty? appreciation of beauty? animal and insect flight? And even the origin of spirituality. And the answers satisfy mind and heart.

Those open to consider a higher source might read W. A. Dembski's book *Intelligent Design* (forward by Michael Behe). Dembski holds a PhD in math from the University of Chicago, a PhD in philosophy from the University of Illinois at Chicago, and degrees in theology and psychology. He has done postdoctoral work at the University of Chicago, MIT, Princeton, and Northwestern University. He received two fellowships from the

---

[889] *Scientific American*, June 5, 1852, p. 289-299.
[890] Norm Scharbough, *Ammunition*, Communiqué Conservative Publishers, PO Box 315, Brownsburg, IN, 1991, p. 177.
[891] Mart de Groot, personal communication May 8, 2018.

National Science Foundation and is a senior fellow of the Discovery Institute Center for Renewal of Science and Culture. He authored *The Design Inference* (Cambridge) and is the editor of *Mere Creation*. He brings his wide-ranging studies into his discussion of the science of Intelligent Design. College- and graduate-level readers will find *The Design Inference* thought-provoking.

From the perspective of his studies he looks at evolution's unsolved problems and at the science of Intelligent Design:

> William A. Dembski, PhD: Indeed the following problems have proven utterly intractable not only for the mutation-selection process proposed [by evolution,] but also for any other undirected natural process proposed to date:
>
> - The origin of life,
> - The origin of the genetic code,
> - The origin of multicellular life,
> - The origin of sexuality,
> - The scarcity of transitional forms in the fossil record,
> - The biological Big Bang that occurred in the Cambrian era,
> - The development of complex organ systems, and
> - The development of irreducibly complex molecular machines.
>
> Dembski continues: These are just a few of the more serious difficulties that confront every theory of evolution that posits only undirected natural processes.[892]

A predictions perspective

Biblical answers to our basic questions can be trusted. They are authenticated by the many predictions (prophecies) in the Bible which have already been fulfilled. Seeing that our Creator does know and reveal the future gives us confidence in His Word. Here are a few examples of fulfilled predictions:

Through the prophet Joel, God had said, "The *sun* shall be turned into darkness, and the moon into blood, before the great and terrible day of the Lord come" (Joel 2:31).

And during His First Advent, Christ added certain details: "Immediately after the tribulation of those days shall the *sun* be darkened, and the *moon* shall not give her light, and the *stars* shall fall from heaven, and the powers of the heavens shall be shaken" (Matthew 24:29).

John the Revelator *added* more: " . . . and . . . there was a *great earthquake;* and the *sun* became black as sackcloth of hair, and the *moon* became as blood; and the *stars* of heaven fell unto the earth, even as a fig tree casteth her untimely figs, when she is shaken of a mighty wind" (Revelation 6:12-13).

In these four prophecies four notable signs heralded the approaching nearness of Christ's Second Coming to earth: *a great earthquake*, followed by *a darkening of the sun and moon*, followed by *a falling of the stars*. Those predictions were fulfilled, in precisely the order forecast:

The Lisbon Earthquake, November 1, 1755, was one of the most severe and widespread (4 million square miles) ever recorded. It was *felt in Africa, continental Europe, Great Britain, Greenland, America, and the West Indies.*

The great Dark Day, May 19, 1780, was a dense darkness, beginning between 10:00 and 11:00 a.m., and the *sun* was dark, lasting for many hours. A standard reference work published in 1881 said that it was "the most mysterious and as yet unexplained phenomenon of its kind, in nature's diversified range of events, during the last century, . . . a most unaccountable darkening of the whole visible heavens and atmosphere in New England." It was observed at the most easterly regions of New England; westward, to the farthest parts of Connecticut, and at Albany; to the southward, it was observed all along the sea coasts; and to the north, as far as the American settlements extended. It probably far exceeded these boundaries, but the exact limits were never positively known. Birds returned to their nests and cattle returned to their stalls. The Connecticut legislature almost adjourned, the terrified members thinking that the Day of Judgment had come.[893] That night, as prophesied, *the moon was blood red.*

---

[892] William A. Dembski, "Naturalism and Its Cure," *Intelligent Design: The Bridge Between Science and Theology*, InterVarsity Press, Downers Grove, IL, p. 113.

[893] Richard M. Devens, *Our First Century: Being a popular descriptive portraiture of the one hundred great and memorable events of perpetual interest in the history of our country, political, military, mechanical, social, scientific and commercial; embracing also delineations of all the great historic characters celebrated in the annals of the Republic; men of heroism, statesmanship, genius, oratory, adventure and philanthropy*, 1878, pp. 89-90.

The "Falling of the Stars," November 13, 1833, was a meteoric shower so thick, so extensive, so long-lasting, that thousands feared the planet was doomed, "the whole firmament, over all the United States, being then, for [eight] hours in fiery commotion! . . . an incessant play of dazzlingly brilliant luminosities was kept up in the whole heavens. Some of these were of great magnitude and most peculiar form . . . the first appearance was that of fireworks of the most imposing grandeur, covering the entire vault of heaven with myriads of fire-balls resembling sky-rockets. On more attentive inspection, it was seen that the meteors exhibited three distinct varieties, as follows, described by Professor [Denison] Olmsted:[894] [of Yale College, Massachusetts, who was known as "America's greatest meteorologist."[895]]

> Professor Olmsted: First, those consisting of phosphoric lines, . . . the most numerous, everywhere filling the atmosphere, and resembling a shower of fiery snow driven with inconceivable velocity . . . transfixing the beholder with wondering awe.
>
> Second, those consisting of large fire-balls . . . . This kind appeared more like falling stars, giving to many persons the very natural impression that the stars were actually falling from the sky. . . .
>
> Third, those undefined luminous bodies which remained nearly stationary in the heavens for a considerable period of time; these were of various size and form.
>
> One, of large size, remained for some time almost stationary in the zenith, over the Falls of Niagara, emitting streams of light which radiated in all directions. . . . No spectacle so terribly grand and sublime was ever before beheld by man as that of the firmament descending in fiery torrents over the dark and roaring cataract![896]

W. J. Fisher, PhD, writing in *The Telescope*, October 1934, said the phenomenon was "the most magnificent meteor shower on record."[897]

The display, as described in a Professor Silliman's Journal, was seen all over North America. He described it as within the limits of the longitude of 61 degrees in the Atlantic Ocean to 100 degrees in Central Mexico, and from the North American lakes to the southern side of the Island of Jamaica.[898]

It was the all-engrossing theme of conversation and of scientific disquisition, for months. Some people watched in terror and mortal fear, thinking that the end of the world had come. The masses were panic stricken. Impromptu prayer meetings were held in many places. One eye witness was awakened by the most distressing cries—shrieks of horror and cries for mercy.

He opened his door, sword in hand, to see upwards of a hundred people lying prostrate and speechless on the ground. Others uttered bitter moans, imploring God to save the world and themselves. The point from which the meteors seemed to come was the constellation Leo.[899]

On a typical night, when there is no meteor shower, the average observer will see only ten "falling stars" per hour.[900]

Arago computes *that not less than two hundred and forty thousand meteors were at the same time visible above the horizon of Boston!"* [901] *That night more than a billion shooting stars appeared over the United States and Canada alone.*[902]

Because biblical prophecies have often had more than one fulfillment (the first, local, the second, more extensive), some Bible-believing Christians expect some cosmology prophecies could be fulfilled again immediately before Christ's promised Second Coming.

---

[894] Richard M. Devens, *Our First Century*, p. 329.
[895] J. N. Loughborough, "The Second Advent Message," *The Great Second Advent Movement: Its Rise and Progress*, Review and Herald Publishing Association, Washington, DC, 1909, p. 96, and *Our First Century*, pp. 330-331.
[896] *Our First Century*, pp. 330-331.
[897] Willard J. Fisher, PhD, "The Ancient Leonids," *The Telescope*, Bond Astronomical Club, Cambridge, MA, October 1934, p. 83.
[898] *Our First Century*, pp. 329-330.
[899] *Our First Century*, p.332.
[900] Peter M. Millman, "The Falling of the Stars," *The Telescope*, Bond Astronomical Club, Cambridge, MA, May-June 1940, p. 60.
[901] *Our First Century*, p. 330.
[902] Peter M. Millman, "The Falling of the Stars," *The Telescope*, Bond Astronomical Club, Cambridge, MA, May-June 1940, p. 57.

IX
Conclusion

Chapter 39
Dark Edges

Certain key figures in the mid- and late-1900s adopted evolution's philosophy, with far-reaching results. Many are unaware of the role evolution played in their actions. Thankfully these pivotal figures are unrepresentative. This chapter needs to be included if for no other reason than to contribute to a comprehensive overview, including evolution's tragic figures.

In order to have a superior stock, Darwin believed in inbreeding. He married his maternal father's granddaughter, who was also his mother's niece. They had ten children. Mary died shortly after birth. Anne died at age 10. Charles Jr. was born retarded and died at 19 months. Henrietta had a serious breakdown at age 15. Three of his six other sons were ill so often that Charles regarded them as semi-invalids.[903]

But, as we shall see in this chapter, even more harm resulted from the part of the title of Darwin's book that is almost never cited: *On The Origin of Species By Means of Natural Selection, or the Preservation of Favored Races in the Struggle for Life*. Although Darwin's "Favored Races" may have included groups of all kinds of organisms, it also included humans.

To repeat for emphasis, when most people refer to Darwin's book, they do not include the entire name: *On The Origin of Species By Means of Natural Selection, or the Preservation of Favored Races in the Struggle for Life*. The Theory of Evolution is not just an academic discussion. In today's world, his book is not politically correct. Darwin was racist. What did he write?

> Charles Darwin: At some future period, not very distant as measured by centuries, the civilized races of man will almost certainly exterminate and replace the savage races throughout the world.[904]

Even evolutionist Stephen Jay Gould, PhD, acknowledges the dark edges of evolution:

> Gould: [Ernst Haeckel's] greatest influence was ultimately in another tragic direction—national socialism. His evolutionary racism; his call to the German people for racial purity and unflinching devotion to a "just" state; his belief that harsh, inexorable laws of evolution ruled human civilization and nature alike, conferring upon favored races the right to dominate others; the irrational mysticism that had always stood in strange communion with his brave words about objective science—all contributed to the rise of Nazism. The Monist League that he had founded and led, though it included a wing of pacifists and leftists, made a comfortable transition to active support for Hitler.[905]

> Richard Milner: He [the German evolutionist Dr. Ernst Haeckel] convinced masses of his countrymen they must accept their evolutionary destiny as a "master race" and "outcompete" inferior peoples, since it was right and natural that only the "fittest" should survive. His version of Darwinism was incorporated in Adolf Hitler's [book] *Mein Kampf* (1925), which means "My Struggle," taken from Haeckel's German translation of Darwin's phrase, "the struggle for existence."[906]

Hitler believed the Germans were the superior race that deserved to rule the world. According to Sir Arthur Keith, "The German Fuhrer . . . has consciously sought to make the practice of Germany conform to the theory of evolution."[907]

Hitler "stressed and singled out the idea of biological evolution as the most forceful weapon against traditional religion."[908] In 1936, "the

---

[903] Ian T. Taylor, "Charles Darwin, MA," *In the Minds of Men*, TFE Publishing, Toronto, Canada, 1984, p. 127.
[904] Charles Darwin, "On the Affinities and Genealogy of Man," *The Descent of Man*, A. L. Burt Company, New York, 1874, p. 178.
[905] Stephen Jay Gould, PhD, "Evolutionary Triumph, 1859—1900," *Ontogeny and Phylogeny*, The Belknap Press of Harvard University Press, Cambridge, MA, 1977, pp. 77-78.
[906] Richard Milner, "Ernst Haeckel (1834-1919) German Evolutionist," *Encyclopedia of Evolution*, Facts on File, New York, NY, 1990, p. 207.
[907] Sir Arthur Keith, "An Evolutionary Interpretation of the Second World War," *Evolution and Ethics*, G. P. Putnam's Sons, NY, 1946 and 1947, p. 230.
[908] Daniel Gasman, *Scientific Origins of Modern Socialism: Social Darwinism in Ernst Haeckel and the German Monist League*, 1971, p. 188.

Reichsgericht (the German 'Supreme Court') refused to recognize Jews living in Germany as 'persons' in the legal sense."[909] This fact opened the way for The Holocaust to kill at least six million Jews during World War II.[910]

Hitler killed the Jews to speed up the process of evolution. "I have the right to exterminate an inferior race that breed like the vermin," he said.[911]

> Robert E. D. Clark, PhD: Adolf Hitler's mind was captivated by evolutionary teaching—probably since the time he was a boy. Evolutionary ideas—quite undisguised—lie at the basis of all that is worst in [his book] *Mein Kampf*—and in his public speeches. . . . In a speech at Nuremberg, in 1933, he argued that a higher race would always conquer a lower.[912]

> *The Evolution Conspiracy:* Darwinian Theory combined with Marxist Ideology formed the cornerstone of the "superman" doctrine—the survival of the fittest or the duty of the strong to trample the weak. . . . Marx admired Darwin, and the two corresponded.[913]

> Stephen Jay Gould, PhD: Agassiz endeared himself to the proponents of slavery when, as Europe's leading national historian, he chose to settle in America and to maintain that blacks represent a separate and lower species. . . . Biological arguments for racism may have been common before 1850, but they increased by orders of magnitude following the acceptance of evolutionary theory.[914]

> Charles Kingsley: The Black People of Australia, exactly the same race as the African Negro, cannot take in the Gospel. . . . All attempts to bring them to a knowledge of the true God have as yet failed utterly. . . . Poor brutes in human shape . . . they must perish off the face of the earth like brute beasts.[915]

The tragic results of evolutionary thinking in the early 1900s are exemplified in the story of Ota Benga:

> Dennis R. Petersen: Darwin's book, *The Descent of Man*, prompted some anti-Christian intellectuals to seek fossils proving man rose from apes. Others believed that "half-man, half-ape" creatures still flourished in remote parts of the world. In the early 20th century, such beliefs led to the wanton killing of Australian Aborigines and the sad story of an African Pygmy man named Ota Benga. An evolutionist researcher captured Ota Benga in 1904 in the Congo. In his own language, his name meant "friend." He had a wife and two children. Chained and caged like an animal, he was taken to the St. Louis World's Fair where evolutionist scientists exhibited him as "the closest transitional link to man." Two years later, they took him to the Bronx Zoo in New York and displayed him as one of the "ancient ancestors of man" along with a few chimpanzees, a gorilla named Dinah, and an orangutan called Dohung. The zoo's evolutionist director made speeches boasting that this exceptional "transitional form" was in his zoo. They caged Ota Benga as if he were an ordinary animal. Demoralized and distraught by his treatment, Ota Benga eventually committed suicide.[916]

Why were Australian Aborigines treated like animals and almost entirely obliterated as a population? Although mass killings have existed throughout all human history, we can ask what influence evolution had on modern mass mortalities. Why (from 1975 to 1979 did Pol Pot execute more than a million people—more than one-third of the population of Cambodia)? Why did Joseph Stalin kill 60 to 100 million of his own people (including those in his own prison camps, who were allowed to die from neglect)? Why did Mao Tse-tung murder six million people in China? Why did Adolf Hitler turn against humanity, call Jews "cockroaches" who needed to be "ex-

---

[909] Ernst Fraenkel, "The Normative State," *The Dual State: A Contribution to the Theory of Dictatorship*, Octagon Books, NY, 1941, p. 95.
[910] "The Holocaust," *Wikipedia, The Free Encyclopedia*. Wilhelm Höttl, an SS officer and a Doctor of History, testified at the Nuremberg Trials and Eichmann's trial that at a meeting he had with Eichmann in Budapest in late August 1944, "Eichmann . . . told me that, according to his information, some 6,000,000 (six million) Jews had perished until then—4,000,000 (four million) in extermination camps and the remaining 2,000,000 (two million) through shooting by the Operations Units and other causes, such as disease, etc."
[911] Adolf Hitler, quoted in *Creation Magazine*, Vol. 18, No. 1, p. 9.
[912] Robert E. D. Clark, PhD, "Good Squib," *Darwin, Before and After*, Moody Press, Chicago, IL, 1967, p. 115.
[913] Caryl Matrisciana and Roger Oakland, "Preparation for the Delusion," *The Evolution Conspiracy*, Harvest House Publishers, Eugene, OR, 1991, p. 65-66.
[914] Stephen Jay Gould, PhD, "Pervasive Influence," *Ontogeny and Phylogeny*, The Belknap Press of Harvard University Press, Cambridge, Massachusetts; London, England, 1977. pp. 127-128.

[915] Charles Kingsley, 1888, Sermons on Natural Subjects, Sermon XLI, pp. 414-417; cited in *Creation Ex Nihilo*, February 2000, p. 51.
[916] Dennis R. Petersen, "Why Is This Subject the Most Important Battleground of History?" *Unlocking the Mysteries of Creation: the Explorer's Guide to the Awesome Works of God*, Bridge-Logos Publishers, Alachua, FL, 2002, p. 142.

terminated," and massacre at least six million?[917]

Chairman [Mao] listed Darwin and Huxley as his two favorite authors.[918]

> In a biography by E. Yaroslavsky: At a very early age, while still a pupil at the ecclesiastical school, Comrade Stalin developed a critical mind and revolutionary sentiments. He began to read Darwin and became an atheist.[919]

Certain tenets of the theory of evolution were dangerous. For Stalin, Mao, and Hitler, three powerful men, Darwin's theory removed all morality—there was now no such thing as right and wrong. With natural selection, might made right.

Karl Marx based his philosophy of communism on evolution: "Darwin's book is very important and serves me as a basis in natural science for the class struggle in history."[920]

> *Holt Biology—Visualizing Life:* You are an animal and share a common heritage with earthworms.[921]

However, the *Holt Biology* view may have a trickle-down effect.

Evolution during the modern era and evolution's aftermath

In 1957, the Russians launched Sputnik, the first satellite to orbit the earth, and Americans panicked. Americans were taught that the Soviets were ahead of the United States because they promoted intense high school studies in advanced science.[922] Two years later, in 1959, the 100th anniversary of Darwin's book *On The Origin Of Species*, the United States Department of Health Education and Welfare promoted a new, expanded emphasis on evolution. The National Science Foundation provided the Biological Sciences Curriculum Study (BSCS) with the funds to provide new textbooks to public schools.[923]

American textbooks were rewritten in the late 1950s and early 1960s to include more evolution. It was called, "The Cold War Reconstruction of American Science Education" *with an emphasis on the role of the biological sciences in the curriculum.*

One wonders whether this Reconstruction may have led some students to see in the evolution emphasis a "proof" that the old morality and God and His laws were dead, *leaving them free to experiment with a new morality. Was it wholly coincidental that in the 1960s* prayer was removed from the public-school system[924] and SAT (college admissions Scholastic Aptitude/Assessment Test) scores plummeted,[925] and statistics were showing significant and rapid increases in:

- Sexually transmitted diseases in 10-14-year-olds (started to skyrocket to 385 percent[926])
- Unmarried birth rates[927]
- Divorce rates[928]
- Child abuse[929]
- Illegal drug use[930]
- Violent crimes[931]
- Suicide rates in 15-24-year olds.[932]

---

[917] *The Dangers of Evolution*, Creation Science Evangelism, 29 Cummings Road, Pensacola, FL, 2007, see also: the Guinness Book of World Records, 1994, p. 460.
[918] *Creation*, Vol. 18, No. 1, p. 9.
[919] E. Yaroslavsky, *Landmarks in the Life of Stalin*, Foreign Languages Publishing 033House, Moscow, 1940, pp. 8-12; see also: Paul G. Humber, "Stalin's Brutal Faith," *Impact #172*, Institute for Creation Research, El Cajon, CA, 1987, p. 1.
[920] Karl Marx, as quoted by Conway Zirkle, *Evolution, Marxian Biology, and the Social Scene*, Philadelphia: University of Philadelphia Press, 1959, p. 86.
[921] George B. Johnson, "Animal Kingdom," *Holt Biology—Visualizing Life*, Holt, Rinehart and Winston, Inc., Austin, New York, Orlando, Chicago, 1994, p. 453.
[922] "Crisis in Education," *Life*, March 24, 1958, pp. 25-35.
[923] Arnold B. Grobman, *The Changing Classroom: The Role of the Biological Sciences Curriculum Study*, Doubleday & Company, Inc., Garden City, NY, 1969, pp. 26, 170-172, 200-204.
[924] Allaboutpopularissues.org, "Prayer in Public School," 2002.
[925] College Entrance Exam Board.
[926] Centers for Disease Control and Prevention and Department of Human Resources.
[927] Statistical Abstracts of the United States.
[928] U.S. National Center for Health Statistics: Vital Statistics of the United States.
[929] U.S. Department of Health and Human Services and Child Maltreatment: Reports from the States to the National Child Abuse and Neglect Data System.
[930] National Institute on Drug Abuse, 2000.
[931] Statistical Abstracts of the United States and the Department of Commerce Census Bureau.
[932] National Center for Health Statistics.

Chapter 40
Final Thoughts

In the introduction of this book is a statement by evolutionary biologist and geneticist Richard C. Lewontin, PhD:

> Dr. Lewontin: We take the side of science *in spite* of the patent absurdity of some of its constructs, *in spite* of its failure to fulfill many of its extravagant promises of health and life, *in spite* of the tolerance of the scientific community for unsubstantiated just-so stories,[933] because we have a prior commitment, *a commitment to materialism* [also known as naturalism]. It is not that the methods and institutions of science somehow compel us to accept a material explanation of the phenomenal world, but, on the contrary, that we are forced by our *a priori* [presumptive, taken for granted] adherence to material causes to create an apparatus of investigation and a set of concepts that produce material explanations, no matter how counter-intuitive, no matter how mystifying to the uninitiated. Moreover, that materialism is an absolute, for we cannot allow a Divine Foot in the door.[934]

My question is (and this is a question you can ask any evolutionary scientist or atheist): Is Dr. Lewontin's scientific worldview expressed here based on science? Or should it, instead, be considered a philosophy?

Some things I just don't understand

I don't understand why people today smoke tobacco. In the nineteenth century, when patients were being bled to death and drugged to death, tobacco was prescribed for any problems one might have with his or her lungs—"the vapor to be produced by smoking a cigar. The patient should frequently draw in the breath freely so that the internal surface of the air vessels may be exposed to the action of the vapors."[935]

You can tell from this statement that not much was known then about the anatomy of the lungs. Today we know a lot more. We know that smoking causes lung cancer and emphysema. We know that it is a dangerous, poisonous habit that costs a small fortune, and stinks up people's homes, cars, clothes, and breath. I can think of no good that comes from smoking tobacco. What early science once thought was good, science now says is bad.

Some people smoke because they think it looks "cool." Unfortunately, because they become addicted to nicotine, their lives are controlled by a weed. And that's sad.

I don't understand why people today use illegal drugs. In the nineteenth century, physicians prescribed calomel, opium, prussic acid, lunar caustic, antimony, arsenic, heroin, mercury, chloroform, and strychnine. Today, science tells us these drugs, along with methamphetamines and misuse of opoids, are dangerous. Not only do illegal drugs cost a small fortune, but they also cause crime, broken relationships, loss of health, incarceration, and premature death. What early science once thought was good, science now says is bad.

People get started using harmful drugs because of peer pressure, and they continue because of serious addictions. And that's sad.

I don't understand why people gamble, start wars, commit murder, cheat on their spouses, misrepresent truth, use foul language, involve themselves with human trafficking and prostitution, and abuse their animals, children, or spouses; all of which can ruin lives. And that's sad.

Likewise, I just don't understand why people believe in materialism. Why don't evolutionists have a good explanation for irreducible complexity, including sexual reproduction? How can anyone look at a newborn baby, with all of its potential, and believe that a miracle like that happened by chance? How can anyone look at

---

[933] In science, a *just-so story* is an unverifiable, unfalsifiable, fictional explanation. Unfalsifiable means there is no possible physical experiment or observation that can be devised to prove or disprove a particular hypothesis or theory. But some still believe the story.
[934] Richard C. Lewontin, PhD, quoted by Joe White et al, *Darwin's Demise*, Master Books, Inc., Green Forest, AR, 2001, p. 139; *see also:* Richard Lewontin, "Billions and Billions of Demons," *The New York Review*, January 9, 1997, p. 31, (emphasis added).
[935] Richard A. Schaefer, "How much more could they bear?" Legacy Publishing Association, Loma Linda, CA, 2005, p. 75.

the lack of intermediates in the fossil record, all of the hoaxes and fraud and scientific assumptions and legitimate, but corrected errors that have occurred over the decades, and believe that man originated from a lightning strike on warm primordial soup? And how can the scientific community look at the geologic column with its continental and worldwide, water-laid, fossil bearing strata, and deny a global Flood when rates of erosion tell us that the continents should have eroded away at least 100 times during the estimated billions of years of their existence.

Many Christians have been taught that the Holy Bible is the inerrant Word of God. The evidence shows that any worldview other than creationism, including theistic evolution (that God used evolution to create the various forms of life), undermines one's faith in the teachings of Scripture. When people start down that path they hit a slippery slope that eventually undermines their faith in God. Because what kind of God would rely on "tooth and claw" to create life as we know it! They were not part of His original action. He said, "An enemy hath done this" (Matthew 13:28).

Evolutionists reject creationism because it is not based on naturalism. Creationism is based at its origin on miracles, and evolutionists claim that they do not believe in miracles. It appears to me that evolutionism and materialism are handicapped by bias and dogma, which forbids its proponents to even consider alternatives (because they "can't allow a divine foot in the door").

> Jonathan Wells, PhD: Like all other scientific theories, Darwinian evolution must be continually compared with the evidence. If it does not fit the evidence, it must be reevaluated or abandoned—otherwise it is not science, but myth.[936]

Evolution explains the universe and its origin in a naturalistic, materialistic, humanistic frame of reference that recognizes no possibility of any other cause. Except for Theistic Evolution, Naturalistic Evolution offers an anti-God paradigm that does not acknowledge any evidence that might support the biblical account of how life started on earth. This worldview allows no hope of a future beyond this life. A belief in evolution's "Time plus Chance plus Natural Selection" and perhaps some "intelligence" inherent in Matter itself, leads many to a depressing sense that life is meaningless and pointless, and that what originated from the hand of mere matter just doesn't matter.

And that makes me *really* sad.

Support for recent creation of life from science

Evolution's model says life on earth began 3.8 billion years ago. These pages have been an introduction to some of the scientific evidence supporting the report from Genesis to Revelation that God created a habitat for humans and life on earth just a few thousand years ago—not billions, though He might have created our galaxy, solar system, and planet countless years ago, according to Genesis 1:1-2. Is there in-depth evidence for this from scientists across many scientific disciplines?

Yes.

During my 10 years of researching the literature for this book, my confidence in the Genesis story has deepened; that story arcs across hundreds of verses from Genesis to the last book of the sacred Scriptures. My interest in science has also deepened.

Following is a summary of some of the scientific findings that especially impressed me as supporting recent creation of life on earth.

Mitochondrial Eve is about as old as Genesis Eve.[937] In 1987, researchers at UC Berkeley reported in *Nature* (after comparing mitochondrial DNA globally) that the "mother" of humanity lived about 200,000 years ago.[938] *In 1998,* the journal *Science* reported that mitochondrial Eve "would be a mere 6,000 years old." The *Science* headline said this raised "troubling questions about the dating of evolutionary events."[939] (Pages vii and 114 discuss other

---

[936] Jonathan Wells, PhD, "Introduction," *Icons of Evolution: Science or Myth?* Regnery Publications, Washington, DC, 2000, p. 5.

[937] Ann Gibbons, "Calibrating the Mitochondrial Clock," *Science*, Vol. 279, January 2, 1998, p. 29; Walt Brown, PhD, *In the Beginning*, pp. 229-231.

[938] Rebecca L. Cann and others, "Mitochondrial DNA and Human Evolution," *Nature*, Vol. 325, January 1, 1987, pp. 31-36, (01 January 1987) doi:10.1038/325031aO

[939] Ann Gibbons, "Calibrating the Mitochondrial Clock," *Science*, Vol. 279, January 2, 1998, p. 29; Walt Brown, PhD, *In the Beginning*, pp. 229-231.

age-related dilemmas that evolution faces in radiocarbon and radiometric dating.)

Evolution's assumptions and explanations invite questions in "candid moments." The naturalistic materialist beliefs of evolution exclude a Creator. Still, there are scientists today who see, in nature's complexities that are beyond human understanding, the hand of One *outside* the material universe.[940] A Designer. Scripture says He became also one of us[941] to recreate[942] those who believe in Him.[943] (See chapter 37).

There are many rock layer formations preserved. And rates of erosion are rapid. Yet, there is so little continental erosion (e.g., Grand Canyon's flat layers) that it suggests the earth is young. Not old.[944] In the evolution model, earth and its dry continents have existed for billions of years. Field researchers across the globe have noted erosion rates on the continents ranging between 3 and 63 feet per 1000 years—faster erosion at higher altitudes. The height of continents averages about 2000 feet above sea level. Even at just one millimeter per 1000 years, the continents would have been eroded to sea level in only 623 million years.[945] And in a minimum two and a half billion years assumed for the existence of the earth's continents, a bare minimum one-millimeter erosion rate should have leveled the continents to sea level four times.[946]

Why are the continents and geologic column still here? Slow up-thrust of mountains is offered as the explanation. However, generally mountain building comes from the bottom up. Erosion, including erosion of fossils, comes from the top down. Yet the fossils and their geologic column are still here. Creationists would say life on earth is thousands, not billions of years old, and that much of the mountain up-thrust occurred in cataclysmic earth upheavals during and after the great Flood: "all the fountains of the great deep were broken up;" "the subterranean waters"—"the springs of the great deep burst forth" (Genesis 7:11)—possibly geysers erupting from deep earthquake activity during the Flood. (See pp. 14 and 140.)

If carbon-14 is up to 37 percent out of equilibrium, life on earth cannot be ancient.[947] Carbon-14 is produced from once-living (organic) material. Yet, carbon-14 being out of equilibrium tells us that at least the atmosphere where carbon-14 is made around our earth is not so old. The evolutionary scientist who developed carbon-14 dating methods, Willard F. Libby, PhD, determined that if a new earth were to be created, it would take about 30,000 years for carbon-14 to reach equilibrium;[948] where the amount entering from the atmosphere matched the decay rate. And if carbon-14 is 37 percent out of equilibrium, that would mean that life on this earth was created far less than 30,000 years ago. On the other hand, I don't see any theological problem in believing that the universe itself is old, as Genesis 1:1-2 may imply. The rest of Judeo-Christian Scripture, however, supports the view that life on this earth was created recently. (See p. 177.)

---

[940]"For since the creation of the world God's invisible qualities—his eternal power and divine nature—have been clearly seen, being understood from what he has made" Romans 1:20 *New International Version*.
[941]"1-In the beginning was the Word, and the Word was with God [the Father], and the Word was God [the Son]. 3-All things were made by Him, and without Him was not any thing made that was made. 14-And the Word was made flesh, and dwelt among us, (and we beheld His glory, the glory as of the only begotten [Son] of the Father,) full of grace and truth. 17-Grace and truth came by Jesus Christ" John 1:1-17 KJV.
[942]"Therefore, if anyone is in Christ, he is a new creation" 2 Corinthians 5:17 NIV.
[943]"For God so loved the world, that He gave His only begotten Son, that whosoever believeth in Him should not perish, but have everlasting life" John 3:16 KJV.
[944]Peter J. Wyllie, "The Great Globe Itself," *Propædia: Outline of Knowledge, Guide to the Britannica, Encyclopædia Britannica*, Encyclopædia Britannica, Inc., William Benton, Chicago, London, Toronto, Geneva, Sydney, Tokyo, Manila, Seoul, 1979, p. 77.
[945]Ariel A. Roth, PhD, "Some geologic questions about geologic time," *Origins: Linking Science and Scripture*, Review and Herald Publishing Association, Hagerstown, MD, 1998, pp. 262-274.
[946]Ariel A. Roth, PhD, "Some geologic questions about geologic time," *Origins: Linking Science and Scripture*, Review and Herald Publishing Association, Hagerstown, MD, 1998, pp. 262-274.
[947]Peter J. Wyllie, "The Great Globe Itself," *Propædia: Outline of Knowledge, Guide to the Britannica, Encyclopædia Britannica*, Encyclopædia Britannica, Inc., William Benton, Chicago, London, Toronto, Geneva, Sydney, Tokyo, Manila, Seoul, 1979, p. 77.
[948]W. F. Libby, *Radiocarbon Dating*, University of Chicago Press, Chicago, IL, 1955, p. 7.

The reliability and trustworthiness of radiometric dating is questionable.[949] Radiometric methods of dating the same fossil or rock sometimes disagree by as much as hundreds of millions of years. This disturbs evolutionists. However, carbon-14 dates with results that deviate substantially from what is expected may be discarded and never published. (See pp. 175-178.)

Red blood cells, etc. in many dinosaur fossils reveal they are not millions of years old.[950] Dinosaurs are said to have lived 65 to 70 million years ago. But everything from unfossilized soft tissue, collagen fibers, and *nucleated red blood cells,* to heme iron, hemoglobin, and flexibly elastic blood vessels have been found in dinosaur bones—*Tyrannosaurus rex,* hadrosaurs, and duck-billed dinosaurs.[951]

Creationists agree: researchers at Geoscience Research Institute, using chemical tests including sequencing, mass spectrometry, and immunofluorescence, confirmed in 2015 that soft, flexible blood vessels were found in the fossil hadrosaur Brachylophosaurus. "This should raise questions over whether these fossils are as old as widely thought."[952] (See pp. 121-123.)

Carbon-14 is found in dinosaur tissue.[953] Some dinosaur bone samples have been carbon-14 dated as 20,000 years old.

Materials from Cretaceous layers (thought to be 100,000 million years old—layers where dinosaur bones are found) were carbon-14 dated at 34,000 years. (See p. 178.)

And measurable carbon-14 was found in the bones of nine dinosaurs from Texas to Alaska and one from China. The dinosaurs would need to have been millions of years younger to yield measurable amounts of carbon-14.[954] (See p. 122-123.)

No carbon-14 can be detected in material older than 100,000 years. Carbon-14 is in coal, oil, and diamonds. (See p. 175.)

Diamonds are pure carbon. Why is the same level of carbon-14 found in deep-earth diamonds as is found in fossils from the top to the bottom of the fossil record (such as in the Grand Canyon)? (See pp. 175-176.)

Coal and oil. Carbon-14 is found in coal. Textbooks teach that some coal formed 250 million years ago. After about 5,730 years, half the amount of carbon-14 in a sample has decayed, so after 10 half-lives, coal should have no carbon-14 left to measure. Implied: Coal is young. (See chapter 32.)

To the creation-minded geophysicist,[955] carbon-14 now in coal is the carbon-14 that was in dead carbon-rich trees and plants. The global catastrophic Flood (and its geologic activity) a few thousand years ago created layers and layers of varied sediments as its waters receded. These heavy layers of organic carbon became coal and oil. (See pp. 175-176.)

Recent studies have shown that (under sedimentary and lava pressure and with volcanic heat) coal and oil can form (under water) in 1-6 years (and does not need 250 million years). (See 6-page summary for scientists of a book by Dr. Andrew Snelling: *Recent Rapid Formation of Coal and Oil,* at https://answersingenesis.org/

---

[949] Robert E. Lee, "Radiocarbon: Ages in Error," *Anthropological Journal of Canada,* Anthropological Association of Canada, Ottawa, Vol. 19(3), 1981, pp. 9-29, (emphasis in the original); see also: Stuckenrath 1977:188.
[950] M. Schweitzer and T. Staedter, "The Real Jurassic Park," *Earth,* June 1997, pp. 55-57.
[951] John Noble Wilford, *New York Times,* "Tissue Find Offers New Look Into Dinosaurs Lives," March 24, 2005; see also: *Science,* March 25, 2005; see also: M. Schweitzer and T. Staedter, "The Real Jurassic Park," *Earth,* June 1997, pp. 55-57; Derek Isaacs, "Hell Creek," *Dragons or Dinosaurs? Creation or Evolution?* Bridge-Logos, Alachua, FL, 2010, pp. 141-142; and as reported also in the *Journal of Vertebrate Paleontology,* Smithsonian magazine, *Cosmos,* the *Proceedings of the National Academy of Sciences,* and *New Scientist.*
[952] Geoscience Research Institute, "More Dinosaur Blood Vessels," *Geoscience Newsletter,* Number 44, January 2016.
[953] Reginald Daly, "Origin of sedimentary mountains, opposing theories," *Earth's Most Challenging Mysteries,* The Craig Press, Nutly, NJ, 1972, p. 280.

[954] Hugh Miller, Hugh Owen, Robert Bennett, Jean De Pontcharra, Maciej Giertych, Joe Taylor, Marie Clair Van Oosterwych, Otis Kline, Doug Wilder, Beatrice Dunkel, Abstract, "BG02-A012, A Comparison of δ13C & pMC Values for Ten Cretaceous to Jurassic Dinosaur Bones from Texas to Alaska USA, China, and Europe;" see also: Paul Giem, MD, on You Tube, "The Missing Presentation," March 30, 2013. https://youtu.be/s_53hGlasuk.
[955] John Baumgardner, PhD, *Thousands . . . Not Billions,* DVD, The Institute for Creation Research, An ICR Special Edition, in association with TEN31 Productions, 2005.

Final Thoughts

geology/catastrophism/how-fast-can-oil-form/).

I hope that the perspective presented in this book will not only confirm the already persuaded, but also it will be welcome news to the open-minded that creation science makes much more sense than previously realized. It can be accepted and even exonerated by the "reasonable doubts" regarding evolution that have been brought to light in these pages.

I believe in God not only because I exist, but also because I have human consciousness. And I don't believe there is any way all the interrelated systems in my body could have evolved by chance. My skeletal system, circulatory system, digestive system, immune system, endocrine system, lymph system, and reproductive system, and most wonderfully, a central nervous system with a brain—all these intricate systems working together in an interdependent relationship demand they all came into existence at the same time, as mature systems. Even more wonderful, this brain and body have been invited to share a loving and eternal relationship with my Creator and fellow created intelligences.

One of the most profound questions I have ever heard is, "Which came first, the chicken or the egg?" Creationists can answer that question. Evolutionists cannot. Nor can they answer how the opposite sexes evolved *in tandem*.

> Michael R. Rose, PhD: The evolution of sex is one of the major unsolved problems of biology. Even those with enough hubris to publish on the topic often freely admit that they have little idea of how sex originated or is maintained. It is enough to give heart to creationists.[956]

> Geoffrey Simmons, MD: Like the chicken-and-egg riddle, there could not have been an egg or a sperm without a male and a female. . . . If evolution were the explanation, millions of intermediate species as well as numerous genetic mistakes should have been found in the fossil record.[957]

Actually, the scientific community announced in mid-2010 that the chicken came first:

> Associated Press: British scientists believe they have found the answer to an ages-old question: Which came first, the chicken or the egg?
>
> Scientists cracked the puzzle after discovering that the formation of eggs is possible only thanks to a protein found in chicken's ovaries. That means eggs have to be formed in chickens first.
>
> The protein—called ovocledidin-17 (OC-17)—speeds up the development of the shell. Researchers from Sheffield and Warwick universities in England laid out their findings in the paper "Structural Control of Crystal Nuclei by an Eggshell Protein."
>
> They used a supercomputer to zoom in on the formation of an egg and realized the protein is vital in kick-starting the crystallization process. It works by converting calcium carbonate into the calcite crystals that make up the egg shell.
>
> Dr. Colin Freeman, from Sheffield University's Department of Engineering Materials, said, "It had long been suspected that the egg came first—but now we have the scientific proof that shows that in fact the chicken came first. The protein had been identified before and it was linked to egg formation, but by examining it closely we have been able to see how it controls the process."
>
> Freeman said, "It's very interesting to find that different types of avian species seem to have a variation of the protein that does the same job."[958]

This is a good example of how what looks like good science can answer a profound question while at the same time ignoring an equally profound question: Where did the chicken come from? But it also may be a good example of some fun fake news from the British "research" team. If so, insiders reading all the details surrounding this impressive scientific discovery must have enjoyed reporting it (or is it merely funny to my British-roots editor?) Surely one does not need to be a biologist to recognize no research here answered where that proteinaceous chicken had come from.

So, do I lean toward creationism? No! I stake my future on it. Evolution usually eliminates God. It explains how life evolved, but it cannot

---

[956] Michael Rose, PhD, "Slap and Tickle in the Primeval Soup," *New Scientist*, Vol. 112, New Science Publications, London, October 30, 1986, p. 55.
[957] Reproduction: Macroscopic," *What Darwin Didn't Know*, Copyright © 2004 by Geoffrey Simmons, MD. Published by Harvest House Publishers, Eugene, Oregon, 97402; www.harvesthousepublishers.com. Used with permission, p. 57.
[958] Associated Press, July 14, 2010

explain how life started. And the theory of natural selection only explains variation (sometimes called microevolution), not macroevolution (how a simple cell evolved into more complex organisms and eventually into man).

Furthermore, there is no such thing as a simple cell. According to the *Encyclopædia Britannica*, a "simple" cell contains enough information to fill a hundred million pages of the *Encyclopædia Britannica*.[959]

> Lee Strobel: Looking at the doctrine of Darwinism, which undergirded my atheism for so many years, it didn't take me long to conclude that it was simply too far-fetched to be credible. I realized that if I were to embrace Darwinism and its underlying premise of naturalism, I would have to believe that:
>
> - Nothing produces everything;
> - Non-life produces life;
> - Randomness produces fine-tuning;
> - Chaos produces information [in the cell, e.g., in DNA];
> - Unconsciousness produces consciousness; [and]
> - Non-reason produces reason.
>
> Based on this, I was forced to conclude that Darwinism would require a blind leap of faith that I was not willing to make. Simply put, the central pillars of evolutionary theory quickly rotted away when exposed to scrutiny.[960]

My research led to some questions for which I have not found answers.

1. Why are the stratified layers of the geologic column in the Grand Canyon, which supposedly took millions of years to form, obviously flat and devoid of wind and water erosion in comparison to earth's present surface, unless by the Genesis Flood and its cataclysmic aftermath of settling layers of sediment?
2. Why did the plants and animals that are known today as "living fossils," (supposedly from hundreds of millions of years ago), not evolve into something else during that time? Scripture says that life on earth was very recent.
3. How could random changes modify cold-blooded animals into warm-blooded animals? Scripture says each creature was created to reproduce "after its" [basic] "kind."
4. How could the multiple interlocking and exquisitely complex chemical steps in blood clotting have developed by natural selection?
5. If amphibians evolved before mammals, why do some amphibians have five times more DNA than mammals and some amoebae have 1,000 times more DNA?[961]

There may be satisfactory answers to some of such questions. But I have not found them.

Michael J. Denton, MD, PhD, has his own series of questions based on the accepted evolutionary sequence:

> Why aren't reptiles today developing feathers? . . . Why aren't invertebrates evolving into vertebrates? Why aren't reptiles evolving into mammals? Shouldn't evolution be ongoing?[962]

A fossilized cockroach is pictured in *National Geographic*:

> *National Geographic*: "[The] fossil imprint . . . shows that roaches have changed but little since their world debut more than 320 million years ago.

National Geographic reported animal life found in amber, looking like today's insects, including a beautifully preserved ant (declared to be 100 million years old), a praying mantis (dated at 40 million years old), and an ordinary house fly and termite (reported to be "millions of years" old).[963]

A coelacanth [fish] reported to have been extinct for 70 million years and used as an index fossil to assign ages to geologic layers, was caught live by fishermen off the coast of Madagascar in 1938. Since then, over 30 specimens have been caught alive. . . . It is often featured as a living link with the past.[964]

Greenling damselflies, tiny insects with 22 mm

---

[959] "Life," *The New Encyclopædia Britannica*, Encyclopædia Britannica, Inc., Chicago, London, Toronto, Geneva, Sydney, Tokyo, Manila, Seoul, 15th ed., Vol. 10, 1979, p. 894.
[960] Lee Strobel, "The Cumulative Case for a Creator," *The Case for a Creator*, Zondervan, Grand Rapids, MI, 2004, pp. 277.
[961] "RNAs AND PROTEINS may communicate regulatory information IN PARALLEL," *Scientific American*, October 2004, p. 62.
[962] Michael J. Denton, PhD, "Evidence Against the Theory of Evolution," *The Case Against Darwin*, Refuge Books, Burlington, Massachusetts, 2002, p. 50.
[963] Paul A. Zahl, PhD, "Golden Window on the Past," *National Geographic*, September, 1977, pp. 422-435.
[964] Dennis R. Petersen, "Some Famous Fossil 'Connections' To Evolution," *Unlocking the Mysteries of Creation: The Explorer's Guide to the Awesome Works of God*, Bridge-Logos Publishers, Alachua, FL, 2002, p. 104.

[0.8661 inch] wingspans, supposedly went extinct 250 million years ago, but there they are, still flying in Australia.[965]

> Warren L. Johns, JD: Wollemi pine trees, conventionally dated at 150 million years before the present, were long considered extinct. To everyone's surprise, hardy Wollemi pines have been discovered alive and well in Australia, far from extinct.[966]

All these "ancient" finds are only a few thousand years old, from the Biblical perspective.

So, do I believe in God? Of course I do. I find the evidence as plain as the sun at noon on a cloudless day.

That doesn't mean there are not clouds or even violent storms that obscure the sun at times. Come to think about it, the sun shining on the earth in good weather and bad is a good metaphor for the great controversy between Christ and the author of evil.

All good comes from God and all evil originates from the archangel who rebelled against Him (Revelation 12:7-9).

An interesting reference written by the ancient Sumerians in cuneiform script, the wedge-shaped characters on clay tablets of the earliest records ever unearthed, reminds us of the biblical story of Eve (Genesis chapter 3):

> Siegfried H. Horn, PhD: "The maiden ate that which was forbidden, the maiden, the mother of sin, committed evil; the mother of sin had a painful experience."[967] This brief statement seems to refer to the event told in Genesis 3, according to which Eve, Adam's wife, brought a curse upon herself, her husband, and her offspring by listening to the alluring voice of the serpent and eating from the forbidden fruit of the tree of knowledge of good and evil. By doing this she became not only the first [rebel on earth] but as our Sumerian poem appropriately calls her, "the mother of sin." This fall into sin brought in its wake not only the loss of innocence and of Paradise but also painful childbearing, hard labor to make a living—truly a "painful experience"—and finally death.[968]

Before the Fall of Adam and Eve, there was no pain, suffering, or death. The animals of Eden ate plants. They did not eat each other. No animal food chain. Eventually, because those living on earth became so inhumane toward each other, and toward all living things, God the Creator became heartsick, and judged that it was time to put a stop to man's inhumanity. So, He offered rescue to whoever would come to Him, and He would have spared any who entered Noah's ark to avoid the Flood. But all except Noah's family rejected His 120 years of warnings, and were destroyed by the global Flood.

One of the strengths of scientists (both evolutionists and creationists) is their philosophical belief in the importance of self-correction. During my ten years of study for this book, I became convinced that the theory of evolution has been proven false. Some of the details leading to my conclusions may in time be unverified. However, I pray that my readers will with an open mind reconsider the sometimes-imaginative explanations and the major assumptions in the theory of evolution and in its "missing links" (and links documented as *rare* or even *man-made*), versus the multiplying scientific evidence which supports creationism.

Evidence for Biblical creation of life recently

- The Law of Biogenesis: that life comes only from life.
- Organisms of irreducible complexity—the eye, the ear, the central nervous system, and wondrous brain, and most interdependently, the sexes and reproductive system.
- Geology—erosion, mountain building, and global evidence for rapid deposition of the geologic column.
- The First and Second Laws of Thermodynamics.
- Carbon-14 on earth not yet reaching equilibrium.
- Carbon-14 found in coal, dinosaurs, and

---

[965] David Coppedge, "Speaking of Science," *Creation Matters*, January/February 2010, p. 5; cited by Warren L. Johns, JD, "Paging Sherlock Holmes, Missing Links?" *Genesis File*, www.GenesisFile.com, Lightning Source, LaVergne, TN, 2010, p. 42.

[966] *Creation ex nihilo [out of nothing]*, December 2000, p. 6; cited by Warren L. Johns, JD, "Superstitious Nonsense: Life from Spontaneous Generation?" *Genesis File*, www.GenesisFile.com, Lightning Source, LaVergne, TN, 2010, p. 43.

[967] Siegfried H. Horn, PhD, quoting Alfred Jeremias, *Das Alte Testament im Lichte des Alten Orients*, 4th edition, Leipzig, 1930, p. 99.

[968] Siegfried H. Horn, PhD, "Light Shed on the Pre-patriarchal Period," *Records of the Past Illuminate the Bible*, Review and Herald Publishing Association, Washington, DC, 1963 and 1975, p. 8.

even diamonds (the hardest and least contaminated substance on earth).
- DNA.
- Consciousness, morality, love, beauty, and the appreciation of beauty.
- The maximum ages of civilization, languages, trees, and deserts.

Evidence against evolution of life billions of years ago

- The fossil record (the lack of the transitional fossils and of the much referred to "common ancestor").
- The conversion of settled evolutionists to rational and ardent creationists.
- Evolutionary predictions.

Predictions

If evolution—rather than Creator God—started life, then a number of predictions could be made, that:
1. Early rocks would exhibit evidence of warm primordial soup with no oxygen.
2. The fossil record would show tens of thousands of missing links in an evolutionary progression from amoeba to man.
3. There would be literally millions (a continuum) of intermediates between all species of plants and animals living today.
4. The fossil record would show more types of plants and animals in the more recent, upper layers of the geologic column than in the older, lower layers.
5. Different methods of radiometric dating would generally verify and confirm each other.
6. Serious evolutionary scientists would be able to create life in the laboratory using a variety of contents to replicate "warm primordial soup" and sparks of various intensities to simulate the postulated lightning strike.
7. Vestigial organs would commonly exist.
8. Miracles would not occur.

These predictions have not been proven true.
The *Judeo-Christian Scriptures* report that creation was followed by a global Flood a few thousand years ago. If evolution started life, then the above could be expected or predicted; yet, these expectations have not been met.

On the other hand, the Sacred *Scriptures* of Jews and Christians *report* that *Creation was later followed by a global Flood.* Therefore, it leads the researching scientist to *expect* that *nature would prove the following to be true:*

1. Flood legends abound.
2. Early rocks show evidence of an oxygen-rich atmosphere and no primordial soup.
3. All plants and animals resist change into different basic kinds of plants and animals, even by the most modern breeding or genetic experimentation.
4. A global Flood covering earth's luxuriant vegetation and abundant animal life caused great reserves of coal, oil, and natural gas.
5. Huge deposits of coal, oil, and natural gas, known as fossil fuels, have been found sandwiched between layers of water-laid, sedimentary rock around the world.
6. There appears to be a greater variety of basic kinds lower down in the geologic column than live here today, known even by evolutionists as the Cambrian Explosion. This includes subtropical plants and animals buried under ice at the North and South Poles.
7. Inherited behavioral patterns in some insects are essential not only for their own survival but also for the survival of each of the one-of-a-kind plants that are specific to a particular insect—a specific plant on which the very life of that kind of insect symbiotically depends. According to the laws of probability, this mutual independence between co-dependent insect-plant pairs could not have evolved simultaneously.
8. Irreducible complexity in anatomy and physiology, including sexual reproduction, is acknowledged as explainable only by having been designed, and is unexplainable by naturalism (time plus chance).
9. The world's calendars are built on a seven-day weekly cycle (Genesis chapters 1 and 2 and Exodus 20:8-11).

Evolution teaches that *death*, through survival of the fittest and natural selection, brought *man* into the world. The Bible teaches that man,

Final Thoughts

through sin (selfishness, self-centeredness, *me first* at another's expense), brought death into the world. (Romans 5:12). These worldviews are at opposite poles.

Creationists have a hope. If creationists are wrong, they have nothing to lose. If evolutionists are wrong, they have everything to lose. If creationists are wrong, life will someday end and that will be the end of everything for them. If evolutionists are wrong, their eternal destiny may be in serious jeopardy. One reason I would rather be a creationist than an evolutionist is that I want to be on the side of truth at the end of our age. As a creationist, I cherish hope for a better world to come. Evolutionists hope for positive mutations to continue humanity. For them, this life is all there is.

A very personal experience worth sharing:

> John M. Cimbala, PhD: I was raised in a Christian home, believing in God and his creation. However, I was taught evolution while attending high school, and began to doubt the authority of the Bible. If evolution is true, I reasoned, the Bible cannot also be true. I eventually rejected the entire Bible and believed that we descended from lower creatures; there was no afterlife and no purpose in life but to enjoy the short time we have on this earth. My college years at Penn State were spent as an atheist, or at best as an agnostic.
>
> Fortunately, and by the grace of God, I began to read articles and listen to tapes about scientific evidence for creation. Over a period of a couple of years, it became apparent to me that . . . scientific data from the fossil record, geology, etc. could be better explained by a recent creation, followed by a global Flood. Suddenly I realized that the Bible might actually be true! It wasn't until I could believe the first page of the Bible that I could believe the rest of it. Once I accepted the fact that there is a Creator God, it was an easy step for me to accept His plan of salvation through Jesus Christ as well. I became a follower of Christ during my first year of graduate school at Caltech.
>
> Since then, I have devoted much time to studying the evidence for creation and a global Flood. The more I study, the more convinced I become that there is a loving God, who created this universe and all living things. God revealed some details about His creation in the Book of Genesis, which I now believe literally. . . . [969]

My view

Evolution usually defines science as first excluding the possibility of there being a creator; it defines science as the search for solely natural explanations for origins and change. It defines searching in the Supernatural as unscientific. Yet there is much in evolution's search methods that lends itself neither to laboratory nor field observation or experimentation.

For reasons outlined in these pages, I believe that "we walk by faith [a trusting relationship with God], not by sight [merely what we can observe]" (II Corinthians 5:7). In His Word—the Bible, and its embodiment, His Son Jesus—and in all He has made, I find my personal, caring Creator.

Creationism is one of the oldest explanations of origins. Critics of the biblical history of earth (creation by God; global Flood . . .) may not have been exposed to the Bible and Judeo-Christian historians enough to consider such history seriously. A more open and objective worldview is willing to consider the evidence even if it would "allow a Divine foot in the door."

However, in recent years there has been a revival in what is known as the "New Creationism" or "neo-Creationism."

Jesus and those who wrote the Bible regarded it not only as God-inspired and life-changing but also as historically true. And what about the Flood (Genesis chapters 6-9)? Does the New Testament record the Flood as literal history? Yes—by the testimony of Jesus (Matthew 24:37-39; Luke 17:26-27), and the author of the book of Hebrews (Hebrews 11:7), and the apostle Peter (II Peter 2:5), and the Christian church they founded.

Peter, in his letters to the young churches, said he was writing (II Peter 3:1-6) "… you must understand that in the last days scoffers will come . . . and laugh at the truth. . . . "

Paul cautioned a new young pastor, "Timothy, guard what has been entrusted to your care. Keep out of foolish and contradictory arguments over what is falsely called knowledge, or you will

---

[969] Edited by John Ashton, PhD, John M. Cimbala, PhD, "John M. Cimbala, PhD," *In Six Days: Why Fifty Scientists Choose to Believe in Creation*, Master Books, Inc., Green Forest, AR, 2000, pp. 200, 201.

miss the most important thing in life—to know God" (I Timothy 6:20-21).

Scientific truths are continually changing. The words of the Bible are continually changing *those who read and incorporate them.*

In many Scriptures the Bible plainly tells us that evil—such as thorns, thistles, poisonous snake venom, the viciousness of animals and man—is a direct result of humanity's rebellion against the Creator. And "all creation is suffering" under the effects of turning away from Him (Romans 8:19-22).

So, we can expect to find many unGod-like things in nature too, like the food chain of violence with the larger eating the defenseless. These are bearing "false witness" to the kind of benevolent being God is.

Such evils did not exist before Adam and Eve rebelled against God. In rejecting His warning and His daily presence, they also rejected His continuous protective blessing on the creation (Genesis 3:1-8). All can see the results.

Despite this false testimony against the Creator, when we believe His written Word, we see the true testimony, and we learn to trust Him, even though there are questions we cannot answer. But God has not left Himself "without a witness" (Acts 14:17) in the beauty, complexity, and intelligent design of His created wonders. He gave us His created works as a witness to His existence and power and generous love. I once heard it said that flowers are God's love letters to us.

It is good to believe in God when we find evidence in the sciences. Sometimes it is not easy to reveal such faith, especially when we are in a wholly secular environment or with those who have not yet become acquainted with their Creator through His written and living Word (John chapter 1).

And soon, God's world will be re-created to perfect Garden of Eden conditions by the New Adam (I Corinthians 15:45-57). Scripture calls it the New Earth (Isaiah 65:17; Revelation 21:1) and tells us about it. No violence. No death. Peaceful as a meadow:

> The wolf also shall dwell with the lamb, and the leopard shall lie down with the kid; and the calf and the young lion and the fatling together; and a little child shall lead them. And the cow and the bear shall feed; their young ones shall lie down together: and the lion shall eat straw like the ox (Isaiah 11:6, 7).

We cannot even imagine how wonderful life will be:

> But as it is written, Eye hath not seen, nor ear heard, neither hath entered into the heart of man, the things which God hath prepared for them that love him (I Corinthians 2:9).

And how do we experience such wonders? What follows I believe is the most important text in Scripture, and the most important words in this book:

> For God so loved the world, that he gave his only begotten Son, that whosoever believeth in him should not perish, but have everlasting life (John 3:16).

Question: Why did I compile this book? Could it be that I care about my human brothers and sisters, and my heirs, and want to share the ultimate future with them?

Glossary

Abiogenesis: the theory that life spontaneously generated from non-living material. See **Biogenesis.**

Anthropology: the study of human societies and cultures and their development.

A priori: presumptive, taken for granted.

Archaeology: the study of human history and prehistory through the excavation of sites and the analysis of artifacts and other physical remains.

Archaeopteryx: the most famous fossils in the world, thought to illustrate a transition between classes such as reptiles and birds.

Archeoraptor: claimed by *National Geographic* to be a missing link; later acknowledged to be a fake fossil, a man-made construct.

Ardipithecus ramidus: a hominin species of fossils found that had not been walking upright like a human, but whose teeth and skull were somewhat human-like.

Big Bang: a scientific theory that the universe started very rapidly from a point and then made stars and groups of stars called galaxies.

Biogenesis: the theory based on observation that life comes only from life. See **Abiogenesis.**

Black box: a term scientists use to describe any system, machine, device, or theoretical construct that they find interesting but inexplicable.

Bombardier beetle: a beetle that can explosively spray hot noxious fluid at its enemies by a mechanism that is a good example of irreducible complexity.

Cambrian: the first layer of the geologic column that has most of the modern phyla of animals, immediately above the Precambrian layers.

Cambrian explosion: the numerous *phyla* of the same group that suddenly appeared in the lowest known fossiliferous rocks, and appeared without transitional intermediates; the inception of modern multicellular life with nearly every major kind of animal anatomy appearing in the fossil record for the first time. The observed phenomenon that in the Cambrian most of the numerous modern *phylae*—for example, jelly fish, arthropods (trilobites) and chordates, and some extinct ones—can be found together without obvious ancestral forms preceding them. This was not expected by evolutionary theory and was a puzzle for Darwin.

Carbon-14: a radioactive isotope of carbon, with 8 neutrons in its nucleus instead of the usual 6 in carbon-12 or 7 in carbon-13, which are stable and not radioactive. Its presence in organic materials is the basis of radiocarbon dating of archaeological, geological, and hydrogeological samples. It is used to date plant and animal fossils that are only thousands of years old.

Catastrophism: the view that geological formations such as the Grand Canyon formed rapidly by sudden, short-lived events, possibly worldwide in scope, in contrast to uniformitarianism. See **Uniformitarianism.**

Circular reasoning: A logical flaw where one uses proposition A to prove proposition B, then uses proposition B to prove proposition A. An example could be the common practice of dating fossils by the rocks they are found in, then later dating rocks and their layers by the fossils found in them.

Common ancestor: believed by evolutionists to be the most recent organism from which all organisms now living on Earth have descended. May also refer to an ancestor of two or more different groups of organisms, as the common ancestor of dogs and wolves.

Creation Design: that God designed and created an engineering wonder—the universe and planet earth, as exhibited by evidence such as irreducible complexity.

Creationists: those who believe that the universe and life originated from specific acts by the Divine Creator.

Dark Ages: A term no longer used by historians—now referred to as the Middle Ages or Medieval Times, in Europe 500-1500 AD following the collapse of the Western Roman Empire. Dark Ages suggests a prevailing ignorance and barbarism—and it was a time troubled by the loss of classical learning—yet, there were also continuous forces for culture and enlightenment.

Dinosaur: Any of certain usually large reptiles discovered in the fossil record, not corresponding to well-know modern reptiles (the *tuatara* is a modern representative of the group). From the Greek *deinos,* meaning "terrible" or "fearfully great," and *sauros,* meaning "lizard" or "reptile;" thought to have died out 65 million years ago.

DNA: Deoxyribonucleic acid; a self-replicating material present in nearly all living organisms as the main constituent of chromosomes; the carrier of genetic information.

Dragons: large, fierce reptiles of legend that roughly match the physical description that modern science now calls dinosaurs. Dragons are mentioned in the Bible 34 times.

Empirical evidence: knowledge or source of knowledge acquired by means of the senses, particularly by observation and experimentation.

Escherichia coli: The *E. coli* bacterium is commonly found in the gut of warm-blooded organisms.

Evolutionary progression: thought by evolutionists to illustrate a multitude of transitional forms from amoeba to man.

Evolutionary theory: the teaching that all plant and animal life forms advanced from simple to complex, also called Darwinism, neo-Darwinian synthesis, Universal Common Ancestry.

Diatomaceous earth: layers of tiny microscopic diatoms (a common type of phyto-plankton); microscopic hard-shelled algae.

Firmament: The starry heavens. The blue sky. Also its clouds. From creation week (Genesis 1:6-8) until the Global Flood, a canopy of water-vapor or thick clouds was stored above the earth. Blue sky and stars could be viewed through its *closed* but see-through "windows," perhaps like today's blue skies with clouds. During the Global Flood, the stored water fell like great waterfalls from the *opened* "windows of heaven" to earth; and, with the "fountains of waters" (great geysers) from the earth itself, their water covered the planet (Genesis 7:11 to 8:2). God said the rainbow is a reminder of His promise there would never be another Global Flood (Genesis 9:9-13).

Fossils: any ancient plant, animal, or human trace, without or with mineralization (such as mineralized "petrified" wood).

Fossil record: fossils preserved in sedimentary rock that show a sequence of life forms from simple in the lower, older layers to more complex in the upper, more recent layers, especially in the vertebrates.

Fruit Fly: *Drosophila melanogaster*, the workhorse of experimental geneticists.

Galapagos Islands: where Charles Darwin collected 14 different varieties of finches; an example of microevolution, as opposed to macroevolution.

Genes: basic physical and functional units of heredity that are transferred from parent to offspring.

Geologic column: a graphic (actual) and theoretical representation of the layers of rock that make up the earth's crust; example of part of the column: the Grand Canyon. The column is the support pillar of evolution for the age of fossils.

Gradualism: the hypothesis that evolution proceeds chiefly by the accumulation of gradual changes. Sometimes called "actualism."

Grand Canyon: A meandering Arizona channel, approximately one mile deep by 277 miles long by 18 miles wide, cut down through numerous sedimentary layers of rock of differing color and texture.

Hanson Ranch: In Eastern Wyoming, where advanced scientific equipment is used by professors and students to study and map the finds in ultra-3-D. Search online for "mapping out the truth/answers in genesis."

Hominids: From the perspective of evolution, the family Hominidae, the great apes, is said to include orangutans, gorillas, chimpanzees, and humans.

Human genome: the complete set of nucleic acid sequence for humans, encoded as DNA within the 23 chromosome pairs in cell nuclei and in a small DNA molecule found within individual mitochondria.

Humanism: attaches prime importance to human nature and human experience. An alternative to religious beliefs about the divine or supernatural, or the hereafter. Stresses the potential value and goodness of humans, emphasizes common human needs, and seeks solely rational ways by which the autonomous self can solve human problems. Each human creates his/her own set of ethical beliefs.

Hyperbaric Oxygen Therapy: The practice of providing 100 percent oxygen therapy under pressure for treatment of certain medical problems, such as the bends in deep-sea divers or poorly healing wounds.

Hypothesis: a supposition or proposed explanation made on the basis of limited evidence as a starting point for further investigation.

Ice Age: a period of long-term reduction in the temperature of Earth's surface and atmosphere, resulting in the presence or expansion of continental and polar ice sheets and alpine glaciers.

Intelligent Design: the theory that life, or some other feature of the universe or the universe itself, cannot have arisen by chance and was designed and created by some Intelligent Entity.

Intelligent Design Movement: promotes the idea that life and the universe appears designed, and is too complex to have evolved without the intervention of an intelligent entity, often thought of as a Supernatural Being.

Invertebrates: animals without backbones.

Irreducible complexity: can be seen as a system or device that has a number of different components that all work together to accomplish a task as a single system composed of several interacting parts, in which the removal of any one of those parts causes the system to cease functioning. Example: the simplest generic mousetrap ceases to function if any one part is removed. All are interdependent.

Java Man: the common name for human fossils promoted as an evolutionary transitional form between apes and humans; named after the island of Java in Indonesia, where they were discovered.

KNM-ER 1470: a famous hominid fossil which looks much more modern than its age according to the standard geologic time scale; it presents a serious challenge to all currently held theories of human evolution.

Krakatoa: the volcanic island between Java and Sumatra which exploded in 1883 and lowered the worldwide temperature for five years.

Lucy: several hundred pieces of bone fossils representing 40 percent of the skeleton of a female of the species australopithecus afarensis.

Ma'adim Vallis: a canyon larger than the Grand Canyon, believed by scientists to have been carved from the Martian surface within a few weeks.

Macroevolution: transmutation or evolution from one Linneaen species into another. For example, some evolutionists say dinosaurs evolved into birds. Megaevolution . . . transmutation or evolution from one larger group such as a family, class, or *phylum* to another. (Required for the overarching theory of evolution, but with much less evidence than that for lower levels of macro-evolution—the word was apparently created by George Gaylord Simpson, an evolutionist.

Materialism: a worldview based on the philosophical belief or doctrine that nothing exists except matter and its movements and modifications. Everything in the world of nature and the universe is material, not spiritual; a naturalistic understanding of reality.

Microevolution: minor changes or diversification strictly within Linnaean species—variation.

Mid-Atlantic Ridge: lava deposit reaching from the volcanic island of Iceland and the volcanic rim of Greenland southward through the Atlantic, forming a massive oceanic range nearly 7,000 miles long.

Missing links: intermediates or transitional forms of life that are missing from the fossil record and that, if found, would support Darwinism by linking a lower life form to a higher one, making the transition look like one led to the next.

Mitochondrial Eve: the idea that all human females were descended from an original female, as evidenced by their mitochondria.

Modus operandi: a particular way of doing something.

Morphology: the branch of biology that deals with the form of living organisms, and with relationships between their structures, rather than the study of their DNA.

Mount St. Helens: a stratovolcano in the Pacific Northwest of the United States, which erupted in 1980.

Mutations: thought by evolutionists to cause large, spontaneous, beneficial genetic changes.

Naturalistic/Materialistic Evolution: the belief that life evolved by itself and no Creator God was involved.

Natural Selection: survival of the fittest, the theory that organisms better adapted to their environment tend to survive and produce more offspring, and that the plants and animals that inherit superior genes and traits will pass on these genes and traits to the next generation. Does not deal with the origin of life.

Neanderthal Man: found to be a subspecies of modern human with the original specimen having arthritis.

Nebraska Man: once considered strong evidence for evolution; was later found to be the water-worn tooth of an extinct pig.

Noachian Flood: the biblical or Genesis Flood of Noah; the global Flood, a watery cataclysm.

Orce Man: promoted as the oldest example of man in Eurasia; later identified as a four-month-old donkey-skull fragment.

Paleoanthropology: combination of the disciplines of paleontology and anthropology.

Paleontology: the study of fossil remains of plants and animals, where they are found in the geologic column, and their relationship to modern plants and animals.

Peat-bog theory: believed by some to explain the creation of coal through the accumulation of large amounts of carbon-rich ferns, mosses, trees, etc. in moist low grounds. From a creation perspective, coal was made by floating mats that sank during the global flood. From an evolutionary perspective, this would have happened gradually, slowly over many millennia.

Peppered Moth: (Biston betularia) thought by evolutionists to be the best example of Darwinian selection, showing an inherited change in coloration of the moths from light to dark, and later back to light. Demonstrates microevolution (variations within moths), but not macroevolution or megaevolution.

Peristalsis: muscular action which keeps food moving through the digestive system.

Pillow lava: lava formed only under water, found as high as 15,000 feet on Mount Ararat.

Piltdown Man: heralded as a transitional, missing link between modern humans and ape ancestors; recognized today as a palaeoanthropological hoax.

Plate tectonics: the theory that Earth's outer shell is divided into several plates that glide over its mantle, the rocky inner layer above the core.

Polystratic trees: fossil trees that extend through several layers or strata of (usually sedimentary) rock.

Precambrian layer: the lowest level of the geologic column and the largest span of time in evolution's geologic time scale.

Presuppositional bias: Philosophical starting point—presupposing certain things as taken for granted, not to be questioned.

Primate: a member of the most developed and intelligent group of mammals (16 families, 72 genera); including humans.

Primitive atmosphere: First atmosphere of earth, thought by some to be methane, ammonia, water vapor, and hydrogen, but shown not to be.

Punctuated equilibrium: (also called punctuated equilibria) is a theory in evolutionary biology which proposes that once species appear in the fossil record they will become stable, showing little evolutionary change for most of their geological history. This state is called stasis. When significant evolutionary change occurs, the theory proposes that it is generally restricted to rare and geologically rapid events. It is the opposite of gradualism.

Quotation Mining: a practice of quoting someone out of context to mean something that the person did not intend and in fact opposed. Quotation mining can be a misleading practice. But the charge of quotation mining can also be leveled to unfairly discredit someone using quotations to legitimately support a contested point.

Radiocarbon dating: a method of dating that measures the ratio of carbon-14 to ordinary carbon in a specimen. Originally the assumption was made that the original ratio of carbon-14 to ordinary carbon was constant in the atmosphere, and therefore in plants and animals that got carbon-14 from the atmosphere. Later it was discovered that there were variations of that ratio in the atmosphere and the ratio was instead compared to that of "known" age samples. How well the age of those "known" age samples is actually known can be questioned.

Radiometric dating, *also known as* **Radioisotope dating:** dating methods that use radioactive isotopes. Other than carbon-14 dating, they are used to date igneous rocks: rocks that were once molten, then cooled and hardened—such as basalt, magma, tuff, or volcanic lava.

Recent life/creation: the "world" view, based on scientific evidence and implied in the Bible, that God created the various forms of life on earth a few thousand years ago.

Science: the intellectual and practical activity encompassing the systematic study of the structure and behavior of the physical and natural realm through observation and experiment (using the human senses and scientific tools). Science is commonly defined as excluding the supernatural realm; creation scientists are open to seeing natural phenomena in ways that lead to experiements studying natural features as well as noting evidences of the Creator's handiwork.

Scientific method: Observing data, forming hypotheses, making predictions, and constructing experiments to test these predictions.

Scientific research: observation and experiment, duplication of valid observation and experiments by other scientists, publication in peer-reviewed publications, seeking professional consensus.

Sedimentary rock: rocks formed from sediment deposited by wind or (much more commonly) water.

SETI: Search for Extraterrestrial Intelligence.

Species: seen by most scientists as a group of animals that can breed with others in the group (interbreed) and produce fertile offspring.

Spontaneous generation of life: an obsolete theory that the formation of living organisms occurred without descent from similar organisms.

Surtsey: a volcanic island that arose from 1963 to 1967; within days or weeks, it became a geological formation that, from a standard geological perspective, would have required thousands of years.

Theistic Evolution: the belief that God used the process of evolution.

Transitional fossils: "missing links"—hypothetical fossils intermediate between two living forms, especially between humans and apes.

Unconformities: missing pages of Earth history in a particular geographical area; a gap in geologic time; a break in the sedimentary fossil record in the geologic column. From an evolutionary viewpoint, the difference in age between the base of the strata above the unconformity and the top of the unit below the unconformity is interpreted as ranging up to millions, even hundreds of millions of years. Some (as in large parts of the Grand Canyon) show no erosion of the sediment below the unconformity, across what are said to be long ages.

Uniformitarianism: the belief that change has always been uniformly gradual over billions of years; claims that "the past is key to the present." Sometimes described as gradualism, in which slow incremental changes, such as erosion, created all the Earth's geological features.

Vertebrates: animals with backbones.

Vestigial organs: organs thought by evolutionists to be useless leftovers from the process of evolution; 180 were identified in 1890; today science has found that none is "vestigial"; all are useful for human function and health.

Young-earth creationism: the belief that planet Earth was created perhaps 6,000 to 10,000 years ago, in six literal days of creative activity (Revelation 14:6-7).

Young-life creationism: the belief that planet Earth (which may be billions of years old) was made fit for plant, animal, and human habitation a few thousand years ago, perhaps 6,000 to 10,000 years, in six literal days of creative activity which are described in Genesis 1 and 2, as well as Revelation 14:6-7.

Young-universe creationism: the belief that planet Earth and the entire universe was created a few thousand years ago, perhaps 6,000 to 10,000 years, in six literal days of creative activity which are described in Genesis 1 and 2, as well as Revelation 14:6-7.

Warm primordial soup: a theory, which is against the currently available evidence, that certain chemicals on earth combined to form amino acids and ribonucleotides, which then made protein and RNA, which then evolved into life. Essentially the same as abiogenesis.

Worldview: a particular philosophy of life or conception of the world; a framework of ideas and attitudes about the universe, ourselves, and life; a comprehensive system of beliefs.

Index

Abiogenesis · · · · · · · · · iii, 55, 58, 65, 225, 229
Ackerman, Paul D. · · · · · · · · · · · · · · · 126, 144
Actualism · · · · · · · · · iii, 5, 120, 125, 226, 229
Alberts, Bruce · 65
Anders, William · 7
Anthropology · · · · · 72, 74, 77, 182, 191, 225, 228
A priori · v, 215, 225
Archaeology · · · · · · · · · · · · · · · · · · iv, 189, 225
Archaeopteryx · · · · · · · · · · · · · 86-87, 118, 195, 225
Archeoraptor · 197, 225
Ardipithecus ramidus · · · · · · · · · · · · · · · · · 76, 225
Austin, Steven A. · 170
Australopithecines · · · · · · · · · · · · · · · 75, 182, 192
Axelrod, Daniel I. · 95
Baker, Sylvia · 107
Bates, Gary · 90
Baumgardner, John R. · · · · · · · · · 137, 141, 175-176
Behe, Michael J. · · · · · · · · · · · · · · · · 63, 67, 205
Berner, Robert · 147
Big Bang · · · · · · · · · · · · · · · 2, 203-204, 206, 225
Biogenesis · · · · · · · · · · iii, 29, 53, 56, 221, 225
Birdsell, J. B. · 182
Black box · 63, 67, 225
Bohlin, Ray · 22
Bombardier beetle · · · · · · · · · · · · · · · · · 66, 225
Borman, Frank · 7
Bowman, Sheridan · 178
Brand, Leonard R. · · · · · · · · 15, 55, 117, 126, 171
Brandstater, Bernard · · · · · · · · · · · · · · · · 15, 43
Brown, Walt, Jr. · 146
Bryan, William Jennings · · · · · · · · · · · · · · · · 58
Cambrian · · · · · · iii, 36, 87, 95-97, 105, 141, 180
206, 222, 225
Cambrian explosion · · · · iii, 36, 87, 95-97, 222, 225
Canby, Tom · 169
Carbon-14 · · · · · · iv, vi, 17, 121-123, 162, 175-179
184, 217-218, 222, 225, 228
Carbon dating assumptions · · · · · · · · · · · · · · 176
Carr, Bruce L. · 75
Carroll, Robert L. · 84
Catastrophe · · · 28, 97, 117-119, 125, 127, 133-135
137, 141, 154, 171, 175, 191

Catastrophic [global] Flood   126, 142, 150, 176, 218
Catastrophism · · · · 5, 115, 134, 137, 141, 144, 225
Chadwick, Arthur V. · · · · · · · · · · · · · · 27-28, 119
Chien, Paul K. · 95
Circular reasoning · · · · · · · 26, 120-121, 181, 225
Clark, Austin H. · 77
Clark, Harold W. · · · · · · · · · · · · · · · · 74, 77, 143
Clark, Robert E. D. · · · · · · · · · · · · · · · · · · · 212
Coal formation · 135
Coffin, Harold G. · · · 9, 13, 15, 49, 53-54, 135-136
139, 145, 170
Cohen, I. L. · · · · · · · · · · · · · · · · viii, 21, 105, 113
Common ancestor · · · · · · · · 3-4, 10, 19, 76, 83-84
92, 111, 225
Creation Design · · · · · · · · · · · · · · · · viii, 222, 225
Creationists · iii-vi, viii, 10-11, 13, 17, 19, 27-28, 30
32, 37, 53-54, 58, 60, 72, 83, 85-86, 90, 92, 95-96
108, 113, 120-121, 125, 129, 127, 136, 141, 148
155, 181, 189, 198, 204, 217-219, 221-223, 225
Crick, Francis H. C. · · · · · · · · · · · inside cover, iv, 67
Comfort, Ray · 83, 108
Cowles, David L. · 55
Czarnecki, Mark · 85
Czerkas, Stephen · 196
Damadian, Raymond · · · · · · · · · · · · · · · · · · 30
Dark Ages · 14, 29, 225
Darwin, Charles · · · · viii, 3-4, 13, 24, 28, 35, 39-40
56, 64, 83, 85, 87, 89, 91, 107, 110, 195, 211, 226
Dawkins, Richard · · · · · · · · · · · · · · · · iv, 57-58, 96
Dawson, J. W. · 144
Denton, Michael J. · · · · · · iv, 40, 60, 63, 67, 84, 91
93, 105, 108, 220
Desertification · 17
Devens, Richard M. · · · · · · · · · · · · · · · · 206-207
de Vries, Hugo · 21
Diatomaceous earth · · · · · · · · · · · · · · · · 126, 226
Dinosaur · · · · vi, 27, 86-87, 119, 121-123, 126-127
145, 178, 187-191, 193, 195-198, 218, 222, 225
DNA (deoxyribonucleic acid) · · inside cover, iii-iv, vii
43-45, 57, 63-65, 67-68, 74, 100
113-114, 197-198, 216, 220, 222, 226-227
Doheny Scientific Expedition · · · · · · · · · · · · 189

Index

Dolphin, Lambert · · · · · · · · · · · · · · · · · · 147
Dose, Klaus · 55
Dragons · · · · · · · · · · · · 122, 187, 190, 193, 226
Dubois, Eugene · 71
Ehrlich, Paul R. · 93
Eldredge, Niles · · · · · · · · iv, viii, 37, 85, 90, 121
Empirical Evidence · · · · · · · · · · 15, 49, 58, 226
Escherichia coli (E. coli) · · · · · · · · · · · · · 25, 226
Evolutionary progression · · · · · · · · · · · 91, 226
Evolutionary theory · · · · v, 8, 14, 19, 43, 46, 53, 56
    67, 85, 90, 96, 147, 196, 198, 212, 220, 225-226
Feduccia, Alan · · · · · · · · · · · · · · · 195, 197-198
Ferrell, Vance · 90
Firmament · · · 7, 117, 131, 147-148, 204, 207, 226
Fisher, W. J. · 207
Fossil record · · · · · · · · iii, viii, 5, 13, 36-38, 71, 77
    83-85, 87, 89, 91-92, 95-97, 105, 117, 119-121
      125, 128-129, 135, 141, 145, 175-176, 198
        206, 216, 218-219, 222-223, 225-229
Fossils · · · · · · · iv, 2-5, 14-15, 27-28, 36-38, 72
    74-78, 83-85, 87, 89-92, 95-97, 106, 117-123
      125-129, 131-132, 134, 136, 141, 143-145
        148, 150, 175-176, 180, 182, 188, 191
    195-197, 212, 217-218, 220, 222, 225-227, 229
Frank-Kamenetskii, Maxim D. · · · · · · · · · · · 22
Freeman, Colin L. · · · · · · · · · · · · · · · · · · 219
Fruit Fly · 24, 226
Galapagos Islands · · · · · · · · · · · · · · · · 3, 226
Gee, Henry · 77
Geologic column · · · · iii-iv, 2-5, 9, 13, 15, 38, 59-61
      92-93, 95, 117-121, 125, 127-129, 135
       139, 141, 145, 176, 180-181, 216-217, 220
         222, 225-226, 228-229
Gibson, L. James · · · · · · · · · · 23, 55, 120, 135-136
Giem, Paul A. L. · · · · · · · · · · · · · · · · · · · 180
Gish, Duane T. · · · · · · · · · · · · · 86, 136, 149, 182
Gitt, Werner · 46
Goldschmidt, Richard B. · · · · iv, 15, 19, 25, 86, 96
Gould, Stephen Jay · · · iv, viii, 23, 30, 83, 85-86, 89
        91, 95, 97, 105-106, 111, 211-212
Gradualism · · · · · · · · · · · 37, 89, 125, 226, 228-229
Grand Canyon · · · · · v, vii, 113, 120, 125, 134, 137
      139-143, 146, 148-150, 183-184, 189
        217-218, 220, 225-227, 229
Gregory, H. E. · 143

Grocott, Stephen · · · · · · · · · · · · · · · · 78, 136
Haeckel, Ernst · · · · · · · · · 28, 56, 110-112, 211
Hanson Ranch · · · · · · · · · · · · · · · · · · 27, 226
Hapgood, Charles H. · · · · · · · · · · · · · · · · 191
Harper, G. H. · 54
Hartnett, John · · · · · · · · · · · · · · · · · · · vii, 50
Hasel, Gerhard F. · · · · · · · · · · · · · · · · 8-9, 135
Hazen, Craig J. · 44
Hitching, Francis · 21
Hominids · · · · · · · · · · · · · · 12, 76-77, 192, 226
Honest moments · · · · · · · · · · · · · · · · · · · 97
Hooker, Dolph Earl · · · · · · · · · · · · · · · · · 131
Hooton, Earnest A. · · · · · · · · · · · · · · · · · 106
Horner, Jack · · · · · · · · · · · · · · · · · · · 122-123
Hubble, Edwin · 44
Hubble Space Telescope · · · · · · · · · · · · · · 203
Hull, David L. · 8
Human consciousness · · · · · · · · · · · · · 26, 219
Human genome · · · · · · · · · · · · · · · · · 67, 226
Humanism · · · · · · · · · · · · · · · · · 53, 134, 226
Huxley, Thomas Henry · · · · · · · · iv, 28, 54, 213,
Hyperbaric oxygen therapy · · · · · · · · · · 147, 226
Hypothesis · · · · 2, 21, 35, 49-50, 57, 66,-67, 78-79
      91, 93, 96, 108, 127-128, 226
Ice Age · · · · · · · · · · · · · · · · · 53, 139, 170, 226
Intelligent Design/Movement · · · iii, 1-2, 10, 19-20
      32, 57, 61, 65-68, 205-206, 224, 227
Invertebrates · · · · 3, 95, 97, 106, 128, 143, 220, 227
Irreducible complexity · · · · · iii, 10, 23, 63, 65-66
     99-100, 105, 108, 215, 221-222, 225, 227
Irwin, Ross · 149
Isaacs, Derek · · · · · · · 106, 122-123, 183, 187, 190
Java Man · 71-72, 227
Johanson, Donald C. · · · · · · · · · · · · · · · · · 74
Johns, Warren L. · · · · · · · · · · · · 8-9, 25, 32, 35
      61, 67, 106, 133, 221
Johnson, Phillip E. · · · · · · · · · · · · · 32, 77, 97, 195
Kemp, Tom S. · 91
Kenyon, Dean H. · · · · · · · · · · · · · · · · · iv, 44
Kingsley, Charles. · · · · · · · · · · · · · · · · · · 212
Kitts, David B. · 84
KNM-ER 1470 · · · · · · · · · · · · · · · · · · 181, 227
Knoll, Andrew · 57
Koestler, Arthur · · · · · · · · · · · · · · · · · · 21, 24
Krakatoa · 169, 227

Landis, Gary	147
Leakey, Mary	iv, 181, 192
Leakey, Richard	181-182
Lee, Robert E.	218
Lemaitre, Georges Edward	2, 203
Lemoine, Paul	79
Lemonick, Michael D.	92
Lewontin, Richard C.	iv-v, 215
Libby, Willard F.	iv, 17, 176-177, 217
Lingenfelter, Richard	177
Lovell, Jim	7
Lovtrup, Soren	67
Lubenow, Marvin	181
Lucy	74-76, 227
Lull, Richard S.	43
Ma'adim Vallis	149, 227
MacInnes, Austin	45
Macroevolution or Megaevolution	iii, 4, 13, 15, 19, 21, 25, 50, 68, 220, 226-228
Mann, Alan	182
Marsh, Frank Lewis	15, 204
Marsh, O. C.	37, 188
Martin, Robert	107
Materialism	v, 66-67, 92, 204, 215-216, 227
Materialistic science	55
Mayor, Adrienne	106
Mayr, Ernst W.	iv, 4, 21, 36, 46, 61, 83, 109
Meyer, Stephen C.	iv, 55, 60, 66-67, 87, 92, 96
Microevolution	iii, 15, 19, 25, 59, 92, 155, 220, 226-228
Mid-Atlantic Ridge	170, 227
Miller, J. R.	76
Miller, Stanley L.	58-61
Milner, Richard	24, 40, 211
Milton, Richard	89
Missing link	iii, viii, 35, 66, 72-73, 76, 83-86, 89, 91, 195-197, 221-222, 225, 227-229
Mitchell, Colin W.	133, 135
Mitochondrial Eve	vii, 114, 216, 227
Modus operandi	43, 227
Monod, Jacques	8
Moon, Irwin A.	20, 68
More, Louis Trenchard	iv, 54
Morphology	23, 75-76, 90, 182, 192, 198, 227
Morris, Henry M.	11, 23, 28, 77, 83, 128-129, 135, 144, 153
Mount Everest clams	148
Mount St. Helens	169, 179-180, 227
Muggeridge, Malcolm	79
Mutations	iii-iv, vii, 15, 21-26, 39, 44-45, 84, 87, 92, 99, 107, 114, 223, 227
National Academy of Sciences	49, 55, 57-58, 63, 65, 71
Naturalistic/Materialistic Evolution	1, 2, 216, 227
Natural selection	iii-iv, 4, 8, 13, 15-16, 21, 23, 25, 31-33, 35-36, 39-40, 44-45, 51, 54, 68, 87, 90-92, 97, 107, 211, 213, 216, 220, 223, 227
Neanderthal Man	73-74, 227
Nebraska Man	72-73, 97, 227
Nelson, Gareth J.	v, 49, 84
Neo-Darwinian synthesis	4, 16, 21, 226
Newel, N. D.	142
Newman, Horatio H.	iv, 97
Noachian Flood	117, 147, 228
Noble, Elmer	45
Nobel, Glenn	45
Olmsted, Denison	207
Olson, Storrs L.	196
Orce Man	77, 228
Origin of love	26
O'Rourke, J. E.	121
Osborn, Henry Fairfield	72-73
Oxnard, Charles E.	75
Paleoanthropology	54, 182, 197, 228
Paleontology	iv, vi, 2, 17, 27, 29, 36, 43, 50, 54, 72, 77, 84, 90-92, 106, 137, 189, 197, 228
Parker, Gary E.	21, 33, 83, 86
Patterson, Colin	89-90
Pauling, Linus	49
Peat-bog theory	135, 228
Peet, Steven D.	133
Peking Man	74
Peppered Moth	33, 228
Peristalsis	40, 228
Petersen, Dennis R.	72, 191, 212
Peth, Howard A.	84
Phillips, David	74
Pillow lava	151, 228
Pilbeam, David	54

Index

Piltdown Man · · · · · · · · · · · · · · · · 72, 97, 228
Pitman, Michael · · · · · · · · · · · · · · 33, 91, 97
Pitman, Sean D. · 63
Plate tectonics · · · · · · · · · · · · · · iv, 132, 170, 228
Platnick, Norman · · · · · · · · · · · · · · · · · · v, 49
Pliny the Elder · 190
Polymerization · · · · · · · · · · · · · · · · · · · 55, 57
Polystratic trees · · · · · · · · · · · · · · · · · · 143, 228
Precambrian layer · · · · · · · · · · · 95, 125, 225, 228
Presuppositional bias · · · · · · · · · · · · · · · · v, 228
Primate · · · · · · · · · · · · · · 3, 71, 73, 83, 109, 228
Primitive atmosphere · · · · · · · · · · · · · · · 59, 228
Probability · · · · · · · · · · · · · · 21, 24, 41, 105, 222
Progressive creation · · · · · · · · · · · · · · · · · · 1, 8
Punctuated equilibrium · · · · · · · · · 86-87, 91, 228
Quotation mining · · · · · · · · · · · · · · · 13, 97, 228
Radiocarbon dating · · · · · · · · · · 175, 178, 225, 228
Radiometric dating · · · · · · · · · · iii-iv, 2, 5, 11, 43
   59-60, 119-122, 142, 175-176
   178-181, 183-184, 217-218, 222, 228
Rates of erosion · · · · vii, 13-14, 125, 142, 216-217
Red blood cell · · · · · · · · · · · 20, 100, 121-122, 218
Raup, David M. · · · · · · · · · · · · · · iv, 36, 38, 84, 90
Read, David C. · · · · · · · · · 2, 8, 17, 21, 23, 25, 72
   74, 85, 96, 111, 121-122, 125, 135, 170
Reasonable doubt · · · · · vi-vii, 1, 5, 11, 13, 32, 53
   60, 129, 178, 219
Recent life/creation · · 1, 30, 60, 140, 216, 223, 228
Reverse cone of diversity · · · · · · · · · · · · · · · · 95
Ridley, Mark · 15, 67
Riegle, David D. · 106
Roberts, Hill · 105
Rose, Michael R. · 219
Rostand, Jean · 41
Roth, Ariel A. · · · · · · · · · · · vi, 1, 4, 23, 38, 55
   57, 60, 84, 95-96, 127, 140-143
Ruse, Michael · · · · · · · · · · · · · · · · · · iv, 26, 54
Sagan, Carl · · · · · · · · · · · · · · · · iv, vii, 4, 41, 86
Sandage, Allan Rex · · · · · · · · · · · · · · · · · iv, 44
Sanford, John C. · · · · · · · · · · · · · · iv, 44-45, 68
Sarfati, Jonathan D. · · · · · · · · · · · · · · · · · 37, 53
Schad, Gerhard · 45
Schweitzer, Mary H. · · · · · · · · · · · · · · · · 121-123
Scientific method · · · · · · · · · 3, 27, 29, 49-50, 178, 229
Scientific research · 229

Scientists who are creationists *(see Creationists)*
Scoffers · 11, 134, 224
Sedimentary rock · · 59, 76, 117-119, 125, 143-144
   150, 177, 183, 222, 226, 229
SETI *(Search for Extraterrestrial Intelligence)* vii, 229
Shapiro, Robert · · · · · · · · · · · · · · · · · · · iv, 67
Shawver, Lisa J. · 105
Simmons, Geoffrey · · · · · · · · · · 99-102, 107, 219
Simons, Lewis M. · 197
Simpson, George Gaylord · · · · · · · iv, 24, 37, 227
Sloan, Christopher P. · · · · · · · · · · · · · · · · · · 196
Snelling, Andrew · · · · · · · · · · 150, 170, 184, 218
Spetner, Lee · 22
Spontaneous generation of life · iii, 13, 53, 55-56, 229
Stahl, Barbara J. · 92
Stansfield, William D. · · · · · · · · · · · · iv, 95, 136
Stern, Jack · 74
Strickland, Monroe W. · · · · · · · · · · · · · · · · · 113
Strobel, Lee · · · · · 31, 59, 65-66, 71, 111, 197, 220
Sunderland, Luther D. · · · · · · · · · · · · · · · 37, 89
Surtsey · 171, 229
Swift, Dennis · 188
Tardieu, Christine · 74
Taylor, Ian T. · 118-119
Theistic evolution · · · · · · · · · · vi, 1, 7-8, 216, 229
Thompson, William R. · · · · · · · · · · · · · · · · · 38
Transitional fossils · · · · 38, 83, 85, 89-90, 106, 229
Tsiaras, Alexander · 101
Tuttle, Russell H. · 192
Unconformities · 229
Uniformitarianism · · · · · · · · · iii, 5, 53, 118-120, 125
   127, 134, 137, 153, 169, 171, 225, 229
Urey, Harold C. · · · · · · · · · · · · · · · · · · · 3, 58-61
Velikovsky, Immanuel · · · · · · · · · · · · · · · · · 118
Vertebrates · · · 3, 106, 110-111, 128, 220, 226, 229
Vestigial organs · · · · · · · · · · · · · · iv, 109-110, 222
Virchow, Rudolph · · · · · · · · · · · · · · · · · 29, 71-72
von Braun, Wernher · · · · · · · · · · · · · · 7, 30, 100
von Fange, Erich A. · · · · · · · · · · · · · 127, 135, 177
Wald, George · 56, 93
Wanser, Keith H. · 144
Warm primordial soup · · · · · · · · · · 2, 216, 222, 229
Webster, Clyde L. Jr. · · 24-25, 32, 49, 57, 92, 136
Wells, Jonathan · · · · · · · · · · · 33, 59, 87, 197, 216
Werner, Carl · 60

West, Ronald R. · 120
Whitcomb, John C. · · · · · · · · · · · · · · 144, 153
White, Timothy D. · · · · · · · · · · · · · · · · · 192
Whorton, Mark · · · · · · · · · · · · · · · · · · 65, 105
Williams, Alex · · · · · · · · · · · · · · · · · · · vii, 50
Woodmorappe, John · · · · · · · · · · · · 154-155
Wootton, Robin J. · · · · · · · · · · · · · · · · · · 38

Wyllie, Peter J. · · · · · · · · · · · · · · · · · · · 11, 14
Xing, Xu · 197
Yokel, W. H. · 108
Young-earth creationism · · · · · · · · · · · · · 229

CPSIA information can be obtained
at www.ICGtesting.com
Printed in the USA
LVHW060821181119
637664LV00013B/5794/P